WHALES
DOLPHINS
AND
SEALS

*With special reference
to the New Zealand region*

Illustrations:

Pinnipeds —   M. Gaskin
Cetaceans —   M. W. Cawthorn, Fisheries
          Research Board of Canada

HEINEMANN EDUCATIONAL BOOKS

WHALES DOLPHINS AND SEALS

With special reference to the New Zealand Region

D. E. Gaskin

SBN 0 435 62285 4

Heinemann Educational Books Ltd
London Auckland Melbourne
Ibadan Hong Kong New Delhi
Singapore Toronto Nairobi

First Published 1972
© D. E. Gaskin

Published by Heinemann Educational Books
Set by Rennies-Illustrations Ltd, Auckland, N.Z.
Printed offset by Lee Fung Printers Ltd, Hong Kong

Contents

To my mother. In memory.

—

. . . and involved us in a gale which effectually
stopped our progress for a week.. It was our first taste
of the gentle zephyrs which waft their sweetness over
New Zealand, after sweeping over the vast, bleak iceberg-
studded expanse of the Antarctic Ocean. Our poor
Kanakas were terribly frightened . . . the cold was very
trying, not only to them, but to us, who had been so long
in the tropics that our blood was almost turned to water.

Frank T. Bullen
The Cruise of the Cachalot

Preface

The term 'marine mammal' includes the whales, dolphins, porpoises, seals and sea lions, sirenians (sea-cows) and sea otters. No book has previously been written on the marine mammal fauna of the western South Pacific. Since so little is known of the tropical fauna of this region I have restricted the detailed descriptions to cover species positively known to occur in the temperate, subantarctic and antarctic zones. Sea otters, sirenians and true porpoises (Phocoenidae) are not known to occur in the region so defined. Even our knowledge of the distributions, life histories and biology of the whales, dolphins, seals and of sea lions is relatively incomplete. More species, especially dolphins, are almost certainly waiting to be reported in the northern part of this region.

Large-scale commercial whaling is coming to a close all over the world because of over-exploitation of the whale populations. However, seals, sea lions, dolphins, and even some medium-sized whales have been exhibited in seaquaria in many places in the last few years, including New Zealand and Australia. The result has been an unprecedented upsurge of public interest in these animals.

There is a great deal of regional information in the literature on marine mammals, but most is scattered in specialist publications and periodicals not readily accessible to the student or the interested public. I have tried to summarise in this book all the information available up to the time of writing, together with a brief discussion of all the important aspects of whale, dolphin and seal research that has been undertaken both in the western South Pacific area and in other parts of the world. I have attempted to do this in such a fashion that the book will be of use to the specialist and layman alike. The bibliography has been made as comprehensive as possible.

D.E.G.
University of Guelph, May 1971.

1. Origins and Evolution of the Major Groups of Marine Mammals

A student interested in the evolution of marine mammals must keep the following point clearly in mind: those features which are fishlike — clammy skin, streamlined body, a tail, forelimbs modified into flippers, and in some, the presence of a dorsal fin — are all adaptations to the marine environment in which these animals live. Such characteristics have evolved separately in several different groups of marine mammals; as a result the relationships are complex and the lines of phylogenetic descent very difficult to ascertain. The student might well be confused by earlier works on the evolution of these animals, since there was a tendency at one time to reason that various anatomical differences represented lines of phylogenetic descent. We now know that a number of anatomical peculiarities, such as the lobulate kidney in whales, can be associated with adaptation to marine life and increase in body size, and that in general, parallelism in functional adaptations is rife in the marine groups of mammals. Anatomical similarities or differences need not be related to phylogeny (McLaren 1960). Because views on the evolution of these groups have changed so rapidly in recent years, I have cited only more recent authorities in the sections that follow.

The Evolution of Seals: The Pinnipedia

It is natural to ask from which group of mammals the Pinnipedia are derived; unfortunately this is not an easy question to answer, although we can at least say that none of the three living lines, Otariinae, Odobeninae and Phocidae, can be considered ancestral to the others (Scheffer 1958). The relationships of these three seal lines remain conjectural. Monophyletic origin of the Pinnipedia versus biphyletic origin (with the Otariinae and Odobeninae in one line and the Phocidae in the other) is still hotly debated.

In recent work on the classification of these animals there has been a tendency to work with basicranial structures (McLaren 1960, Mitchell 1968a), since these are considered least likely to be affected by aquatic environmental modifications. Both these authors, like King (1964), lean strongly towards theories of biphyletic origins for Pinnipedia. Notable authors putting forward evidence for monophyletic origin of the Pinnipedia have been Simpson (1945), Davies (1958) and most recently Ling (1965b). Simpson believed that the seals arose from basically canoid ancestors, while McLaren and King suggested that the otarine seals appear to have arisen from a common stock with the Ursidae (bears), and the phocine seals from a common stock with the Lutrinae (otters). Ling (1965b), studying the histology of sweat and sebaceous glands in the two major seal lines, concluded that the arrangement of these structures suggested a sequence from the Otariidae to the Phocidae. However this conclusion was challenged by Mitchell (1967), who argued that the conditions observed were parallelisms associated with increasing adaptation to an aquatic environment with progressive loss of hair in both groups.

Postulation of common ancestry of the three major pinniped lines involves one in the difficulties of explaining

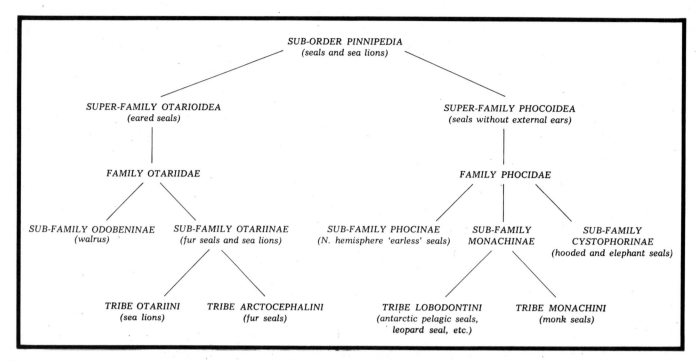

Figure 1.

Classification of living seals and sea lions.

the zoogeographical distribution of the various groups, especially the fossil forms. While some authors avoid the issue to some extent by referring to the Pinnipedia as having 'common arctic ancestry', the fact remains that the fossils of the otarine seals seem to occur exclusively around the margins of the North Pacific, while the phocid groups are found on the Atlantic coast of America and in Eurasia. The most primitive living phocids are freshwater animals, leading King (1964) to suggest that the Phocidae arose from otter-like ancestors in Palaearctic lakes and moved into salt water in the Miocene, while the Otariidae arose from bear-like ancestors in the Pacific regions of North America.

The North Pacific marine mammal fossil deposits are among the richest known in the world, and some excellent work carried out in this region has added much to our knowledge of these extinct groups, and allows limited conjecture on lines of descent to living subfamilies and genera. The first otarine fossils are found in Middle Miocene strata, for example the rather aberrant sea lion *Allodesmus*, which apparently resembled an elephant seal in external appearance, even to an inflatable proboscis (Mitchell 1966). Nevertheless, even early species such as this were clearly well adapted to an aquatic existence, and we have no remains directly linking them to completely terrestrial ancestors (King 1964), although such forms are believed to have lived in the Oligocene, between 25 and 40 million years ago.

Another relatively early sea lion was *Desmatophoca*, remains of which are quite common in some areas. This was a less specialised genus than *Allodesmus*, but would clearly be recognised as a sea lion if it were alive today. Other Middle and Late Miocene genera were *Neotherium*, *Pontolis*, and *Atopotarus*; *Pontolis* was still in existence in the Pliocene era, less than 11 million years ago. Another Late Miocene genus was *Pithanotaria*, which appears to have had fairly close relationships with modern *Callorhinus* (King 1964). A point worth making about a number of early otarine seals, such as *Imagotaria*, is that they were obviously too specialised to be the forerunners of modern forms (Mitchell 1968a).

In the Pliocene we find the remains of two more genera, *Dusignathus* and *Pliopedia;* the latter shows tantalising relationships to both modern *Eumetopias* and modern *Odobenus*. In Quaternary strata are remains which differ from modern animals only at the species level (Kellogg 1922). For example in Japan we find a fossil species of *Zalophus*, *Z. kimitensis*, in Australia a fossil *Neophoca*, *N. williamsi*, and in New Zealand a fossil *Arctocephalus*, *A. caninus* (King 1964).

The walruses are also believed to have originated in the North Pacific region (Kellogg 1922), becoming more widely distributed in the Oligocene. The oldest known walrus is *Prorosmarus* from Miocene strata. The range of walruses was obviously, judging from fossil remains, considerably greater in the past than at the present time, for remains of odobenines have been recorded from both southern California (Mitchell 1961, 1962) and from the south eastern United States (Ray 1960). No pre-Pliocene remains have yet been found in Europe (Scheffer 1958). Fay and Ray (1967) discussed the influence of climate on the distribution of walruses, with special reference to the living species, supplementing an earlier paper by Fay (1957).

The phocids, as far as the fossil record shows, date from the Middle Miocene. The oldest known genus is *Leptophoca* from Maryland Miocene strata (True 1906). In Europe a number of other fossil genera have been found, dating from Late Miocene to Pliocene, including *Miophoca, Pontophoca,* and *Prophoca.* Around the coasts of the North Atlantic most of the living genera of Phocidae have also been found as Quaternary fossils; and even the earlier phocids named above can be placed without difficulty in the modern subfamilies (Kellogg 1922).

The Evolution of Whales and Dolphins: The Cetacea

The earliest known group of animals to appear in the fossil record which can be called whales are the Zeuglodonts, of the Archaeoceti. These make up the oldest line of three major whale groups, the other two being the toothed whales or Odontoceti, and the baleen whales or Mystacoceti. The first traces of archaeocetes have been dated to the Middle Eocene, about 40 million years ago, so whale lineage goes back much further than that of the Pinnipedia. Nevertheless, Slijper (1962) argued that the Archaeoceti could not be considered as direct ancestors of either modern baleen whales or modern toothed whales; that it was unlikely that they gave rise to the ancestral forms of either group. The Archaeoceti may be regarded as a less successful independent line which died out perhaps 10 million years ago. These animals were elongate slender animals with long jaws, rather like caricatures of present-day Ziphiidae. Hind limbs were present, but as in modern whales much reduced and probably not visible externally. The group was reviewed in great detail by Kellogg (1936).

One of the earliest known Archaeoceti is *Basilosaurus,* dating from the Middle Eocene. In common with other genera of the group the skeleton shows many affinities with terrestrial-type mammals of the period, although the skull shows some cetacean features. In one genus, *Patriocetus,* the process of telescoping of the skull can be seen to have begun; a process which has resulted in the nostrils of present-day cetaceans being on top of the head. However, most archaeocetes had nasal bones in the usual mammalian position. The neck vertebrae of archaeocetes were free, not fused, and the teeth were still differentiated into functional groups, i.e. incisors, canines and molars, as in terrestrial mammals. Remains of Archaeoceti have been found in many parts of the world.

In the Late Oligocene, several million years before the last archaeocetes disappeared from the earth, the first true toothed whales, the Squalodontidae appeared. This group did not last as long as the Archaeoceti, which were a dominant group for about 25 million years from the Middle Eocene to the Early Miocene, but were still important as late as the Late Miocene, about 14 million years ago. The teeth were still divisible into functional incisors, canines, premolars and molars, yet the two latter series were also becoming very numerous and less specialised, a condition found in most modern Odontoceti. These were carnivorous species with slender snouts, and the serrated edges of the back teeth have been noted to bear a strong superficial resemblance to the teeth of sharks (Mitchell 1966). The squalodonts were a successful group in their time, and were probably important marine predators. A number of remains have been described from New Zealand by Benham (1935, 1937b, c), Marples (1949a), and Dickson (1964).

At the same time as the squalodonts ranged the world ocean two successful families of smaller animals, the Hemisyntrachelidae and the Acrodelphidae, also evolved. The latter family, otherwise known as long-snouted dolphins, bore some relationships to modern Ziphiidae, and are considered to be characteristic of Early Miocene marine faunae (Mitchell 1966). Their remains have been found in Eurasia and both the American continents. In the Early Miocene strata the modern odontocete families Physeteridae, Ziphiidae, Delphinidae and Platanistidae (river dolphins) also appear among the fossils; this appears to have been a time of explosive, very rapid evolution, perhaps associated with a major change in earth climate, although a really steep decline in palaeotemperatures is not recorded until the Late Miocene (Keyes 1968, Jenkins 1968, Beu and Maxwell 1968, and Devereux 1968), too late to assist in explaining such a burst of evolution. However, a study of New Zealand palaeoclimates by Fleming (1962) indicated that the Late Oligocene was one of the relatively cool periods in local history, and presumably this was also true for other parts of the world.

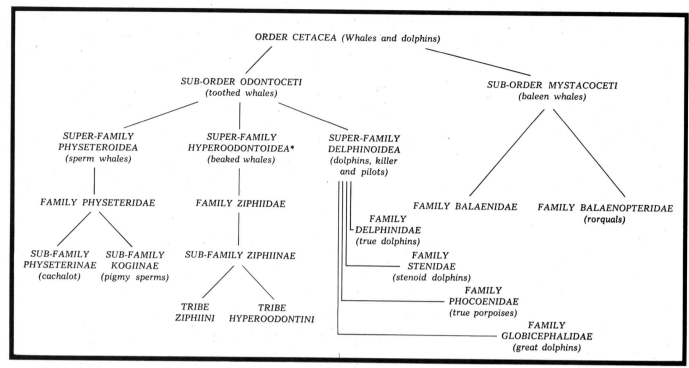

Figure 2.

Classification of living whales and dolphins. *After Moore (1968).

Onset of such relatively unfavourable conditions might have triggered an evolutionary 'burst'. Rapid evolution in cetaceans may well be aided by the fact that the total population size of any species is generally relatively small in biological terms, a few million at the most throughout the world, judging by modern species. Under these circumstances, especially if unfavourable conditions decrease the population size still further, there is the likelihood of a relatively large percentage of mutations being passed to the succeeding generation (Clarke 1954). Association of cetacean species with fairly limited temperature ranges (Gaskin 1968c) suggests that this factor could be an important one in the evolutionary history of the group.

One of the oldest acrodelphid genera known is *Argyrocetus,* from Early Miocene strata of Patagonia. Even within this family evolution appears to have been fairly rapid, since the Middle Miocene *Agabeus* was toothless. At this time we also find the remains of early physeterine or kogiine sperm whales, such as *Aulophyseter* and *Physodon,* and a distinct family of slender, long-beaked, toothless (or with back teeth only) dolphin-like animals called the Eurynodelphidae.

The first baleen whales, their origins obscure, appear in the fossil record for the first time in the Middle Oligocene. The major early baleen whale family was the Cetotheridae, of which many species have been described from a number of countries. Their remains have been found in abundance in some parts of the American Pacific coast (Mitchell 1966), and specimens are known from New Zealand (Benham 1937a, with erroneous nomenclature; 1939, with corrected nomenclature; Marples 1949b, 1956). The specimen figured in this book is a recent find, taken from a road bank near Taihape by Dr Roy Yensen, and Mr Keith Ellworthy, both of Palmerston North, together with a number of other helpers. Most of the twenty or more cetotherian genera so far described contain small species, often less than about twenty-five feet in length. Even the Early Miocene cetotheres from Californian strata are small, some only ten feet in length (Mitchell 1966). In the Middle Miocene deposits in California these primitive baleen whales are quite abundant; Mitchell pointed out their close resemblance to the Gray Whale, the only living member of the family Eschrichtiidae.

The first Balaenidae appear in the Late Miocene strata, along with primitive Balaenopteridae. The rorquals are believed to have evolved first; certainly their fossil record goes back several million years beyond that of the Balaenidae (Sanderson 1956). One genus, *Mesotaras,* appears to be a good link between the two families, since it cannot easily be placed in either. One of the

4

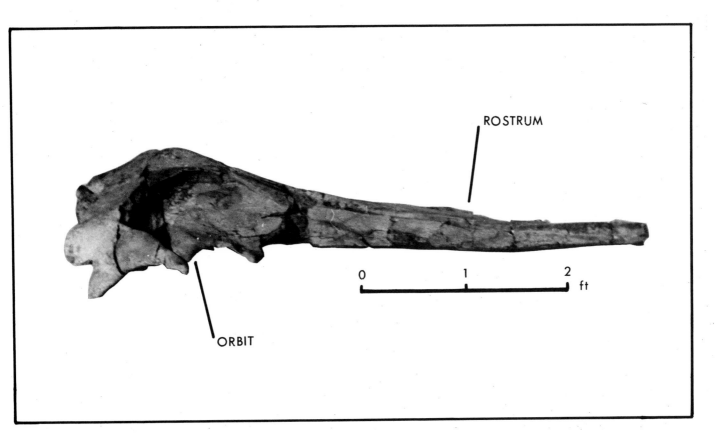

Figure 3.   (K. Ellworthy, Palmerston North, New Zealand.)
Right lateral view of fossil whale skull, apparently that of a
cetothere, found near Taihape, New Zealand. Skull dimensions
5ft. 5in. long, by 2ft. 5in. at widest point.

relatively early rorquals *Plesiocetus* of the Early Pliocene
attained lengths of fifty to sixty feet, although it is very
noticeable that none of the fossil rorquals were nearly as
large as present-day fin and blue whales.

There is near unanimity among specialists that the
ancestors of cetaceans were also the ancestors of the land
mammals known as the Artiodactyla, of which modern
representatives are the camels and rhinoceros. There are
many points of anatomy in common (Kleinenberg 1959,
Slijper 1962, 1966), and the serological typing work of
Boyden and Gemeroy (1950) clearly showed that the
serum proteins of cetaceans had more in common with
Artiodactyla than any other mammalian group. Nevertheless, the fossil record which could confirm the origin
of the cetaceans from terrestrial or freshwater mammals
still has many gaps.

It would be wrong to consider whale speciation as
now ended; the process is still continuing, albeit too
slowly for us to observe noticeable effects within even
several lifetimes. During the Pleistocene and the
Quaternary epoch climatic changes involving both
warming and cooling appear to have led to rorqual and
right whale populations being segregated into northern
and southern hemisphere stocks; while there may be interchange there is little doubt that it is relatively small.
This phenomenon is called antitropical distribution (Davies
1963a). Ivashin (1958) attempted to build a case for sub-
specific status for the humpback populations in the different Antarctic areas based largely on colour pattern
differences, but his evidence has been strongly challenged
by Chittleborough (1965).

The case for similar northern-southern segregated
populations has been argued for the dolphin genus
Lagenorhynchus by Bierman and Slijper (1947, 1948),
although some of their systematic arguments are not
sound (Fraser 1966a). Davies (1960) demonstrated for
another odontocete genus *Globicephala* that two temperate zone populations, *G. melaena melaena* in the
North Atlantic and *G. melaena edwardi* in the
Southern Ocean, are separated by the distribution of
Globicephala macrorhyncha in the Caribbean, tropical
Atlantic and Indian Ocean.

5

2. Anatomy and Physiology of Marine Mammals

Our understanding of the anatomy and physiology of marine mammals has advanced so much in recent years that it becomes a daunting task to summarise the basic information in a few pages. A much more thorough treatment of these subjects will be found in 'Marine Mammals' by Harrison and King (1965).

Few biologists would challenge that whales, dolphins and porpoises are, evolutionarily speaking, the highest forms of life in the sea. These mammals thrive in a medium so physiologically dry that terrestrial mammals die in it for lack of fresh water, and some species can tolerate water temperatures as low as 0°C for indefinite periods. The adaptations of form and function which permit this are of the utmost interest to both pure and medical science. It is now possible to study seals and dolphins in captivity under almost natural conditions. In earlier years most physiological data consisted of highly suspect measurements made on dead animals — freshly dead only if the experimenter was exceptionally lucky. Similarly most early histological studies were based on material often many hours old, so that the tissues were already decomposing before they were fixed and sectioned for examination.

While we still lack complete understanding of many aspects of anatomical and physiological functions, at least the major organ systems and the structure of their tissues can now be described with some degree of accuracy, and some experimental observations are available on the function of most of these systems. There are a number of ways of summarising this information; in this volume I have chosen to discuss very briefly the systems of the body, treating them separately for convenience, but relating one to another where necessary.

Nervous System:
Brain, Spinal Cord and the Problem of Intelligence

The central nervous system of the Pinnipedia is not as highly organised as that of cetaceans (Harrison and King 1965). The external features of the brains of a number of species were described by Fish (1898), and the organ has been described in more detail by later workers (Ogawa 1935; Langworthy, Hesser and Kolb 1938; Breathnach 1960), but we still know relatively little about its histological structure and the functions of the various parts, especially the cortical areas. Langworthy and his co-workers elucidated the functions of most of the motor region, and Alderson, Diamantapoulos and Downman (1960) partially mapped the important auditory centres of the cortex.

The cetacean brain has received far more attention than that of seals. Kojima (1951) found the average brain weight of 16 male sperm whales to be 7.8 kg, and also that the sperm whale has a larger brain per size of body than the baleen whales. Some studies on baleen whale brains have been made by Wilson (1933), Jansen (1951, 1953) and Jansen and Jansen (1953, 1969). These workers examined the brain of blue and fin whales. A somewhat earlier descriptive study of the fin whale brain was made by Langworthy (1935); Riese and Langworthy (1937) also made a comparative study of the brains of both fin and sperm whales. The brain of the humpback whale was described by Breathnach (1955).

Smaller cetaceans have also received attention. The brain of the common dolphin was described by Addison (1915); and that of the bottlenosed dolphin by Langworthy (1931, 1932). Kruger (1959, 1966) has carried out useful work on the brain of the Pacific bottlenosed dolphin. The brain of the Atlantic harbour porpoise *Phocoena* was discussed by Breathnach and Goldby (1954), and the spinal cord of the same species by Hepburn and Waterston (1904) and Pressey and Cobb (1928). More recently Flanigan (1966) studied the spinal cord of the Pacific Striped Dolphin.

Briefly summarising the findings of this work: the cerebellum is well developed, but not much more so than that of the average terrestrial mammal; this may be because although a cetacean requires great muscular and nervous co-ordination, it is subjected to less severe equilibrium problems than a terrestrial mammal, being supported in a fluid medium. The parts of the cerebellum known to be associated with muscular co-ordination are well developed, as might be expected. Olfactory bulbs are absent in cetaceans, but the part of the brain normally associated with olfactory functions is well developed. This may have taken on some other function; it is generally accepted that whales have little or no sense of smell. The auditory cortex is highly developed in cetaceans, reflecting the importance of sound reception for echolocation and communication in these animals. Sight is generally thought to be of secondary importance to whales and dolphins, but the areas of the brain associated with sight are not reduced, and the bottlenosed dolphin has been shown to have considerable powers of visual discrimination (Kellogg and Rice 1966).

The spinal cord shows certain modifications which can be related to the marine habit of cetaceans. There is a moderate cervical enlargement, and virtually no lumbar enlargement. These correlate respectively with the reduction of the complexity of the forelimbs, and the total loss of functional hindlimbs. There is considerable development of motor cells in the spinal cord, apparently in association with the development of the flukes as the major propulsive unit (Harrison and King 1965).

How intelligent are cetaceans? This is still the most controversial question in cetacean studies today. We can make a number of fairly limited statements of fact. For example, the cerebral cortex of cetaceans is highly convoluted (Breathnach 1960), and the complexity of these convolutions resembles in a general way that found in the brain of primates. The cetacean brain is large, especially that of dolphins, but the neurone density is relatively low. But this is something to be expected with a brain of large absolute size, and is not necessarily a measure of cetacean intelligence. Some workers have suggested that this complex development of the cortex is associated with the great degree of muscular co-ordination exhibited by cetaceans, since it seems certain that not all such functions are under the control of the cerebellum alone.

Our problem is made worse by the inevitable man-centred approach of any observer, no matter how hard he tries to be impartial. A layman visiting a dolphin display at an oceanarium comes away convinced of the intelligence of dolphins. Certainly the major emphasis on play in bottlenosed dolphin behaviour is something which many biologists would associate with a high level of intelligence. Yet if one examines the abilities of dolphins very critically, the claim made by some that they may be the near-equals of man seems more doubtful. For example, although dolphins are superbly adapted to their environment this cannot be shown to have anything to do with intelligence. They can tolerate changes in their environment, but so can other lower animals. They have a considerable capacity for learning and have been shown to be susceptible to a high degree of training. Within their morphological and anatomical limits they are great mimics. They can solve simple problems, but do not score particularly highly in this respect. Nor do they rate highly as innovators. Dolphins in captivity sometimes learn to do things for themselves in the context of their new environment; possibly captivity provides stimuli not present in their natural surroundings.

The major problem is that we cannot really define intelligence. The 'intelligence tests' of a decade ago are now no longer regarded as being any more than very fallible, rough guides. Histological criteria based on examination of brain tissue; considerations of brain weight, size and brain to body ratios, and comparisons of behavioural patterns are all unsatisfactory methods of trying to assess intelligence. In conclusion perhaps all we can say for certain is that dolphins, especially *Tursiops truncatus,* exhibit intelligence at least equivalent to that of the lower primates, and perhaps higher. Unfortunately in the oceanarium type of environment it is difficult to differentiate between behaviour which can really be called intelligent, and behaviour which is the result of training in response to visual and auditory stimuli.

The Major Sense Organs:
Eyes and Ears

Little is known of the tactile sense of whales, but the skin is well supplied with nerves in both baleen and toothed whales, especially around the mouth. There are short bristles on the snout of a number of species; it has been suggested that these might be pressure receptors in baleen whales, but tactile in the freshwater dolphins such as *Inia*.

Rather naturally most attention has been focused on the major sense organs, the eyes and ears. All cetacean eyes are constructed on basically similar lines, but the sclerotic outer layer is much thicker in some species than in others; in the sperm whale it is heavily calcified. Lachrymal glands are absent, since it is not necessary to keep the conjunctiva moist as in terrestrial mammals. Seals of course spend much more time above water than cetaceans; in these an oily secretion keeps the surface of the eye lubricated. Since there is no lachrymal duct the seal eye 'weeps' profusely all the time. The lens in cetaceans is much more spherical than in terrestrial mammals; more so in toothed whales than in baleen whales. The whale or seal cornea does not bend light rays like that of the land mammal; the cornea, aqueous humour and vitreous humour all have the same refractive index as water (Matthiessen 1893). A detailed study of the extrinsic eye muscles of whales showed that these also had a number of unique features (Hosokawa 1951). The retractor bulbi muscle system, by which the eye is retracted into the socket, is particularly well developed in cetaceans. The sensory nerves of the eye muscles (from the ophthalmic nerve) run quite separately from the motor nerves (coming from the oculomotor and abducens), not closely adjacent to them as in most other mammals. The eyes of seals have one other adaptation which should be mentioned; the pupil is capable of great expansion to facilitate vision at moderate depths. The cornea is astigmatic, but for vision in air the natural longsightedness of the system is compensated for by the reduction of the pupil to a narrow vertical slit. This counteracts the astigmatism of the cornea and permits a reasonably sharp focus (Harrison and King 1965).

Early studies on the anatomy of the ear of marine mammals were carried out by Hunter (1787); Home (1812) and Buchanan (1828), mainly on right whale and narwhal; Rapp (1836) and Stannius (1841) on the harbour porpoise; Murie (1873) on pilot whale; Beauregard (1894) on a number of species; Anderson (1879) on freshwater dolphins; Boenninghaus (1903) on har-

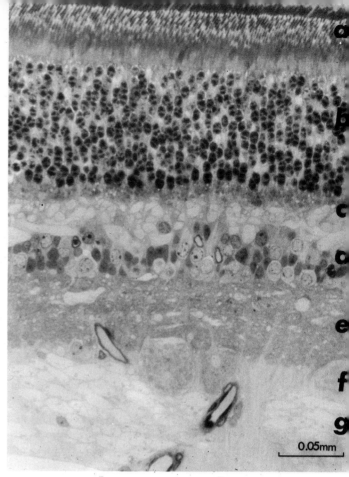

Figure 4. (B. Peers, M.Sc. thesis, University of Guelph.)
The retina of *Tursiops truncatus* resembles that of a nocturnal terrestrial mammal. Morphologically, the retina is adapted for vision under conditions of low illumination. a. visual cell layers (photoreceptors); b. outer nuclear layer; c. outer plexiform layer; d. inner nuclear layer; e. inner plexiform layer; f. ganglion cell layer; g. nerve fibre layer.

bour porpoise; Lillie (1910) on baleen whales; and Anthony and Coupin (1930), Scholander (1940) and Yamada (1953) on various species of beaked whales and the pigmy sperm whale.

Cetacean hearing ability has been intensively studied in recent years by Purves (1955, 1966), Fraser and Purves (1954, 1960), Purves and van Utrecht (1963), Reysenbach de Haan (1957), Kellogg (1960) and Norris (1968, 1969); the last author discussing the evolution of acoustic mechanisms in toothed whales. These workers have greatly advanced our understanding of the physiology of hearing of these mammals, although there is still disagreement on a number of important points. The cetacean sense of hearing has been shown to be acute and responsive to a wide frequency range (Harrison and King 1965). While Reysenbach de Haan asserts that the external auditory meatus (blocked by a wax plug in baleen whales) plays little or no part in the transmission of sound to the inner and middle ear, Fraser and Purves consider that the meatus is as important as in terrestrial mammals. The tympanic membrane of cetaceans is in two parts, an inner part forming a concave drum, and an external part which fits over the basal end of the wax plug. Experimental studies have shown that the wax plug can transmit sounds with little loss of definition along its longitudinal axis. The tympanic bone of whales is a large bean-shaped object, hollow, and only attached to the skull proper by a pair of thin projections in baleen whales and one projection in toothed whales. The insides of the tympanic cavity and associated sacs are filled with foam; it has been postulated by Fraser and Purves that this compound functions both in hearing and the equilibration of pressures during diving. The whole system, according to these workers, is particularly well adapted for the reception of high frequency sounds. There is of course no external 'ear' or pinna in cetaceans; Fraser and Purves believe that auricular muscles around the external auditory meatus can regulate the tension of the tube and assist the animal in assessing the direction from which sounds are coming.

A small pinna is present in the Otariinae among seals, but not in other groups; in the phocids the aperture of the external auditory meatus is minute. Relatively little attention has been given to the ear of seals compared with that of cetaceans, but since there is good evidence that these animals can produce echo-locatory sounds, there is little doubt that their hearing under water is very acute.

Endocrine System:
The Major Glandular Organs

Our knowledge of the general inter-relationships between the endocrine organs of marine mammals, and the part the whole system plays in maintaining physiological equilibrium, actually amounts to very little, although a number of detailed studies have been made on individual organs.

The pituitary/hypophysis has been described and analysed biochemically in a number of species. Useful hormones have been isolated from this organ, especially ACTH (adrenocorticotrophic hormone) which is used extensively in the treatment of serious burns and to ease rheumatoid arthritis (Slijper 1962). The major notable studies in this field have been those of Geiling (1935 on fin and sperm whale); Wislocki and Geiling (1936); Valsø (1938 on blue whale); Jacobsen (1941); Hanstroem (1944); Hennings (1950 on fin whale); Benz, Schuler and Wettstein (1951 on fin whale), and Sverdrup and Arnesen (1952 on fin whale). The cetacean pitui-

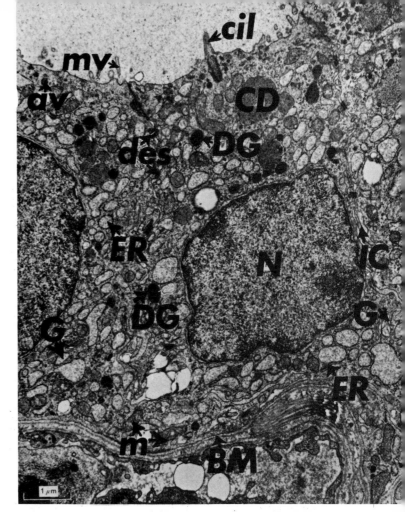

Figure 5. (Dr R. J. Harrison, University of Cambridge.)
Electron micrograph of the thyroid gland follicular cell structure in the common dolphin *Delphinus delphis*. Key to letters: microvilli *mv*; a cilium *cil*; apical vesicles *av*; desmosomes *des*; intercellular channels *IC*; basement membrane *BM*; nucleus *N*; colloid droplets *CD*; dense granules *DG*; endoplasmic reticulum *ER*; Golgi zone *G*; mitochondria *m*. Although the dynamics of synthesis, storage and secretion of the thyroid hormone are reasonably well known, it is not known how it actually leaves the follicular cells.

tary is a dark, roughly heart-shaped structure, with a thin but distinct connective tissue sheet separating the anterior and posterior lobes, lying as in other mammals, under the lower surface of the brain. The pituitary of seal pups is proportionately large compared to that of the adults. It is believed to act in the modification of water metabolism; such modification is necessary because seal milk has a very high fat content. Structure and function of the organ in marine mammals was recently fully reviewed by Harrison (1969b).

The cetacean thyroid gland is a deeply bilobate structure, one lobe lying each side of the larnyx. This gland is dark red in colour, and on each side of the thyroid lobes can be found the smaller pinkish oval parathyroids, which are believed to regulate the amount of calcium in the body, as in other mammals. The thyroid is, of course, well known for its part in iodine metabolism, but in cetaceans this organ, which is relatively large compared to that of terrestrial mammals, is thought to play an important role in the regulation of general body metabolism, especially with regard to control of heat loss (Crile and Quiring 1940). The lobes of the thyroid are themselves lobulate; the microscopic structure was described by Turner (1860) and Harrison, Johnson and Young (1970). Harrison (1960, 1969) discussed the results of a careful analysis of the cycles of thyroid activity in seals. He found that there were two activity peaks; one was found to be associated with the lactation cycle, the other with late foetal development.

The thymus gland is common to all juvenile mammals; it lies as a dark red multi-lobate mass around the pericardium, but begins to degenerate at the onset of adolescence and in normal adults disappears completely. The only detailed anatomical and histological study is that of Turner (1860). More recently Nowell published a brief note on the collection of whale thymus glands for medical research (Nowell 1956).

Jorpes (1950) undertook a study of baleen whale pancreas with a view to commercial production of insulin from these animals; however it was not found to be an economic proposition. The gland itself lies approximately dorsal to the loop of the duodenum; it is diffuse and pinkish. Studies on the histology and digestive enzymes have been made by Neuville (1936), Ishikawa and Tejima (1949), and Quay (1957). As in other mammals, it regulates the sugar metabolism of the body, especially in the liver.

The liver of cetaceans is a large bilobed organ without a gall bladder; it may weigh as much as a ton in the big baleen whales. Seals have a well developed gall bladder, and the liver in these animals is drawn out into six elongate lobes. Both cetacean and seal liver are important sources of vitamin A, and liver oil is a valuable byproduct of modern whaling ships or stations. A review of the total vitamin content of whale liver was given by Braekkan (1948).

As in other mammals the adrenal glands of cetaceans are found on the anterior dorsal aspect of the kidneys. They are small (about 2-3in.) oval organs, generally yellowish and rather flattened, with the outer surface slightly lobulate in appearance. They are not large in proportion to body size, and a study of this was made by Crile and Quiring (1940), who reasoned that this was associated with the generally placid mode of existence, that under normal conditions cetaceans did not need rapid secretion of large amounts of adrenalin into the blood stream. Again, the interested reader may refer to Harrison (1969b) for a full review of the marine mammal adrenal glands.

The mammary glands of the big baleen whales are very large organs, several feet in length and as much as a yard in diameter. Their thickness varies according to the condition of the gland. An actively lactating gland can be 12in. deep. The tissue is pink when resting, yellowish pink when active, lobulate, and each lobule has a collecting duct for the milk which eventually, through larger and larger vessels, reaches the main duct to the nipple. The latter are two in number in cetaceans, each set in a groove anterior-lateral to the female genital opening. Whale calves are 'force-fed'; the milk leaves the gland under some pressure. The structure of baleen whale mammary glands has been studied in considerable detail by van Utrecht (1968). Otariids, together with bearded seals and monk seals (Harrison and King 1965) have four teats; other phocids have only two. They are retractable, and the actual glands are relatively diffuse structures below the blubber.

The milk of marine mammals is very rich in fat compared with that of most terrestrial mammals; the value can be as high as 28 to 45 per cent (Harrison and King 1965). In human milk the fat content is generally much less than 4 per cent. Studies on cetacean milk have been conducted by Ohta and others (1955), Pedersen (1952), Gregory, Kon, Rowland and Thompson (1955), and Lauer and Baker (1969).

Excretory System: Kidney and Osmoregulation

While the kidney is regarded by some as an endocrine organ, it is more usual to consider it along with the ureters and bladder, as part of the main nitrogenous waste excretory system of the body. Actually the structure of the bladder (which is relatively small) and the ureters, is virtually unknown in marine mammals.

The cetacean and pinniped kidney has a remarkable appearance. Instead of the familiar bean-shaped, smooth, oval organ of most terrestrial mammals we find an elongate lobulate structure, somewhat flattened and set against the dorso-lateral body wall. The kidney in this case is composed of a large number of small subspherical bodies called renculi compressed together into the main organ. It must be pointed out that this type of kidney is not confined to marine mammals; it also occurs in elephants, bears, otters and cattle (Slijper 1962).

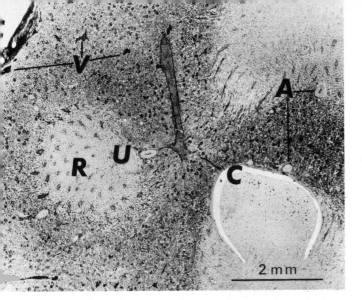

Figure 6.   (J. Dragert, M.Sc. thesis, University of Guelph, Ontario.)
A low power photomicrograph of a section of the kidney of a 3-month old harp seal, *Pagophilus groenlandicus*, stained with Mallory's triple stain.
A fibrous capsule (FC) surrounds the kidney. Note that arteries (A) are not accompanied by veins (V), which demark the periphery of a renculus (R). A branch of the ureter (U) delivers urine from a calyx (C).

The more familiar type of mammalian kidney is made up of an outer cortex and an inner medulla. In the cortex are a number of functional excretory units called lobuli (Freeman and Bracegirdle 1966). These in turn are grouped into functional areas or 'lobi', which centre around collecting ducts. These ducts pass excretory products in solution down through the medulla to the ureter. There is some disagreement on terminology between specialists. Ommanney (1932) and Harrison and Tomlinson (1956) used renculi and lobuli as synonyms, whereas van der Spoel (1963) pointed out that a renculus should be considered homologous with a basic functional area of the 'normal' kidney, and that each renculus in fact contained a number of lobuli.

Cetaceans, and to a lesser extent Pinnipedia, face formidable problems of osmoregulation. Their blood is not isosmotic with the surrounding medium, so that they always have to maintain gradients for different ionic anions and cations against outside concentrations (Fetcher and Fetcher 1942). They are assumed to be able to produce fresh water from the oxidation of fats and from other food sources (Harrison and King 1965); the very long intestine found in cetaceans may be associated with this. Slijper (1962) observed that the freshwater platanistid dolphins had relatively fewer renculi in their kidney than marine forms of similar body weight and size. The cetacean kidney is also presumed to produce urine in very large quantities but so far the evidence to back this assumption is lacking.

Cave and Aumonier (1961) offered some interesting speculation concerning the need for the urine of diving animals to be forced out under slight pressure from the kidney; they contended that the existence of a slight musculature system around the kidney of cetaceans supported their view. Many theories have been offered concerning marine mammal kidney function, generally dealing with the structure as an adaptation to marine life—overlooking the fact that multilobulate kidneys are found in the terrestrial animals listed earlier. Slijper (1962) supported the view that lobulation is associated with increase in body size, though there seem to be some exceptions even to this.

A number of workers have looked for physiological and biochemical adaptations in marine mammal kidney; Fetcher (1939, 1940), Fetcher and Fetcher (1942), Eichelberger, Leiter and Geiling (1940). However, such studies as have been made on the osmotic pressure and composition of marine mammal urine have not shown these to be significantly different from results obtained for land mammals. Van der Spoel did find an important anatomical difference between the venous system of the harbour porpoise and that of the harbour seal. In the latter there is no discrete renal vein from the kidney; instead there is a plexus of blood vessels surrounding the organ, called the 'stellate plexus'. In the harbour porpoise such a system is also present, but it is supplementary to and does not replace the normal renal venous system.

The Digestive System:
Feeding Habits, Diet and Digestion

Detailed discussions of the feeding habits of different species of whales, dolphins and seals are given in section III, but the general patterns of diet of these mammals can be summarised as follows. Toothed whales feed on squid, octopus and fish, rarely on crustaceans; baleen whales on euphausid shrimps and sometimes on schools of small fish; seals feed on crustaceans, fish, squid and octopus. Normal teeth are lost in the baleen whales and replaced with filtering whalebone plates. The dentition of modern toothed whales is highly modified from the pattern found in other mammals, all the teeth being of similar shape and size, usually elongate and pointed. These teeth are used only to hold food temporarily in the mouth before it is swallowed whole; only very large items are likely to be bitten into pieces. Seals on the other hand habitually bite and tear their food, and have teeth much more like those of other mammals.

The general anatomy of the alimentary canal is similar in both whales and seals, although there are a number of

11

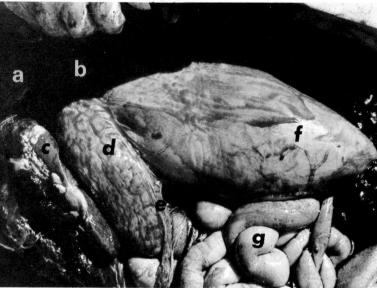

Figure 7. (G. J. D. Smith, M.Sc. thesis, University of Guelph.)

Photograph of stomach and associated viscera of a harbour porpoise *Phocoena phocoena*. a. lung; b. diaphragm; c. liver; d main stomach; e. omentum; f. forestomach; g. small intestine.

variations in stomach structure which were discussed by Slijper (1946). The tongue of whales is a large spongy structure with little musculature or mobility. The oesophagus, particularly that of baleen whales, is amazingly narrow for the size of the animals, barely 4 inches in the fin whale. The stomach is composed of two or three compartments, like that of cattle. The first compartment appears to have a storage function, as well as being the site of preliminary digestion. Peptic ulcers have been found in the first stomach compartment of captive dolphins (Geraci and Gerstmann 1966). The lining of the first stomach is like that of the oesophagus. The second, middle or main stomach is typically mammalian in structure, with a folded glandular lining that secretes hydrochloric acid and the common digestive enzymes lipase and pepsin (Harrison, Johnson and Young 1970). The third compartment is uniformly lined with mucin-producing glands (Geraci and Gerstmann 1966). There is no division of the remainder of the alimentary canal in these animals into small and large intestine, and a caecum is lacking. However the intestine is noteworthy for its great length, which may be associated with water resorption as well as absorption of food.

Figure 8. (G. J. D. Smith.)

Histology of the forestomach of harbour porpoise, *Phocoena phocoena*. A. mucosa; B. sub-mucosa; C. muscularis externa; a. stratified squamous epithelium; b. lamina propria; c. muscularis mucosa; d. circular muscle; e. longitudinal muscle.

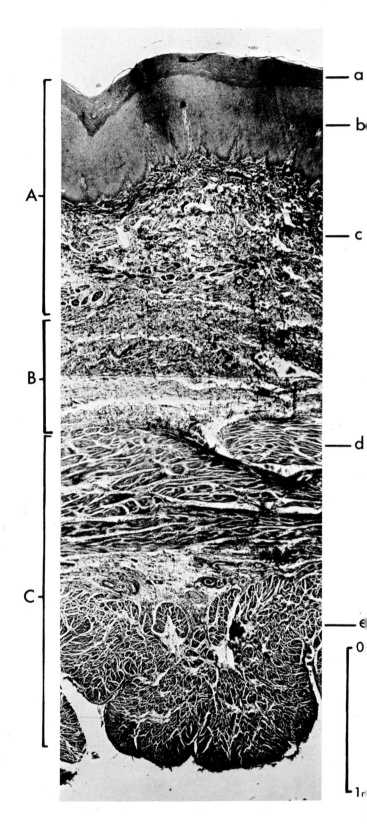

Respiratory and Vascular Systems: Diving Abilities of Whales and Seals

The heart and other major units of the blood vascular system of whales show relatively few divergences from the pattern common to land mammals. The posterior vena cava is subject to considerable variability; single or double vessels appear apparently at random in different species. Detailed studies on the vascular system of cetaceans (both adults and foetuses) have been made by Slijper (1936, 1962), Barnett, Harrison and Tomlinson (1958), and Walmsley (1938).

However there are considerable differences between the peripheral blood system of cetaceans and seals, and that of most other mammals. In various parts of the body this divides into complex networks of anastomosing arterioles, forming large plexuses called 'retia mirabilia'. These were first described in detail by Breschet (1836) and later by Mackay (1886). The stellate plexus around the seal kidney was mentioned in the section on that organ.

It is generally assumed that these vascular retia have a function during diving (Ommanney 1932, Walmsley 1938), but the evidence, reviewed by Barnett, Harrison and Tomlinson (1958), is far from conclusive. Studies by Scholander and Schevill (1955) suggested that some retia act to store heat by taking it from arteries leading to the laminar fin, flukes and flippers, thus regulating to some extent the heat loss that occurs through these vulnerable areas. Modifications of the peripheral blood vascular system of Pinnipedia have been discussed by Barnett and his co-workers, also by Harrison and Tomlinson (1956, 1963), King (1964), Elsner (1969), Irving (1969), Harrison and King (1965). Apart from the existence of the retia another notable feature in these animals is a structure called the 'caval sphincter' which surrounds the posterior vena cava. Experimental work by Murdaugh, Brennan, Pyron and Wood (1962) showed that this seems to regulate the supply of venous blood to the heart during diving.

Studies of the respiratory systems of marine mammals have centered on their enigmatic ability to make deep and prolonged dives when their lung capacity is not much greater, relative to body size, than that of land mammals. Careful studies have shown that the size and numbers of cetacean red corpuscles can exceed twice the corresponding values for terrestrial mammals; however, the affinity of these cells for oxygen is little greater than those of man, for example, and consequently neither the greater size nor concentration of the cells is sufficient to explain the diving capabilities of these animals — Harboe and Schrumpf (1952), Quay (1954). The Japanese scientist Dr T. Tarawa (1951) suggested that other pigments in addition to haemoglobin carry oxygen, but his results cannot be considered very convincing.

As a general rule it is the larger odontocetes of the Physeteridae and Ziphiidae which are capable of the longest sustained dives; among the seals the elephant seals appear to be able to submerge for as much as half an hour (Harrison and King 1965). Examination of sperm whale muscle has shown that the content of myohaemoglobin in this animal is from 8 to 10 times higher than in land mammals. The key to the diving abilities of these animals may well lie in their having blood cells with relatively low affinity for oxygen and muscle pigments with high affinity for oxygen. Thus the blood pigments function largely in rapid transport of oxygen to the tissue pigments, which build up a much larger reserve than could be carried in the blood alone (Irving 1939). Nevertheless, other factors also appear to contribute: the distribution and flow of arterial blood may be restricted to major areas such as brain and muscle during diving and temporarily diverted from less immediately essential areas such as the alimentary canal; the muscles and other body tissues are able to function at lower oxygen tensions than those of most land mammals, and the brain and muscles are less sensitive to carbon dioxide and lactic acid build up than the organs of terrestrial counterparts (Harrison and King 1965). These problems have been further discussed by Lenfant (1969), Elsner (1969), Kooyman and Andersen (1969).

The rate of heart beat of cetaceans decreases markedly when they dive (Irving, Scholander and Grinnell 1941), but not nearly as drastically as that of seals (Irving 1939, Scholander 1940, Harrison and Tomlinson 1963). These workers studied mainly the harbour seal and the grey seal, but Scholander also examined diving birds. In these animals a phenomenon known as 'Bradycardia' occurs when the individual dives. Two ganglia on the vagus nerve control the heart beat rate, which falls from 55-120 beats per minute to 4-15 per minute (Harrison and King 1965), and also cause the caval sphincter on the posterior vena cava to contract, reducing the flow of venous blood back to the heart.

The foam-filled sinus area in the head region of cetaceans may take up nitrogen during dives to depths with considerable pressures. As the air in the lungs becomes compressed it passes into these subsidiary spaces instead of being forced into the tissues. Once air has been forced into cellular tissues under pressure a rapid decrease in external pressure, brought about by the animal surfacing rapidly, can cause bubbles to form; the basis of the very dangerous 'caisson sickness' or the 'bends' which can affect divers who surface too rapidly after breathing compressed air. The sinus spaces and the foam within them seem to prevent cetaceans developing this condition when they surface. Some of the foam, which is also plentiful in the bronchial tracts, is expelled with the 'blow' of whales (Tomilin 1947, Fraser and Purves 1955).

Basic anatomical studies of the lungs, including the histology of the tissues, have been made by Wislocki (1929, 1942), Haynes and Laurie (1937), Bonin and Belanger (1939), Wislocki and Belanger (1940), Murata (1951), Engel (1954), and Baudrimont (1956); and of the nasal passages and trachea by Beauregard and Boulart (1882) and Raven and Gregory (1933).

Dermal Layers:
Skin, Blubber and Thermoregulation

The actual skin or epidermis of cetaceans is very thin, often much less than half an inch, and it overlies an even thinner dermis. The main thickness of the dermal layer is made up by the hypodermis, a fat-rich layer more commonly known as the blubber. Minute dermal projections called papillae intrude into the blubber in rather definite linear patterns, holding the two layers in close proximity, and permitting a small degree of movement of the skin on the firm hypodermal base. This skin mobility is very important in swimming. Skin 'waves' build up as the animal moves; they lie across the line of water flow and reduce turbulence effects, such as occur against the hulls of boats, and instead bring about laminar flow along the skin, with very great reduction in frictional drag.

The blubber can be shown to function as a food store for long migrations (Chittleborough 1965), and when a cetacean is taken out of the water it can act as a very efficient insulator in air, literally boiling the animal in its own metabolic heat if the animal is not cooled in any way. Its value as an insulator in water is generally assumed in the literature, but its actual primary function in this respect must be doubted. Studies on whale metabolism have shown (Kanwisher and Sundnes 1966) that these animals are grossly over-insulated; to such a degree that it is hard to take seriously the idea that the blubber primarily functions as an insulator. These workers suggested that it may play an important part in maintaining hydrostatic buoyancy.

There is quite good evidence to suggest that contrary to popular ideas, large cetaceans in fact experience problems with keeping cool, rather than keeping warm. The thermal conductivity of water is quite high, so the evolution of a vascular blubber layer at least places a barrier between body and medium, even if its function as an insulator in the general sense of the word is doubtful. Heat loss appears to take place largely through the thin flippers, flukes and dorsal fin (Tomilin 1951) and perhaps through the throat grooves of rorquals (Gilmore 1961). Valve systems in the peripheral circulation can restrict or permit full blood flow to these extremities when heat regulation is required (Scholander and Schevill 1955).

For obvious practical reasons there are few reliable body temperature measurements for cetaceans, but the average values obtained seem to lie somewhere between 95° and 97°F (Slijper 1948, Parry 1949c, Tomilin 1951, Kanwisher and Leivestad 1957, Kanwisher and Senft 1960, Belkovich 1961, Morrison 1962).

The anatomy, histology and physiology of the dermal layers of seals have received as much attention as in whales. The hair layer in particular has been intensively studied (Ling 1965a, 1965b, 1968, Ling and Thomas 1967). The blubber layer of the hypodermis is thickest in the elephant seals. The fine under-fur of fur seals traps air bubbles and keeps the skin relatively dry in water. These animals experience over-heating problems on land when the weather is hot and dry. In response to these conditions they can be seen fanning their flippers; these thin structures are major areas for heat loss when a current of air passes over them (Harrison and King 1965). A detailed study of thermoregulation in fur seals was made by Irving, Peyton, Bahn and Peterson (1962). Thermo-regulatory behaviour in the walrus was examined by Fay and Ray (1967). Ray and Smith (1968) studied the thermoregulatory capabilities of the Weddell seal, a phocid with a relatively thin dermal layer. They found that a steep temperature gradient existed in the skin, with the outer few millimetres having a temperature only a degree or so above that of the surrounding medium. This gradient was maintained or altered by a constrictor system in the peripheral blood system. Similar results were obtained for southern elephant seals by Bryden (1964).

Seals tend, like cetaceans, to have a basic metabolic rate somewhat higher than that of most land mammals (Irving and Hart 1957, Hart and Irving 1959, Scholander, Irving and Grinnell 1942). Seal body temperature ranges have been found to be greater than those of cetaceans, with all values obtained to date between 97.7° and 99.5°F (Bartholomew and Wilke 1956, Harrison and King 1965).

Musculo-skeletal System:
Skull, Skeleton and Major Muscles

Since the bones are the parts most readily obtained in good condition from stranded animals, the osteology of most cetacean species was described fairly adequately by the eighteenth and nineteenth century naturalists. The number of specific papers is very great; those relevant to the animals of the New Zealand region are indicated in the appropriate parts of Section III.

There are major differences in the morphology of baleen whale and odontocete skulls. The elongate rostrum, made up of the maxillae and premaxillae, is more sharply delineated from the cranial region in the latter. The baleen whale skull is symmetrical, but that of the odontocete asymmetrical, especially with regard to the nares (nostril) placing. The lower surface of the baleen whale skull bears a long median ridge, which in life runs between the two rows of suspended baleen plates. The maxilla projects ventrally in both groups, passing below the frontal bone in baleen whales, but almost obscuring the frontals dorsally in odontocetes. The parietals form the sides and the supraoccipital the rear of the skull in both groups, as in other mammals.

The skeletal system is quite similar in both groups, although there are some interesting modifications in the primitive pigmy right whale which have not yet been fully explained. The major feature in all cetaceans is of course the great reduction in the hind limbs and pelvic girdle. There is variation from one genus to another in the degree of fusion of the cervical (neck) vertebrae;

Figure 9. (P. W. Arnold, University of Guelph.)
Lateral view of skull of female *Phocoena phocoena* (harbour porpoise), from Letite Passage, New Brunswick, Canada. a. premaxilla; b. maxilla; c. nasal; d. frontal; e. parietal; f. occipital condyle; g. squamosal; h. basioccipital; i. pterygoid; j. jugal (incomplete); k. palatine.

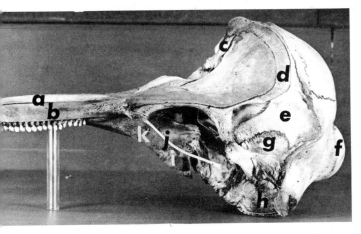

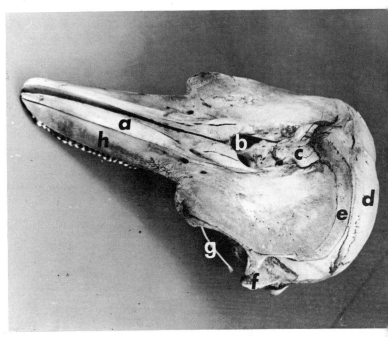

Figure 10. (P. W. Arnold.)
Anterio-dorsal view of skull of female harbour porpoise from Letite Passage, New Brunswick, Canada. a. premaxilla; b. right naris (nostril); c. nasal; d. supraoccipital; e. frontal; f. process of squamosal; g. jugal bone (incomplete).

these are completely fused into a single unit in the right whales.

The skulls of all seal genera (with the exception of the walrus with its immense tusks) show much less modification from the usual mammalian pattern than those of whales and dolphins. The skulls of otariids and phocids can be distinguished because the former have large mastoid processes and small tympanic bulbae, while the latter have small mastoid processes and large tympanic bulbae (King 1964, Harrison and King 1965). There are also differences in dentition; these are discussed on a systematic basis in Section III.

Where the rest of the skeleton is concerned, phocids have a more flexible spine than otariids; the anterior part of the trunk is much more muscular in otariids in association with the more massive structure of the forelimbs and their greater competence and agility on land.

The relation of the skin structure to cetacean locomotion was discussed in the last section; briefly, the arrangement of sub-epidermal papillae permits the skin to move into waves and reduce turbulence and frictional drag, and permits laminar flow. However, the flukes, powered by the longitudinal muscles, provide the major propulsive force.

The flukes have no bone structure other than the

15

Figure 11. (Dr D. K. Caldwell, Marineland of Florida.)
Adult female bottlenosed dolphin *Tursiops truncatus* at Marineland of Florida, showing the flexure of the flukes during the downward stroke of the caudal region during swimming.

medial tapering caudal vertebrae; instead they are constructed with a core of very tough ligamentous tissue which is dense enough to resist change of shape under stress to a very large degree (Felts 1966). Dr Felts reviewed the macro and micro-structure of the flukes and flippers of several cetaceans, and discussed the functional morphology. The forward movement of cetaceans results largely from force transmitted from the upward stroke of the tail flukes, the posterior or trailing margins of which are always angled with respect to the caudal peduncle, and also against the direction of water flow. However, the downward stroke is made with the flukes angled upwards, and this accelerates water passing the body posteriorly, adding to the forward impetus (Parry 1949a, 1949b, Slijper 1961). The overall subtleties of movement still defy detailed mechanical analysis in terms of levers, fulcra and moments. There also appears to be differential output by the dorsal and ventral caudal muscles for the upward and downward strokes respectively, and slight corrective movements of the flippers at the same time compensate for the differences in attitude resulting from the different actions; in this way the posture of the body does not alter significantly (Felts 1966).

The flippers function as paired hydroplanes, although they can just as easily be used independently. These limbs are capable of rotation, of abduction and adduction in the vertical plane, and extension and flexion in the horizontal plane (Felts 1966). The major muscles of the cetacean body have been described in detail by Howell (1930) and Slijper (1936, 1939), but there is still much work to be done on the exact functions of many of these. When the blubber is removed the longissimus dorsi muscles lateral to the spinal column are particularly noticeable, and the flexor and extensor muscles of the caudal region are also very well developed.

Seal locomotion is discussed in a systematic context later, but one major point needs to be mentioned in this general review. Backhouse (1960) contended that the forelimb of otariids was the major unit of propulsion; this statement has been extensively quoted by later authorities; for example King (1962, 1964), and Harrison and King (1965). However, film of *Arctocephalus* swimming under water off the coast of Victoria gives little support for this opinion. The animals can clearly be seen to use the fore flippers for sculling only to maintain equilibrium at slow speed, and for a few strokes during acceleration after a sharp turn; possibly they assist to maintain upright posture during a turn by sculling. Most of the time the seals hold the flippers sloping back with a negative dihedral, and flex the caudal region. Relatively little work has been carried out on seal musculature. Virtually the only papers of note are those by Howell (1928, on *Pusa* and *Zalophus*), and Huber (1934).

Figure 12. (After Slijper, 1936.)
Major muscles of false killer whale *(P.crassidens)*, with left flipper removed: spl., spinalis; s.spl., semi-spinalis; lg.dsi., longissimus dorsi; mltf., multifidus; ilc., iliocostalis; lg., longissimus; ex.cd.lt., extensor caudae lateralis; int.cd.d., intertransversarius caudae dorsalis; flx.cd.lt., flexor caudae lateralis; lv.an., levator ani; obl.int., obliquae interalis; obl.ex., obliquae externalis; scl., scalenus; stn.-thy., sterno-thyreoideus.

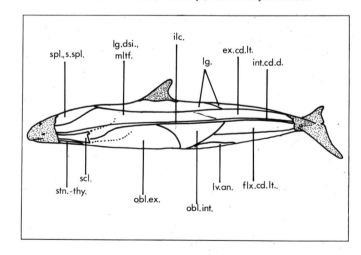

Reproductive Systems

Anyone opening the abdominal cavity of a male cetacean for the first time is at once struck by the relatively huge size of the testes. These are large oval whitish bodies lying just to the side of and to the rear of the kidneys. They are usually quite different in size; the right may be larger than the left, or vice versa. The position of the testes in the body cavity varies from one group of whales to another; in baleen whales and most dolphins they lie immediately posterior to the kidney, in *Mesoplodon* more than their own length posterior to the kidney (Anthony 1922), while in the dolphin *Stenella* they occupy an intermediate position (Matthews 1950). Testis weights in cetaceans vary from a few grammes in immature harbour porpoise to more than 40 kg in the blue whale.

In all baleen whales an annual cycle of testis activity has been noted, as measured by changes in the seminiferous tubules (Chittleborough 1955). Such changes have not been satisfactorily demonstrated in the sperm whale (Aguayo 1963), but recently Best (1969) reported cyclic changes in the size of interstitial cells (Leydig cells), suggesting that the output of the hormone androgen varies throughout the year.

The epididymis is a narrow lobular structure extending in an arc across the ventral surface of the testis; this passes into the twisted and convoluted vas deferens which runs to the penis. The prostate gland is not immediately evident in cetaceans; it surrounds the genital tract to the base of the penis. The penis is only extruded for copulation, normally lying inverted within the genital slit, which is situated well anterior to the anus in the male. The penis is a slender tapering organ with flexible skin which can be everted by blood pressure and some muscular action. There is no os penis (penis bone) in these animals. There is practically no lengthening or thickening of the organ during erection; in anatomical construction it strongly resembles that found in the Artiodactyla (Slijper 1966).

Copulation in cetacean species has been observed only very infrequently in the wild, but more often in seaquaria. It may occur when the male and the female are swimming with their ventral surfaces opposed, or when crossing in an X pattern; in humpbacks it takes place as the animals breach from the surface with the throat grooves pressed against those of the other animal (Slijper 1962). However it is performed, the act is always very short, of 1 to 30 seconds duration.

Detailed accounts of the male reproductive systems of cetaceans have been published by Meek (1918), Ping

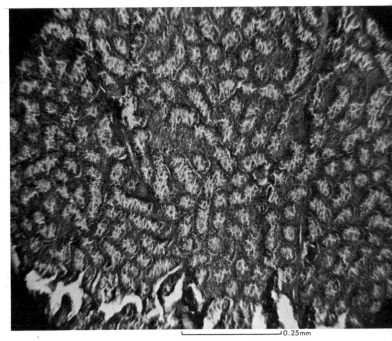

Figure 13. (G. J. D. Smith.)
Transverse section of testis of immature harbour porpoise *Phocoena phocoena* from Letite Passage, New Brunswick, Canada. Much interstitial material between tubules, tubule lumena closed, no spermatozoa in tubules.

Figure 14. (G. J. D. Smith.)
Transverse section of testis of mature harbour porpoise *Phocoena phocoena* from Back Bay, New Brunswick, Canada. Most interstitial tissue gone (b), tubule lumena open, with spermatoza (a).

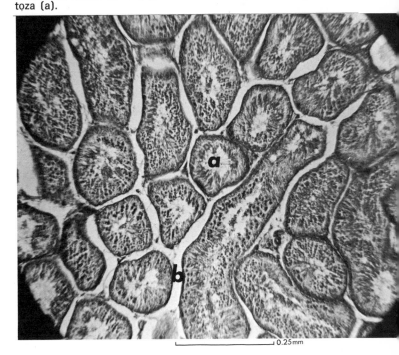

(1926 on harbour porpoise), Ommanney (1932 on fin whale), Slijper (1938), Matthews (1950 on *Stenella*), Preuss (1954), Chittleborough (1955 on humpback), Bassett (1961), Yablokov (1961), and Harrison (1969a).

The testes of otariids are external and enclosed in a scrotal sac as in many terrestrial mammals, while phocids and the walrus have inguinal testes within the blubber, but not deep in the body cavity like cetaceans. The penis is retractable, and normally lies within an epidermal pouch. Copulation takes place in shallow water, or on land or ice (Harrison and King 1965).

Reproductive cycles in seals have been thoroughly reviewed by Harrison, Matthews and Roberts (1952) and Harrison (1969a); in all species examined an annual cycle of testis activity was found.

As in most mammals, the uterus is a bicornuate (two horned) structure in cetaceans, with a muscular cervix at the base of the vagina. However, in the upper part of the vagina in these animals a series of muscular valve-like folds are present (Harrison 1949); these probably serve to retain semen in the upper part of the vagina after copulation. The uterus, which may weigh many hundredweights in a large baleen whale, has a characteristically ridged inner lining. The horns of the uterus, even in the resting condition, are much more massive than the central part of the organ (*corpus uteri*). The distal ends of the horns bend back on themselves and lie with their apices directed caudally. The fallopian tubes are strongly convoluted, and present funnel-shaped openings against the ovaries.

Baleen whale ovaries are relatively elongate structures, those of odontocetes are more shortened and spherical. The egg finds its way into the fallopian tube after being extruded from the ovary (in point of fact from one of the cherry-shaped follicles which rise to the surface from the inner part of the organ). If pregnancy occurs, then the follicle undergoes some drastic physiological changes and develops into a *corpus luteum*, a large body which produces hormones for the duration of gestation. In baleen whales this may attain diameters of up to 20cm (in fin whales, van Utrecht-Cock 1966), and is attached to the ovary only by a slender peduncle. After parturition it regresses into the ovary and becomes greatly reduced in size. It also becomes small and hard, and is then termed a *corpus albicans*.

Ovulations appear to occur in the caudal part of the ovary, and the focus of ovulation shifts anteriorly as the whale ages (Laws 1957, Slijper 1966). The greatest ovary weight of whales is about 26 kg in the blue whale, during pregnancy. Detailed studies of the female reproductive systems of cetaceans have been made by Meek (1918), Ommanney (1932 on fin whale), Pycraft (1932 on common dolphin), Comrie and Adam (1938 on false killer), Slijper (1939 on false killer), Dempsey and Wislocki (1941 on humpback), Matthews (1948 on blue and fin), Harrison (1949 on pilot whale), van Lennep (1950 on blue and fin), Chittleborough (1954 on humpback), Robins (1954 on humpback), Laws (1957, 1961), Stump, Robins and Garde (1960 on humpback), Zemsky (1956 on baleen whales), Rankin (1961 on *Mesoplodon*), Sokolov (1961 on *Lagenorhynchus*), Sergeant (1962a on pilot whale), McCann (1964a on *Mesoplodon*), Harrison, Boice and Brownell (1969), and Harrison (1969a).

The cetacean placenta has been described by Wislocki (1933), Hoedemaker (1935 on harbour porpoise), Wislocki and Enders (1941 on bottlenosed dolphin), Matthews (1948 on blue, fin and humpback), Morton and Mulholland (1961 on pilot whale), and Slijper (1962, 1966). It is a relatively thin structure, usually present in the left horn of the uterus, where the greatest number of egg implantations occur, but part also extends into the right horn. The umbilicus has a zone of weakness which breaks after birth normally without any assistance from the mother or another animal. Birth almost always occurs tail first (McBride and Kritzler 1951, Dunstan 1957). The sexual cycles of cetaceans, especially those of commercial importance, have been intensively studied. Details for each species occurring in the New Zealand region are given in section III.

Multiple births are not common, although a significant number do occur (Kimura 1957). Although Kimura did not think polyovuly was a likely cause of multiple births in cetacea, a case was recently confirmed in a sei whale (Best and Bannister 1963).

Multiple births in fur seals have been reported by Peterson and Reeder (1966). The seal vagina lacks the muscular folds characteristic of the cetacean vagina, and the bicornuate uterus has a medial septum in otariids which separates the organ effectively into two functioning halves (Harrison and King 1965). Otherwise the basic uterine structure is not very different from that of other mammals. The structure of the zonary placenta of seals has been described by Harrison and Tomlinson (1963) and by Harrison and King (1965). The female seal almost always eats the placenta when it is delivered, thereby saving a sizeable fraction of her body reserves of iron.

18

3. Population Research on Commercially Important Species of Marine Mammals

No matter how advanced and technical the methods of capture and processing become, whaling and sealing are still basically hunting enterprises, and hunting is a relatively unsophisticated way of utilising a living resource. Nevertheless, different types of hunting can be defined; one category has an underlying philosophy of conservation, the other does not. All hunting can be classified as either rational (managed) or irrational (uncontrolled). Either type of operation can be economic or uneconomic, depending on circumstances.

Consider irrational hunting first. Most aboriginal hunting is irrational, with no thought for the morrow. However, if the hunting effort is low because the hunter population is small, then the exploited animal population may well be able to absorb the losses without the size of its population diminishing. This is not to say that all aborigine hunting need be free from controls: ostriches have been exterminated in some parts of Africa by primitive man, and the addition of the high-powered rifle to the eskimos' armoury of weapons appears to have increased their hunting powers enough to put the walrus in danger of extinction (King 1964). Low effort irrational hunting, or subsistence hunting, is essentially primitive. With the advent of the expansion phase of European civilization, hunting effort for species such as walrus, right whales and fur seals increased very rapidly, and as exploitation became gross over-exploitation, the hunting changed from economic to uneconomic, and the population densities of many marine mammal species declined to alarmingly low levels. It was unfortunate for the northern large whales, especially the Greenland right whale, that until quite late in the industry's history, it was an economic proposition to send out a vessel to catch just one or two whales.

Not all irrational hunting needs to be regulated, even when the hunting effort is very high. For example, some shellfish populations can be heavily exploited until the density falls to such a low level that the fishermen find it uneconomic to carry out further operations on the beds. The beds are then left alone for a few years while the numbers increase again; in the meantime the fishermen catch something else. Such self-regulating industries are unfortunately fairly rare, but when they exist they need only a minimum of management. At one time it was hoped that the southern pelagic whaling industry would prove to be self-regulating, but such was not the case (see the chapter on world whaling).

Over the last few decades governments have been co-operating, not always with notable success, to ensure that international resources of marine mammals are harvested on a rational basis, with the catch not exceeding the recruitment in any one year. Obviously, rationally controlled economic hunting is the ideal, although in some countries the need for protein and oil has been so high that governments have been prepared to subsidise uneconomic whaling industries to obtain the products. In the long run, such policies often prove to be cheaper to a country than trying to find new industries in which to employ the whalers. Consequently in the modern industrial world we find that uneconomic whaling can be tolerated, although obviously any government or company would like their operations to pay for themselves if at all possible.

Sometimes, produce prices drop so low because of market saturation, and labour costs become so high because of steady inflation, that even small scale whaling may not be economic. The New Zealand coastal sperm whaling industry of the mid-1960s went into abeyance not through a dearth of sperm whales in the New Zealand region, but because the greatly reduced world oil price and the distance of the station from the best grounds made it impossible to catch sufficient animals to make the operation pay. While the New Zealand government was prepared to give indirect, and finally, direct assistance for a short period, it was not prepared to subsidise indefinitely an industry that was not really necessary to the country's economy, and did not employ a large number of workers.

If the existence of any whaling industry can be justified in terms of human needs for oil and protein, then ideally that industry should operate on a rational basis with firm management control backed by sound research. It was unfortunate that at the time the obvious over-exploitation of the southern rorqual population first caused concern, reliable methods of estimating population sizes and yields of whale stocks were not available. These were developed later by the scientists of the International Whaling Commission, but by the time their estimates were accepted, the damage had been done.

Before a population of whales or seals can be managed and conserved for indefinite exploitation certain basic data must be available. The total size of the population must be determinable at any given time with a fair degree of accuracy; the annual mortality must be known, and the rate of recruitment of animals of catchable size into the population must be calculated.

Such information is not at all easy to obtain. Rough estimates of total population size can be made by direct census at sea, with ships carrying experienced observers making sample observations. However, this method is expensive, and the reliability of the results is very limited. The real quality of the observations can never really be gauged accurately, since they are subject to effects of changing wind, sea state, light, visibility and observer fatigue, not to mention the varying experience of individuals. Nor are whale distributions random; the inclusion of a concentration area can make a great difference to the apparent density value. At no time do we know whether to accept a mean, or a high estimate (concentration area included in calculations), or a low estimate (concentration area ignored in calculations).

In general it is far better to estimate population sizes from biological data on pregnancy rates, birth rates and mortality related to operational parameters such as catch per unit effort. It is vital to record length, sex and pregnancy data for all specimens taken, since these yield valuable information on population trends. For example, a continuous decrease in the percentage of older animals is a fair indication that the population is being over-exploited. Similarly, a decrease in the mean body length of the catch could also suggest over-exploitation, or at the very least indicate that fishing mortality (probably with gunner selection for large animals) was significant enough to bring about a change in population structure.

These methods of population study hinge on one vital point; reliable determination of age. Much research has been directed towards cetacean age determination, but the results are still not completely satisfactory. In 1929 Mackintosh and Wheeler published evidence that the number of ovulations experienced by a female baleen whale could be estimated from a count of the corpora albicantia in the ovaries. These bodies are the regressed remains of hormone-secreting corpora lutea which function during pregnancy and are derived initially from the follicles. Other major papers on this subject have been written by Wheeler (1930), Chittleborough (1954, 1959c), Laws (1961), Gambell (1968) and Jonsgaard (1969). Results for baleen whales are not consistent; Chittleborough estimated that the mean number of ovulations per year in humpback whales was about 1; Gambell found a value of about 1.1 for sei, while Law estimated 1.43 for fin whales. Nevertheless by correcting to include pre-ovulation years while the animal is immature this method does provide a rough index of age.

The ages of immature baleen whales have been estimated by examination of growth ripples on the baleen plates (Ruud 1940, 1945, 1959, Ruud, Jonsgaard and Ottestad 1950, Ottestad 1950). In practice the amount of wear on the plates limits the use of this method to whales of up to five years of age.

The method most widely used in recent years for estimating baleen whale age is based on counts of laminations in the wax plug found in the external auditory meatus of these animals. The structure plays a part in the transmission of sounds to the middle ear (Purves 1955). This method has not only been used alone (Purves 1955, Purves and Mountford 1959), but has also been correlated with a fair degree of success against estimates of corpora albicantia accumulation rates (Laws and Purves 1956, Nishiwaki, Ichihara and Ohsumi 1958, Ohsumi 1964a). Until very recently there was dispute concerning the number of laminations laid down each year. Chittleborough (1960, 1962) examined the ear plugs of humpbacks which had been marked as yearlings, and concluded that two layers were deposited each year. However Ohsumi (1964a), Mackintosh (1965), Ichihara (1966a) and Roe (1967) have argued convincingly for only one layer being laid down each year. A number of samples were studied by an international group of specialists at a recent symposium on age determination held in London in 1968 under the auspices of the International Whaling Commission. One factor at the root of the problem was found to be that different scientists had been referring to different layers as laminations. Although no agreement could be reached on the interpretation of layers on sei whale ear plugs, fin whale plugs can now be read satisfactorily (IWC Report No. 19, 1969).

Figure 15.
Structures on which marine mammal age determinations are based: A. Vertical section of mandibular tooth of sperm whale. 1. Cementine. Layers are recognisable in this, but they are of limited value. Generally they are closely packed, and much subject to distortion and wear. 2. Dentine. Growth layers in the dentine are generally regarded as giving an index of age. One clear and one opaque layer is deposited each year. In practice, layers are never as distinct as in the diagram.
B. Wax plug from the auditory meatus of humpback whale. The tapering plug fits snugly within a membranous structure associated with the tympanum, called the 'glove finger'. 3. Longitudinal grooves and ridges are often characteristic of the structure. 4. Growth layers can generally be seen in the concave base of the plug. For accurate work the plug must be sectioned vertically so that the layers can be counted.

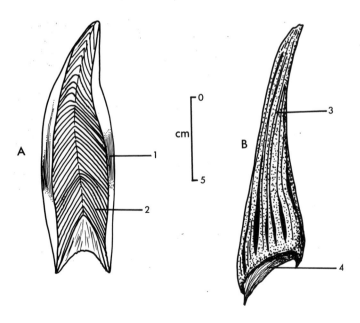

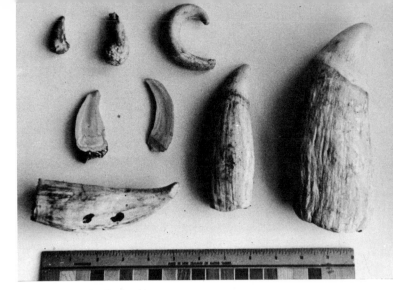

Figure 16. (D. E. Gaskin.)
Teeth of the sperm whale, *Physeter catodon*. Upper row: maxillary teeth, from upper jaw. First and second are typical, third one unusually curled. Below these are two sectioned maxillary teeth; the pulp cavities of these are closed, so they are of only limited value for age determination. Right: mandibular (lower jaw) tooth from 55ft male. Centre is a mandibular tooth from a 43ft male, and bottom, a tooth from the same animal showing caries. These developed on a surface flattened because two teeth had erupted and grown abnormally close together.

Laws (1953b) reported that distinct laminations could be seen in the teeth of elephant seals, and that these appeared to correlate with age. Since then modifications of his techniques have been applied to a number of species of odontocete whale (Sergeant 1959, 1962a), and especially to sperm whales (Nishiwaki, Hibiya and Ohsumi 1958; Nishiwaki, Ohsuumi and Kasuya 1961; Ohsumi, Kasuya and Nishiwaki (1963). Similar laminations have also been found in the lower jaw of the sperm whale but unfortunately the practical difficulties of obtaining large numbers of jaw slices on whaling vessels are far greater than the problem of obtaining teeth. At the symposium on age determination in 1968 the international group decided that most evidence pointed to an accumulation rate of one lamination per year in sperm whales, although this may not be the case in all odontocetes (Sergeant 1959). The accumulation rate of corpora albicantia in sperm whale ovaries has also been correlated with lamination numbers (Best 1968), but the albicantia do not persist as distinctly in sperm whale ovaries as in baleen whale ovaries.

Once a reasonably reliable method of ageing a species is available, and several years of detailed catch statistics have been accumulated, then population analysis can be attempted with some confidence. Actual methods of analysis are beyond the scope of this book, and all require at least some knowledge of algebra and arithmetical functions. A simplified working approach to present methods has been given by Mackintosh (1965), and the reader with a mathematical background could refer to two volumes, *The Dynamics of Exploited Fish Popula-*

tions by Beverton and Holt (1957) and *Handbook of Computations For Biological Statistics of Fish Populations* by Ricker (1958). Papers discussing methods of particular application in whaling research have been published by Hjort, Jahn and Ottestad (1933), Chapman, Allen and Holt (1964), Chapman, Allen, Holt and Gulland (1965), Allen (1966), Boerema, Chapman, Doi, Gambell and Gulland (1969), Bannister (1969), Gambell (1969), and Doi and Ohsumi (1969).

However, reduced to the simplest elements, the problems are as follows. It is necessary to standardize catch statistics on a catch per unit effort basis; this is difficult because catcher horsepower has increased markedly over the years and the addition of sonar and other technical improvements have increased the hunting efficiency of the average whale chaser. It is now customary to make corrections for such differences in the form of estimates based on catcher tonnage.

Not surprisingly, weather conditions also greatly affect catcher efficiency, and efforts must be made to ensure that when statistics are compared, the different sets of information were collected under similar weather conditions. Often, further corrections are necessary if there is a detectable bias by gunners towards one species when another was also present. In the last few years Japanese gunners have shown a bias towards sei whales even when the larger fin whales were present in the catching area, since the sei is preferred for its meat yield even though it is a smaller whale.

When reliable catch per unit effort figures are available they are examined comparatively. If the 'C/E' shows a tendency to decrease steadily, so that the whalers have obviously had to exert more effort for every whale taken with passing time, then it must be assumed that the whale population is either undergoing redistribution, or decreasing in size. Bannister (1969) showed such an increase in effort needed to maintain a catch level in his analysis of the Western Australian sperm whale catch for the mid-1960s. From a value of 3.65 in 1962, catches per effective operational day in 1966 declined to only 1.76.

Natural mortality, which is a largely unknown factor in our knowledge of the life of whales, is estimated indirectly from a plot of observed total mortality of the age classes against catching effort. Then the value for mortality brought about by actual whaling can be estimated as the residue left after the natural mortality is subtracted from the total mortality.

Recruitment rate is estimated from a careful study of the youngest recognisable age class in the catch over a number of years. Complete examination of all females taken in the catch will enable a direct estimate to be made of the pregnancy rate in the population.

The size of the catch that can be sustained indefinitely varies with the size of the population, but within such limits it depends on there being a surplus of recruitment over natural mortality. It is important to realise that in a number of ways we do not fully understand, hunting disturbs the equilibrium of a population. If a population of animals is permitted to find a state of equilibrium with its environment, obviously recruitment and natural mortality will be exactly equal to keep the population size static. From such a population it is not possible to take even a single animal without reducing the total size of the population. One of the most difficult points of basic population dyamics for the layman to grasp is that the largest sustainable catch does *not* come from the largest natural population of whales and seals. As the population approaches its natural maximum, the amount of recruitment surplus over mortality (to us the catchable fraction), becomes less and less. Similarly in a very small population, catching the surplus prevents the total population size from increasing, and so the catchable fraction is bound to stay small. Between these two extremes, where the population is, so to speak, struggling to reach its natural maximum size, there is a midpoint at which the rate of population growth is at a maximum, as is the surplus of recruitment over natural mortality. The ideal aim of rational management is to keep an exploited population near this point, for here the greatest return for catching effort can be reaped, without ever causing the size of the resource population level to decrease. This concept, initially discussed by Hjort, Jahn and Ottestad in 1933 in relation to whaling, was formalized in a paper by Schaefer (1953), and is termed the 'Maximum Sustainable Yield'. This theory is now used as a basis for fisheries and marine mammal management all over the world; perhaps the best example in marine mammal exploitation to demonstrate this principle is the North Pacific Fur Seal Commission, which attempts to regulate carefully the kill of this species using these methods. Unfortunately, efforts to talk the major whaling nations into accepting such rational conservation measures to preserve the southern hemisphere blue, fin and humpback populations were not at all successful. For many years the exploitation rate was considerably greater than the sustainable yield of each population; in the case of each species the final decline in numbers was very rapid.

4. Handling and Studying Marine Mammals in Captivity

Considering that so much of the literature about marine mammals refers to our over-exploitation of their populations, it is encouraging that the development of large seaquaria around the world is leading to a new appraisal of the relationships of man to these animals. Certainly dolphins and porpoises are still taken as food animals in parts of Asia, for example Ceylon and Turkey, but recently the Russian government ended the Black Sea dolphin fishery in their territorial waters, the stated reason being that there was now convincing evidence of the intelligence of these creatures.

The first seaquaria to attempt to keep dolphins, in the United States shortly before the Second World War, in fact pioneered a new form of entertainment operation which has now swept the world. There are now dozens of similar Marinelands all over the world, on every continent except, of course, Antarctica. The development of reliable artificial seawater concentrates has enabled successful large seaquaria to be established even as far from the nearest sea coast as Niagara Falls and Montreal.

To date nearly twenty species of cetacean have been recorded as maintained in captivity for greater or lesser periods of time; the list given below is certainly not exhaustive, and will be obsolete by the time this book is published. Dolphins were in fact kept in tanks as early as 1873, and belugas were brought to England from Newfoundland in 1877. Although some animals were kept alive for a few months in these experimental ventures, most died within a few days.

At present *Tursiops truncatus,* the bottlenosed dolphin, is still the most popular species; most large seaquaria prefer this species if it is obtainable. *Delphinapterus leucas,* the beluga, is held in New York; *Globicephala melaena,* the Atlantic pilot whale, is at Miami; *G. scammoni,* the North Pacific pilot whale, is at the Marineland of the Pacific; *Orcinus orca,* the killer whale, at San Diego and Vancouver; *Lagenorhynchus obliquidens,* the Pacific striped dolphin, at Marineland of the Pacific; *L. obscurus,* the dusky dolphin, at Napier, New Zealand; *Delphinus delphis,* the common dolphin, at Napier and Tauranga; *D. bairdi,* the North Pacific common dolphin, and *Pseudorca crassidens,* the false killer whale, at Hawaii; *Phocoena phocoena,* the harbour porpoise, in the Netherlands; *Stenella plagiodon,* the spotted dolphin, at Miami; the freshwater platanistid, *Inia geoffrensis* or boutu, at Niagara Falls; and a species of *Sousa,* which may be new to science, at Tweed Head Marineland and Marineland of Australia, in New South Wales and Queensland respectively. In addition to the above animals, specimens of *Feresa attenuata,* the slender blackfish, and a single minke whale, *Balaenoptera acutorostrata,* have been kept for short periods in Japan; a *Kogia breviceps,* pigmy sperm whale, was held for a few days at Napier, New Zealand, before it died; and a *Ziphius cavirostris,* or Cuvier's beaked whale, at Marineland of the Pacific, also for a few days only. Both the latter were suffering on arrival from infections which could not be cured, and which had led to their stranding in the first place.

The capture of cetaceans for seaquaria is rarely a simple business. Sometimes swift action permits an

Figure 17. (Dr D. K. Caldwell.)
Trio of bottlenosed dolphins *Tursiops truncatus* performing at Marineland of Florida.

organisation to rescue a stranded specimen; this was possible in the case of some pilot whales on the Pacific coast in 1948 and 1958 (Norris and Prescott 1961), with a dusky dolphin near Napier in 1965 and the pigmy sperm whale and Cuvier's beaked whales mentioned above. Unfortunately such specimens have often stranded because of some weakening disease, and rarely live long. More seriously, unless isolated, they can transmit infections to healthy animals already held in captivity. Cetaceans are prone to primary and secondary respiratory infections in the wild, and in captivity (Ridgway 1965) these are usually fatal. Pilot whales often strand in schools, possibly because the shallow water environment confuses their echo-locatory powers (Slijper 1962). These animals are generally healthy, which explains the relative success the American marineland had with pilot whales obtained in this manner.

In general, seaquaria prefer to select, pursue and capture their animals at sea. There are a number of ways in which this can be done. Set nets can be used to capture semi-estuarine species such as beluga or bottlenosed dolphins, or alternatively seine nets can be used. Dolphins are penned in bays or coves with rock-free shallow shelving bottoms, and either netted actively or are driven into the set nets across the entrance as they dash past the boats towards the open sea.

Deep water species can be taken with a hand-thrown net which enmeshes the head and flippers; this method had been used with success by the Marineland of the Pacific. Another alternative is a tail grab, which was pioneered first in the Mediterranean and has since been perfected independently by the Marineland of New Zealand. In essence this consists of a pair of spring-loaded lazy tongs on the end of a short shaft, which is in turn attached to a length of line. The apparatus is thrown from the boat of the catching boat at the tail stock of a bow-wave-riding dolphin; if the thrower misses the grab can be retrieved for another attempt, still loaded. If the thrower strikes the dolphin's tail stock the grab snaps shut, with a layer of sponge rubber or foam plastic within the arms of the actual tongs cushioning the blow. The dolphin is generally allowed to run out with a length of line until it recovers from the initial shock; in practice the animals rarely run more than 100 ft. or so, since the apparatus is actually quite heavy. As soon as the first panic flight is over, the boat closes with the dolphin and the handlers manoeuvre a canvas sling beneath it so it can be hauled aboard without injury.

The transportation of cetaceans out of water is beset with problems, although these have now with experience been overcome. It has been found that specimens should not be hauled into a boat from a depth of more than 2 or 3 ft., otherwise severe shock can result. Even completely healthy dolphins can suffer fatal skin and eye damage if left exposed to direct sunlight out of water for a few hours. The skin blisters in response to exposure to ultraviolet light, and the eyes develop cataracts. It is possible under these conditions to virtually boil a dolphin within its skin by its own metabolic processes. While the thick insulating layer of blubber beneath the skin may serve a very necessary function of heat conservation in cold water, it also effectively serves to prevent heat loss in an exposed dolphin. This condition can be rapidly accentuated if the dorsal fin, flippers and flukes are covered by accident or ignorance. Most necessary heat loss by radiation takes place through the vascular networks in these flattened slender extremities. The skin must be kept moist at all times, and the eyes protected with wet cloths. Experience has also proved that canvas slings or stretchers are far safer structures for carrying dolphins than ropes passed around the body. Some problems with feeding dolphins in captivity have been shown to results from the animals having damaged larynxes because tight rope slings were used to hold the heads. This organ appears to play an important part in echo-location of food. A dolphin with a damaged larynx conceivably might not even be able to find a dead fish drifting down through the water. Forcible feeding can of course be carried out, but this is very time consuming and causes the animal much distress. In the case of large animals such as pilot whales there is a real need for deep foam plastic cartage beds, since these animals are so large, with body weights of a ton or more, that unsupported by water their own bulk may cause fatal damage to internal organs.

While the most successful seaquaria tend to be those close to or actually on the coast, great improvements in handling techniques in recent years now allow animals to be kept out of water for two or three days without harm, and even flown several thousands of miles. A few years ago two dusky dolphins were flown from Napier to the Tweed Head Marineland in New South Wales.

Almost the first task a trainer has, once a dolphin is safely lodged in a pool, is to get the animal to feed. One of the big breakthroughs in handling any dolphin is to get it to eat dead fish for the first time. Some dolphins persist for months in only eating live fish, which greatly increases the food supply problem of any organisation. Bottlenosed dolphins can fast for about 10 days; after this time if an animal still refuses to eat it is best released. The daily quantity of food eaten by a captive cetacean of course varies greatly from species to species and from individual to individual. Three pilot whales in captivity at Marineland of the Pacific of 17 ft, 13 ft and 12 ft body lengths consumed respectively 100 lb, 80 lb and 40 lb of squid and mackerel per day each (Sergeant 1962a). Common dolphins at Napier take about 12 lb of herring and other small fish species per day when first taken into captivity, but as they become accustomed to their new environment their consumption drops to about 9 lb per day (Messrs F. Robson and A. Dobbins, in letters).

The capture and maintenance of seals and sealions on the whole is easier and requires less elaborate facilities. However, in commercial research it is necessary to be able to capture, brand and release very large numbers of seals in a relatively short period of time, and this raises considerably more problems than does obtaining one or two animals for a zoo or seaquarium. Recently Stirling (1966b), described a simple and effective method for immobilizing large pinnipeds such as Crabeater and Weddell seals. A sack with ropes tied to each corner was drawn over the head of an animal lying on or moving across the ice; once the sack was in place the animals were soon found to become still, and by maintaining tension on the ropes one man

Figure 18. (Marineland of New Zealand, Napier.) Underwater training session at the Marineland of New Zealand of four common dolphins (*Delphinus*).

could control the seal while the other carried out any branding, tagging or similar operation requiring an immobilized animal. Stirling stressed the obvious advantages of handling seals by this method, compared with using a net requiring several men which could injure the animal if it struggled.

In the last decade some work has been carried out on the North Pacific fur seal population and the elephant seals on Macquarie Island with a view to finding safe drugs which can be used to immobilize animals for branding. As yet no completely satisfactory compound has been discovered. Ling and Nicholls (1963) used succinylcholine chloride to temporarily immobilize elephant seals on Macquarie Island, and Peterson (1965) found the same compound moderately useful for field immobilization of North Pacific fur seals. Generally speaking the drugs used fall into two broad categories; first, central nervous system depressants such as thiopental sodium, phencyclidene, propio-promazine and insulin, and second, peripheral nerve inhibitors such as nicotine, gallamine and succinylcholine chloride.

Almost all compounds have similar disadvantages for practical use with seals; just not enough is known about the required dosage. Some of the drugs listed above produce undesirable side effects. Others are safe and effective for calming animals and permitting branding, but have long recovery periods in which drugged animals have to be protected by humans from attacks by other

25

animals, thus cancelling out any saving of manpower that the use of tranquiliser guns might initially appear to give. Some compounds, such as nicotine, have been found to have a therapeutic ratio too low for safety in the case of seals, although useful for large terrestrial mammals. This means that the ratio between the size of the necessary dose for immobilization and the fatal dose is very low. To use such a drug with complete safety the body weight of any individual must be accurately known, and this is obviously nearly always impossible in field conditions. Peterson found that phencyclidene was the most useful and safe of the central nervous system depressants, and concurred with earlier workers such as Ling and Nicholls and also Flyger, Smith, Damm and Peterson (1965) that succinylcholine chloride was the best of the peripheral nerve inhibitors, even though the effective dosage and lethal dosage were quite close together.

Similar work on the immobilization of cetaceans at sea has been strongly criticised by the American Society of Mammalogists' Marine Mammal Committee (Schevill and others, 1967). Recent physiological studies indicate that breathing is not under autonomic nervous system control in cetaceans as in humans; so that anaesthesia can quite quickly lead to suffocation. For this reason Ridgway (1965) recommended that central nervous system depressants not be used on cetaceans unless an effective method of artificial respiration can be devised. However, the same author described successful anaesthesia of bottlenosed dolphins under operating theatre conditions using the compound halothane. This was administered through a vaporizer and was found to have many advantages, especially rapid induction, a deep plane of anaesthesia compared with nitrous oxide, and rapid recovery.

Work such as this has become an essential companion study to the many behavioural and communication studies being carried out on captive dolphins, since these animals are subject to many illnesses similar to those known in other large mammals. Only in the last few years have we begun to accumulate enough knowledge of cetacean and pinniped anatomy and physiology to treat the ailments of captive animals with any degree of confidence (Brown, McIntyre, Delli-Quadri and Schroeder 1960). Dermatosis conditions are common in both captive cetaceans and seals, but especially in the latter. Apparently, lack of certain vitamins and trace elements can have great effect on normal moulting processes. Many deaths of small cetaceans can be attributed to bacterial infections such as erysipelas and pneumonia; both can sometimes be treated with antibiotics, but are often fatal illnesses.

When recovery does occur the survival of a sick dolphin can be as much a result of care by its fellow animals as by the handlers. A dolphin with any respiratory infection, growing weak, can soon lose the ability to surface in order to breathe. Animals in this condition have been observed on more than one occasion to be supported by their companions working in relays, with an animal beside and slightly below the sick individual on each side lifting it to the surface every few seconds (Caldwell and Caldwell 1966). Especially where bottlenosed dolphins are concerned it is, as a rule, a grave mistake to isolate a sick animal, even if antibiotics are being administered as part of a treatment. While it is difficult to generalize about the health problems of captive small cetaceans, the pelagic deep water species such as *Lagenorhynchus* seem to be more prone to common bacterial infections than shallow water species such as *Tursiops* (Ridgway 1965). Almost all individual cetaceans and pinnipeds prove on autopsy to have heavy internal paratitic infestations; including tapeworms and nematodes. There is a suspicion that in the wild state the parasites rarely cause the hosts inconvenience, but in captivity, especially if the diet is not balanced, parasites can become troublesome. Two compounds, thiabendazole and diphenthane 70, have been found to be very effective agents for clearing up infestations of gastro-intestinal parasites in dolphins (Ridgeway 1965).

Training of both seals and small cetaceans in captivity is best established by a repetitive stimulus such as a whistle or a distinct movement, or even a voice command. This basic stimulus can be reinforced with food as a reward for an action performed correctly. However there is no doubt that establishment of a close degree of rapport between handler and animal is equally important. For example, bottlenosed dolphins have been taught to jump from their pool onto the tiles at the side, a situation from which they cannot always extricate themselves without human assistance. More than any other experiment, this demonstrates the great confidence which a captive dolphin can come to have in its handler. A further development in the last few years has been the release of a trained animal in the wild, with subsequent voluntary return on a recall signal (Norris 1965, Norris, Baldwin and Samson 1965).

Research in the last decade on dolphins in captivity has concentrated on problems of echo-location and communication. So many papers have been produced in this area and so much progress made that it is impossible in the available space in this volume to do more than review briefly the major elements. Regarding what has been achieved, one must realise that research of this kind is fraught with problems, not all anticipated when the studies first began. We now know that certain equipment under certain circumstances can produce artificialities in the recordings easily mistaken for dolphin-produced sounds by the unwary. There are also many extraneous sounds to contend with, such as the slap of wavelets against the side of a boat or pool, the surging of water across the face of the hydrophone or microphone, and sounds made by forms of marine life other than mammals; many fish are most vocal at certain times of day. Sounds recorded at different depths may in fact be produced in the same way by the experimental animal, but sound quite different when played back because of hydrostatic pressure distortions. One also has to remember that sounds recorded by the equipment need not have originated in the water; one odd noise noted on tapes made at the Napier seaquarium was particularly baffling, until it was realised that it was the grossly distorted and muffled sound of the trainer shouting orders.

It is probably true to say that in the initial stages of research of this nature the equipment available tends to some extent to dictate what shall or shall not be done. For example while we now know a great deal about the underwater sounds produced by cetaceans and pinnipeds, we know a good deal less about their hearing capacities (Scott Johnson 1967). Fraser and Purves (1954) carried out excellent work on the anatomy of the ear in cetaceans, but there is still much unknown about the functioning of this organ in relation to the marine environment. The technical problems of determining the auditory thresholds of marine mammals are formidable, because of the difficulty of maintaining a sound field in the average holding tank and the near impossibility of judging completely beyond doubt when the experimental animal has detected the sound. Both these aspects have been discussed by Møhl (1964) and Turner (1964), and also Scott Johnson (1967), and to a large extent the above difficulties have been at least partially mastered. Unfortunately, these studies have not progressed very far yet, and the results can only be discussed in technical terms within a limited framework of reliability.

Most work on cetacean sounds has been concentrated on the smaller toothed whales, simply because these are relatively convenient experimental animals. Underwater sounds produced by these animals are undoubtedly associated with the detection of food and establishing

Figure 19.   (D. E. Gaskin.)
In initial stages of training at the Marineland of New Zealand a common dolphin, *Delphinus delphis,* learns to take fish, first from a pole, and then from the trainer's hand.

reference points in the environment, both in the strictly navigational sense and in the more complex context of communication with other animals in a school (Kellogg 1961; Norris, Prescott, Asa-Dorian and Perkins 1961; Norris 1964; Evans, Sutherland and Beil 1964; Dreher and Evans 1964; Busnel, Dziedzic and Andersen 1965; Lilly 1966; Backus and Schevill 1966; Dreher 1966; Busnel 1966; Bateson 1966; Dudok van Heel 1966; Busnel and Dziedzic 1966; Evans 1967; Norris and Evans 1967; and Dreher 1967). The above list of authors and dates is also quite indicative of the rate at which this new field of knowledge is evolving. Seals have been less intensively studied in relation to underwater sound production, although Poulter (1963) reported sonar-type activity by a sea lion, and Evans and Haugen (1963) carried out an experimental examination of the ability of *Zalophus* to echo-locate objects.

Odontocete whales are capable of producing a range of sounds; most of these appear to be based on either a series of clicks, possibly produced in the outer nasal region (Evans and Prescott 1962), or modulated or pure whistles made by the larynx (Evans and Prescott 1962; Fraser and Purves 1954). The clicks appear to be broadband in almost all cases, from low to very high frequency. Backus and Schevill (1966) analysed sounds produced by sperm whales, and found that each animal sent out regular series of clicks, each consisting of six very rapid pulses of sound. They were also able to detect significant differences in the sounds produced by individual animals, so assumed that this could represent a 'signature' helping to fix not only the position in the school but also the identity.

Interestingly enough, these clicks are the only sounds that sperm whales appear to make. It has been suggested for delphinids that the clicks are associated with echo-location and the whistles with communication; but in this case one would be forced, in support of this theory, to say that sperm whales do not communicate. Some qualitative differences in toothed whale clicks have been identified as well as the 'signatures' of individual variation within a species; the clicking rate of a sperm whale is generally considerably lower than that of a bottlenosed dolphin, which can range up to 800 pulses per second. To the human ear clicks in rapid sequences blur into single drawn-out creaking sounds. We do not know if it is the sound or modulations of it which is meaningful to a cetacean, or the high speed individual components. If the rate of clicking is changed, sounds classifiable as 'lowing', 'barks' or 'creaks' are produced.

The whistling noises, often called squeals, are quite different in nature and origin. Frequently they are narrow-band noises, sometimes pure, sometimes with harmonics (Schevill 1964). They may rise, fall, or

Figure 20. (Redrawn from *The Biology of Marine Mammals* edited by H. T. Anderson. Reproduced by courtesy of Dr K. S. Norris and Academic Press Inc. of New York.)
Paired clicks of the Pacific bottlenosed dolphin *Tursiops truncatus gilli* recorded at sea off Santa Catalina Island. Most sonar 'clicks' of this species can be resolved into paired components whose peaks are less than 1msec apart. The rate of repetition of the click train is 40-50/second.

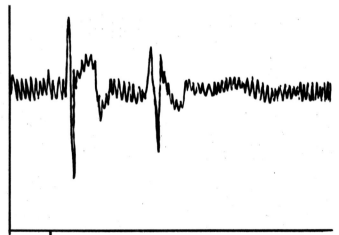

1msec

remain steady for long or short periods. A growing number of papers now contain analyses of the various cetacean sounds, together with reproductions of recorded sonograms, for example Dreher (1961), Kellogg (1961), Vincent (1959), Dziedzic (1952), Lilly and Miller (1961), Lilly (1962, 1963), Evans and Dreher (1962), Evans and Prescott (1962), Dreher and Evans (1964), Evans (1967).

Work of this nature is bound to be of limited application where captive animals are concerned unless family groups of animals can be held in captivity for long periods; only then are the dolphins likely to be in even remotely analagous situations to natural conditions. Some of the work on family groups in captivity is discussed in the section on the bottlenosed dolphin, but there is a long long way to go before we can be sure that we are beginning to elucidate a dolphin 'language', or even if such a thing exists in the terms we can define. After much work on the problems of cetacean communication, Dreher and Evans (1964) could report with a reasonable degree of certainty that a whistle with a steadily rising contour seemed to be associated with search behaviour, that a rising then falling whistle indicated disturbance, and that a whistle with a falling contour definitely indicated that the animal was very disturbed and frightened. Juvenile animals showed somewhat different whistle patterns from adults, including one which was frequently recorded, where a more or less continuous whistle fell very sharply and was abruptly cut off. In a more recent paper Dreher (1966) attempted to analyse dolphin whistles into six contour patterns; but it must be pointed out that some scientists have excellent reasons for not accepting the hypothesis that contour differences imply differences in information content. For example, a sentence could be spoken by two men in two different British dialects, one in an even tone and the other on a rising contour, such as can occur when a Welshman asks a question. The contours could be quite different; the information content exactly the same.

There is little doubt that there is great potential for future work wth captive cetaceans. However, there is a trend towards open ocean work (Norris 1965) which may ultimately yield far more knowledge, since the animals are in a natural environment. Some (Lilly 1961) are not only convinced of the existence of a dolphin language, but also that we shall soon be able to communicate with them and exchange information. However, it must be pointed out that the intelligence and problem-solving ability of dolphins is much more questionable than popular literature would have us believe.

5. *The Development of New Zealand Whaling*

One of the anomalies confronting the student of New Zealand whaling history is that the activities of the mid and late nineteenth century whalers were far more extensively documented than those of twentieth century operations. This has been partly attributed to the somewhat more romantic nature of the earlier period, and the compulsive diary-keeping habits of educated Victorians, but at the pragmatic level it is also explicable in terms of the relative size and importance of the respective operations. In the middle years of the nineteenth century the New Zealand archipelago drew whaling fleets from all over the world. At the height of the old pelagic industry more than 500 vessels worked around these islands. However, at the end of the century New Zealand became a quiet backwater as the focus shifted elsewhere.

The first certain record of a whaling ship visiting New Zealand is that of the *William and Anne,* in 1792 (Dawbin 1954). At this time the American pelagic sperm whaling industry was in the ascendency, as were smaller British, Dutch, French and Portuguese enterprises (Macy 1835, Starbuck 1878, Sanford 1884).

Politically the last decade of the eighteenth century was an unstable one. Relationships between England and the newly independent American colonies were ripe for deterioration into a new war; while the French had lost their claim to Canada the revolutionary government in Washington still had designs on the remaining British possessions in North America.

One factor which limited very early expansion of pelagic whaling in the Far East was discussed by McNab (1914). In the last years of the eighteenth century the powers of the British East India Company extended far beyond the geographical boundaries of the subcontinent; not only did the company govern mercantile trade in the Indian Ocean and the East Indies, it also controlled by permit all whaling between longitudes 51°E and 180°.

However, by 1798 England was at war with Spain, and part of the Spanish naval forces was strategically dispersed about her South American possessions. The deployment of these forces included a squadron manoeuvering within striking range of the Straits of Magellan and Cape Horn. In this year British whaleships in the Pacific were ordered into the relative safety of Australasian waters until the war was over or the threat from Spanish warships had been eliminated.

From 1798 onwards, when the East India Company granted an extension of whaling limits, sperm whaling centres grew up along the east coast of Northland, not for working up whales, but for refitting and refurbishing pelagic vessels. There was a settlement at Doubtless Bay, but activity was concentrated especially at Kororareka and Russell in the Bay of Islands. By the 1830s the former had become a temporary base for more than 100 whaling vessels, American and British, and the behaviour of their crews ashore was the despair of the small but growing population of settlers at Russell. Relationships between

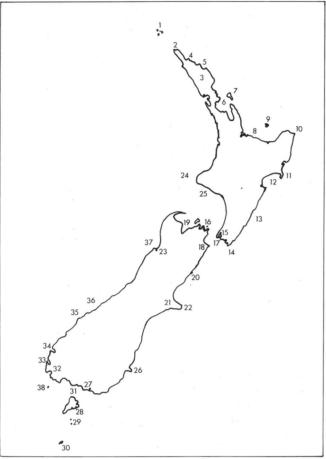

Figure 21.

New Zealand; with major localities mentioned in the text. 1. Three Kings Is. 2. North Cape. 3. Northland Peninsula. 4. Doubtless Bay. 5. Russell, Bay of Islands. 6. Hauraki Gulf. 7. Great Barrier Island. 8. Bay of Plenty and Tauranga. 9. White Island. 10. East Cape. 11. Mahia Peninsula. 12. Hawke Bay and Napier. 13. Wairarapa Coast. 14. Cape Palliser. 15. Wellington. 16. Arapawa Island, with Tory Channel Whaling Station. 17. Cook Strait. 18. Cape Campbell. 19. Nelson, and Tasman Bay. 20. Kaikoura Peninsula. 21. Bank's Peninsula. 22. Akaroa. 23. Westport. 24. Cape Egmont. 25. South Taranaki Bight, and Wanganui. 26. Otago Peninsula. 27. Bluff Harbour. 28. Stewart Island. 29. The Traps. 30. The Snares. 31. Foveaux Strait. 32. Preservation Inlet. 33. Dusky Sound. 34. Doubtful Sound. 35. Milford Sound. 36. Jackson Bay and Open Bay Islands. 37. Cape Foulwind. 38. Solander Island.

no facilities for working the animals for their oil. Despite many problems with hostile local natives he succeeded in ekeing out a bare living by taking the baleen from whales and selling it to ships calling in to Cloudy Bay and the Marlborough Sounds.

This was the beginning of the 'Bay Whaling' industry for right whales which was to dominate New Zealand inshore whaling for the next decade. Guard was first joined and then overshadowed by two men, Barrett and Thoms, who were more successful, and within three years other stations were starting up in Cloudy Bay, Jackson's Bay and Palliser Bay.

The first phase of New Zealand bay whaling was completely dominated by enterprises financed by Sydney and Hobart merchants; this was an immediate offshoot of a thriving but short-lived right whaling industry starting in Tasmanian coastal waters (Crowther 1919, Philip 1935). The phase of Australian ascendency was short; by 1836 Americans had become the most important element in bay whaling (McNab 1913). Dutch, British, French and Portuguese were also involved, the French concentrating around Bank's Peninsula.

Descriptions of New Zealand bay whaling and observations on the habits of right whales have been provided by Polack (1838, *Travels and Adventures . . . between 1831 and 1838);* Dieffenbach (1841, *New Zealand and its Native Population*); Knox (1850, *The Whale and Whaling);* Thomson (1859, *The Story of New Zealand*) Pratt (1877, *Colonial Experiences . . . of Thirty-four Years in New Zealand);* Sherrin and Wallace (1890, *Early History of New Zealand*); Jacobson (1893, *Tales of Bank's Peninsula*); McNab (1913, *Old Whaling Days*); McNab (1914, *From Tasman to Marsden*); and Morrell (1935, *New Zealand*). Unfortunately these works have become almost impossible to obtain except through large central libraries. However, a great deal of information on this period, both published and unpublished, has been assembled into a single most useful volume by L. S. Rickard (1965, *The Whaling Trade in Old New Zealand*).

During the earliest phase of bay whaling, operations were conducted from anchored vessels or suitably sheltered beaches. The whalers erected crude huts beside the array of trypots used for rendering down the blubber, and generally lived in the most primitive conditions. Later in the 1830s some of these enterprises were abandoned. Others became more permanent, with sheltered lookouts, substantial buildings, and a scaffolding and pulley arrangement called the 'shears'. This was used to roll the carcase during the 'flinching' process of removing the blubber, and also to suspend the whale and prevent it from sinking

whalers and Maoris were generally good. As the industry's association with New Zealand developed Maoris became more and more common among whaling ship crews, and it is recorded that they made exceptional 'boatsteers' (harpooners).

During 1827 Captain John Guard arrived in Tory Channel after an unsuccessful sealing expedition. Noting the abundance of right whales he abandoned his sealing and began, with much trouble, to take a few whales. He had very few men and only home-made whaling gear, and if the local topography dictated working on the carcase in water rather than on the beach. Sharp blubber spades were used to remove the oil-rich coating; the blubber was cut into strips or squares for boiling down, the method varying from one station to another.

Whales were killed by the most primitive of techniques. Since females with calves came into shallow water, the whalers would row or sail out to them and fasten on to the calf first. The mother invariably stayed with the dead or injured calf, and was in turn easy to kill. Male right whales generally stayed some distance from the coast, and were hunted by pelagic whaling vessels on the offshore grounds.

Independent operators played a part in this bay whaling industry. European interpreters or 'tonguers' (McNab 1913) with Maori boat crews would kill whales and tow the carcases to the whaling stations. In return they would be given the tongue, from which 5 to 10 barrels of oil could be obtained, and the flinched (flensed) carcase. Meat from the latter would feed a tribe for weeks. Whale meat from humpback whales is still a source of protein for native whalers in Tonga to this day.

In later years the need for the Europeans in this subcontracting enterprise became less, as most Maoris associated with the whalers learned English. Maori whale boats operated on their own, and sold whales to the Te Awaiti station in Tory Channel at £20 each (Dieffenbach 1841).

Middle-men thrived. Whalers were paid in salt, flour, sugar and rum instead of cash; this effectively bound them to the local industry and prevented them buying passages out on vessels returning to England after depositing colonists. Nearly twenty buildings were erected at a settlement in Preservation Inlet; this became an important pelagic whaleship refitting centre, especially for vessels working for sperm on the nearby Solander Ground off Stewart Island. Vessels working around the Chatham Islands, which were themselves an important sperm whaling centre, often called in to Cloudy Bay, another important South Island whaling centre. Kapiti Island, the Mahia Peninsula and Palliser Bay became relatively important North Island centres of whaling activity (Rickard 1965).

Right whaling reached a peak in 1839 and declined rapidly after that. Webb (1871) remarked that the practice of killing calves appeared to be closely related to the decline of the right whale population in New Zealand waters. By 1870 the bay whaling industry was only of very minor importance (Hocken 1871).

Bay whaling as a major industry really came to an end in New Zealand by the middle years of the 1840s; yet Maori boat crews still operated out of six localities in the Bay of Islands until 1900, from the Mahia Peninsula until 1910, and from Te Kaha until as late as 1930. These men hunted humpbacks using traditional long boats and hand irons, and worked up the animals using spades, shears and trypots. All the paraphernalia of the nineteenth century whaling industry remained in use in this anachronistic relic enterprise more than sixty years after it had been abandoned in the rest of the world. However it is important to note that whaling was only a sideline to the Maori families involved; this was a profitable seasonal operation to supplement income from farming and other sources. Nevertheless, with the exception of this very small-scale industry, New Zealand whaling was of no importance by 1870.

It is tempting to separate New Zealand whaling since 1798 into chronological phases of species exploitation; sperm whale, right whale, humpback whale and sperm whale again. However, for a number of reasons the result is not very satisfactory. The intensive sperm whale phase in the New Zealand region lasted from 1798 to approximately 1875. Bay whaling for right whales was a major industry only from 1830 to 1845, being completely overlapped by the sperm whaling industry. Even during the 15 peak years of bay whaling much mixed whaling was common in the area, with pelagic vessels taking right whales and sperm whales according to season. Both phases were almost completely over by the end of the 1870s, not only in New Zealand waters but in other parts of the world as well (Andrews 1911, Dunbabin 1925, Dakin 1934, Dorsett 1954).

While the origins of humpback whale exploitation from small but permanent shore stations on the New Zealand coast can be traced back to 1890 it is best to discuss first the brief but distinct 'Norwegian' phase, even if out of chronological sequence by about twenty years.

This attempt to set up Norwegian controlled operations represented an unsuccessful offshoot of a new northern hemisphere based, steam-powered whaling industry which was expanding its activities at considerable profit in the Falkland Islands Dependencies as well as the North Atlantic and the Arctic Ocean. The history of the abortive western South Pacific enterprises was detailed by Risting (1922).

In 1911 the 'Laboremus' company of southern Norway fitted out the sealing vessel *Mimosa* as a small floating factory ship, and at the end of the year sent her to Tasmania with a single whale chaser to hunt rorquals, humpbacks, and right whales if they could be found. The expedition had no success in Tasmanian waters, and little more near New Zealand.

A second attempt to exploit the resources of this region was made by the 'New Zealand Whaling Company' of Larvik · the following season, which despatched the floating factory *Rakiura* with four steel chasers to the western South Pacific. This expedition, together with a chartered vessel the *Prince George,* worked in the Bay of Islands between July and October 1912 mainly taking humpbacks (Lillie 1915). The *Rakiura* and her chasers searched all round the archipelago, even venturing down into Antarctic latitudes. The expedition was not a success and returned home by 1913.

The third and last foray by a foreign enterprise was financed by the 'Australia' company, which sent the small *Loch Tay* with two steel chasers in a search for whales between the east coast of Australia and Campbell Island. This expedition had no more success than its predecessors, and between January and April 1913 restricted its activities to a small sperm whale fishery based at Bluff in Southland. The operation was saved from financial ruin only by a large ambergris find (Dakin 1934).

In 1890 Mr H. F. Cook began humpback whaling at Whangamumu in Northland, most of the station staff being Maoris. Between 1890 and 1909 the whales were captured in a unique fashion (Ommaney 1933). Humpbacks migrating past the Northland coast frequently moved very close inshore. Cook and his men ran a stout cable across a 50 yard wide neck of the Whangamumu Channel and suspended from this a series of large coarse nets, formerly made of rope and later of steel links. As soon as a humpback became entangled in a net the whalers rowed in close and killed it with hand lances.

The Whangamumu operations were modernised in 1910 when Cook purchased a steam chaser and put this into service. This chaser allowed the station to take from 27 to 74 humpbacks each year from the northern migration peak in May-July and the southern migration in October, for 20 years. After the depression of the 1930s, the station did not reopen, operating for the last time in 1931 (see Appendices, table 1). Two motor launches supplemented the steam chaser for a short time after 1930. Operations at Whangamumu have been described by Lillie (1915), Sleeman (1921) and Ommaney (1933). In 1911-12 Cook sent the steam chaser down to Campbell Island to investigate persistent stories of large seasonal concentrations of right whales there. While a description of this particular enterprise has not been published there is mention in a diary of another small station set up on Campbell Island at the same

time (*Diary of Two Whaling Seasons* 1911, 1912; Dominion Museum unpublished achives), of a finback taken by 'Cook's Station'.

A small station operated for right and humpback whales at Kaikoura between 1917 and 1922 (Dawbin 1956a, and unpublished Marine Department records), never taking more than 20 animals a year. This station was never a great success, partly because of the poor anchorage at South Bay, and partly because finance was lacking.

Mr J. Perano began a station at Tory Channel, based on Te Awaiti and Fishing Bay, in 1909, opening what was to be New Zealand's longest-lived shore whaling station, operating without a break from 1915 to 1964.

Between late April and early August humpback whales pass through Cook Strait on a northward migration to their breeding grounds in the tropics (Dawbin 1956a). To exploit this resource the Perano family first used traditional methods and gear, including hand lances. However, in 1915 Mr Perano introduced three fast 34 ft motor launches, each capable of 30 to 40 knots for extended periods. These launches had a light whale gun in the bows with a 1¼ in. bore, firing a 14 lb harpoon. The harpoons carried a small grenade, which usually only stunned the whale. Compressed air was then pumped into the thorax to prevent the whale sinking. The *coup de grâce* was administered by 1½ lb of gelignite in the hollow iron head of a long lance; the charge being detonated electrically. In calm or moderate seas these vessels were able to range as far afield as Cape Terawhiti on the Wellington side of Cook Strait. Humpbacks coming up the coast of the South Island towards the Tory Channel entrance were spotted from a sheltered lookout on the headland opposite the whaling station. In the early days another Perano brother had a small station on this side of Tory Channel, with considerable rivalry resulting. The Perano brothers Joseph and Gilbert who took over control of the main station at the death of their father, continued to run the company until its final close down in 1964, although the humpback phase ended in 1962.

Between 1956 and 1962 a series of companies operated a whaling station on Great Barrier Island in the Hauraki Gulf. This station took from 29 to 135 humpbacks and Bryde's whales each year, except in its final disastrous year 1962 when only eight were captured.

The growth of the factory ship industry in the Southern Ocean again after its disastrous losses in the Second World War, and the corresponding growth of the Australian humpback shore whaling industry, together contributed to a gross over-taxing of the species in Antarctic Areas IV and V. The sum total catches of Australian, New Zealand and pelagic operations annually exceeded the recruitment of young animals into the breeding population by a considerable margin. By 1961 the humpback population in Area V was collapsing. In 1960 the Tory Channel company reached an all-time peak of 226 humpbacks; the same year in which Great Barrier took its greatest total of 135 animals. In 1961 the Tory Channel catch slumped to 55, the lowest total since 1937, and Great Barrier took only 26. In 1962 the Tory Channel catch was down to 27, and that of Great Barrier to only 8. The latter company then went into liquidation.

In late 1962 the Perano brothers carried out a survey from a number of points along the Cook Strait coasts and decided that there might be enough sperm whales moving off the east coast of New Zealand to support their industry's continuation providing they could develop deep-sea capability. To this end they negotiated the purchase of the steam chaser *Orca* of 156 tons, from an Australian company, and approached the Marine Department with a request for a scientific survey of the sperm whale population size, composition and movements in the Cook Strait region. The author was appointed to carry out this task in October 1962.

Until 1963 sperm whales formed an insignificant fraction of the catch taken by shore whaling stations in New Zealand. None were ever taken by Whangamumu, according to Marine Department records, or Great Barrier, and Tory Channel whalers took only 17 between 1915 and 1962; one in 1938, two in 1940, nine in 1947, two in 1957 and three in 1962. This dearth of catches represents lack of interest rather than rarity, and that the main concentrations to be found off the east coast of New Zealand lay well beyond the normal operating range of any of the stations. It is probable that the sperm whale is, and always has been, one of the commonest whales of commercial size in this part of the South Pacific.

The Marine Department sperm whale survey ran from October 1962 until December 1964. The Royal New Zealand Air Force and the Civil Aviation authority provided planes for whale survey work from November 1962 onwards in the Cook Strait region, and auxiliary spotting reports were supplied by pilots and navigators of the National Airways authority. Lighthouse keepers were requested to keep regular lookouts at selected points around the coast, and selected coastal, island and trans-Tasman ships also returned regular sighting logsheets. The weather station personnel on Campbell Island and

Figure 22.　(M. W. Cawthorn, Fisheries Research Board of Canada.)

The Tory Channel whaling station on Arapawa Island during the 1964 whaling season. Between the station and the whale chaser *Orca* can be seen one of the tiny motorboats used, prior to 1963, for catching humpbacks. Cook Strait lies beyond the channel entrance, and further away still is the hilly country to the southwest of Wellington.

the Kermadec Islands also participated in this scheme. The United States Navy offered facilities for observers on their weather station picket vessels steaming through the summer months from Dunedin to latitude 60°S and back.

However, the basic 'legwork' was carried out by the three launches *Paea, Manga* and *Mako* of the Fisheries Protection Squadron of the Royal New Zealand Navy. In the first phase of the survey these vessels were commanded by Lt. John Beauchamp, Senior Officer, Lt. S. F. Teagle and Lt. F. Arnott, and in the latter phase by Lt. Cmdr. D. Davies, Senior Officer, Lt. Teagle and Lt. L. Merton. The first cruise using these vessels began on 14th February 1963, and the survey continued until 18th November 1963; eight cruises being made in all, with a total of 165 days operations (Gaskin 1963, 1964a). Part of the survey duties of these vessels included the task of marking as many whales as possible with stainless steel 'Discovery' marks fired from shotguns. It was necessary to fire these at a range of not more than 90 ft in order to achieve complete penetration through sperm whale blubber. The tips of these marks were coated with

33

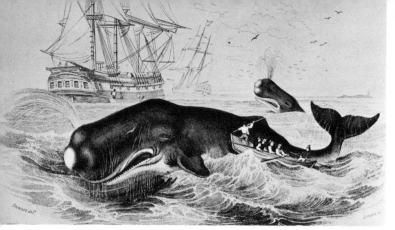

Figure 23. (Alexander Turnbull Library, Wellington.)
A lithograph of sperm whaling around the New Zealand coast towards the end of the eighteenth century.

Figure 24. (Royal New Zealand Air Force.)
The Russian whaling factory ship *Slava* (right), towing a group of fin and sei whales, meets the supply tanker *Kherson* in Pegasus Bay, New Zealand, during March 1964.

penicillin to minimize the chances of infection of the small wound left in the skin. A brief report on the objectives of the New Zealand whale survey was given by the author in an earlier publication (Gaskin 1965).

The *Orca* began operations in 28 April 1963, taking two sperm whales not far from the mouth of Wellington Harbour. The season was not a good one for the company. They began operations still in the hope that they would be able to take humpbacks as well as sperm whales; and it soon became evident that the winter months were the time when the density of sperm whales on the whaling grounds was at its lowest (Gaskin 1968b). The humpback migration showed no miraculous recovery of numbers; in fact only 11 were seen of which 9 were taken. The season lasted for eight months, until 11 December 1963, when the chaser had to suspend operations a few weeks early because of engine trouble. The total catch for the period consisted of 114 sperm whales, nine humpbacks and one fin whale.

The Marine Department then granted permission for a short summer season, in order to obtain detailed biological observations from catches on the whaling grounds throughout this crucial period. The summer season opened on 11 January 1964 and lasted until 17 April 1964. Catches were much better, the company closing operations in the autumn with a total of 77 sperm whales. It was decided to open catching again in October 1964, for another eight-month season. However, in 1964 two other factors intervened. The morale of the whaling company staff was greatly depressed by the appearance as early as March 1964 of the Soviet factory ship *Slava* with 15 chasers on the southern part of the Cook Strait whaling grounds. This expedition in fact took mainly sei whales from a migration stream coming up the coast unusually close inshore, but a number of sperm whales were also caught. The international catch statistics, when available for that year, also showed a vast increase in sperm whale catches in the South Pacific by the Japanese factory ship *Nisshin Maru No. 3*, which had taken 1,219 compared with only 100 in the previous season. The low

oil yield, only four tons per animal, and the statistically incredible number of whales recorded for the minimum size limit indicated that many under-sized animals were being taken. Previously the 38 ft minimum length regulation for capture of sperm whales by pelagic vessels and the fact that Tory Channel was a cold-water whaling station taking mainly male sperm whales, accorded the species a large measure of protection in the region.

However, a second factor proved to be more serious. When sperm whaling operations began in April the price of sperm whale oil on the international market was in the region of £85 per ton. During the period of operation this price began to slump steadily downwards, until in October 1964 it was less than £60 per ton. The *Orca* came out for her third season on 19 October 1964, and worked until the Christmas recess on 23 December, taking 57 sperm whales and 4 sei. By December the oil price had sunk to an all-time low of barely £40 per ton. Over the Christmas vacation the directors of the company held urgent conferences with their staff and with officials of the Treasury and the Marine Department. The Government decided with great reluctance that under the circumstances it could no longer assist the company, which had come to the end of its liquid assets. These had earlier been run down to a dangerously low level by the costs of modernising the operations. Shortly after Christmas 1964 the directors of the company announced to the press that they would not be operating in 1965, and that the chaser, the plant and other assets would be put on the market. So New Zealand whaling finally came to an end on 23 December 1964, after a final frantic burst of activity. Today, Tory Channel whaling station is a tourist's curiosity, to be seen from the inter-island ferries as they commute between Wellington and Picton. It looks surprisingly intact from the outside, but, in actuality, is a gutted shell. Ironically, a few months after the final decision was made to close down, the world market price of sperm whale oil began to rise again towards its previous high level.

6. The Development of World Whaling

To understand fully the factors responsible for the dramatic shifts of emphasis in New Zealand whaling throughout its history, it is really necessary to be familiar with the changes of much greater magnitude which have come about in world whaling since the pursuit of cetaceans first became an industry.

Coastal whaling has been carried out by the Japanese for centuries (Fraser 1937, Sanderson 1956, Omura 1953), but only in recent years have their activities become significant on a world scale. In the West the whaling industry really began in the Bay of Biscay, where at one time the black or northern right whale, *Eubalaena glacialis*, was seasonally found in great numbers. This is not to say that the Basques of the Bayonne district and the nearby towns of Biarritz and St Sebastian were the first Europeans to hunt whales, but they are the first people of whom we have records who made whaling into something which justified the name 'industry' in the modern sense of the word. However, there is little doubt that as long ago as 1100 B.C. the Phoenicians operated shore-based whaling for sperm whales in the eastern Mediterranean (Sanderson 1956).

Altogether about a dozen coastal settlements were involved in coastal whaling in the Bay of Biscay, each killing perhaps less than a dozen whales each year. There is no record at all of when they began to whale; Slijper (1962) suggested that they learned the art from the essentially barbarian Norsemen and coupled it with more southern European mercantile instincts. Sanderson (1956) traced the origins of Basque whaling through old documents; the earliest record appears to predate A.D. 1150, by which time whaling was already a thriving enterprise all along the Basque coast. The Basque towns were active in whaling from the twelfth to the sixteenth century, but late in the thirteenth century they began pelagic whaling using three-masted caravels; these vessels developed in size and complexity during the long period of Basque whaling ascendency, although the actual whaling was carried out from small whale boats, called shallops. In the latter years of the industry the invention of the try-works by Captain Francois Sopite (c. 1600) allowed the animals to be completely processed for their blubber at sea while still fresh. While the Basque whaling industry was purely coastal in the early years, it is known that the Basques reached Newfoundland as early as 1538, and perhaps even earlier. Sanderson noted that the Carta Catalan de Mecia de Viladestes of 1413 in the National Library of Paris showed a basque caravel, with a whale, in a position far to the northwest of Iceland.

Early in the seventeenth century the Basques lost their whaling supremacy to the Dutch and the English; by this time they were only a minor power in a very rapidly expanding Europe, with England, Holland, Spain and Portugal struggling for military and mercantile supremacy. The Dutch discovered Spitsbergen in 1596, which proved to be one of the richest whaling grounds yet found, and by the early years of the 1600s the Dutch and English (with the aid of a number of renegade Basque whalers) were able to exclude Basque enterprises not only from these waters, but also from Newfoundland, which had just been taken by the British.

Whaling at Spitsbergen went on spasmodically. The Muscovy Company of England, under several names, sent the *Lionesse* and *Amitie* to the island in 1610, and these had some success in Deere Sound. In the following year the *Elizabeth* and *Mary Margaret* followed; the latter was sunk on the way back to England. In 1612, six vessels, four of them British, worked around the island taking 17 whales at a substantial profit. In 1613 the number of vessels working Spitsbergen waters rose to 27 and continued to increase for some years. The Dutch built a small fort on Spitsbergen to give naval protection to their vessels going home to Amsterdam and to curtail the activities of British and French privateers. After 1625 the whaling at Spitsbergen was almost entirely a Dutch affair with the object of the hunt being the Greenland right whale, *Eubalaena mysticetus*. Even by 1621 there were 52 Dutch vessels working in these waters, ranging farther north to the ice limits in summer, since the whales were becoming rarer close to Spitsbergen. In the last years of the 1630s the Germans of the then Hanseatic League began to show an interest in Spitsbergen as a whaling base. During the rest of that century they sent dozens of ships northwards.

The Spitsbergen industry, especially the Dutch section, had a series of setbacks in the latter half of the seventeenth century, culminating in virtual destruction during the second English-Dutch war of 1665-67. However, by the time the third such war broke out in 1672, the Dutch were supreme again, and they dominated northern whaling from this time on until 1799 when the fleet was annihilated in the new wave of strife that was later to be known as the Napoleonic Wars.

Arctic whaling continued in a number of forms until about 1910, when the last Dundee whaler is reported to have come back 'clean' from the Davis Strait region. British whaling ships found their way into Davis Strait in the late spring of 1773, though the Dutch had been working there for many years before this. German activity was also strong. While Dundee and Aberdeen became the major Scottish whaling ports, Kingston-upon-Hull became the major English centre, with a fleet of 64 ships by 1816. Hull also produced the first steam whaler, which went into the northern ice in 1857. By 1873 the whole Dundee whaling fleet was steam-powered, for the whalers were quick to see the advantages it could give them in the northern ice-bound regions.

By the 1870s it was becoming harder and harder to find either Biscayan right whales or Greenland right whales, and in the 1880s the Scottish whalers took to hunting, first, white whales (belugas) and then bottlenose whales. The hunting of the latter species began as a shore-based operation in the Faeroes, Orkneys, Shetlands and the Hebrides. The Norwegians were, by this time, beginning to assert themselves in northern whaling, and brought the Scottish monopoly to an end by

Figure 25.   (M. V. Brewington, Director, Kendall **Whaling Museum**, Sharon, Massachusetts.)
Attacking a right whale and 'cutting in', Currier & Ives, New York.

fitting out ships specially to take bottlenoses. By 1888 there were over 60 ships taking part in the industry, killing several thousand bottlenoses each year. However, it is important to note that both these species were taken only after the right whales were reduced to very small numbers, and white whaling and bottlenose whaling can be regarded as phases of the same industry.

The history of the Basque whale fisheries has been documented by Markham (1881) and Jenkins (1921); the Arctic whale fisheries from the English point of view have been described by Scoresby (1820, 1849), Lubbock (1937) and Sheppard (1911), the latter concentrating on the evolution of the whaling industry at Kingston-upon-Hull. Munroe (1853) provided details of catches in the Arctic fishery from 1772 to 1852, and Oesau (1955) gave a complete account of the operations of German whalers in the Arctic from the times of the Hanseatic League onwards to the steam whaling days. The activities of the Russians in early Arctic whaling were described by Weberman (1914), while Kuekenthal (1886) discussed the development of the white whale fishery. The development and decline of Scottish whaling has been considered by Southwell (1904), Haldane (1905, 1908, 1910), D'Arcy Thompson (1918), Harmer (1928) and Pyper (1929), while the history of the smaller Irish whale fishery was traced by Scharff (1910).

The whaling industry of the New England coast grew from co-operative ventures between local Amerindians and white colonists. The Indians had, in fact, been hunting right whales in small numbers long before Europeans came to that part of America. This shore whaling enterprise thrived between 1625 and 1720. However, the sperm whaling industry of New England did not begin until one of these animals was taken more or less by accident by Cpt. Christopher Hussey of Nantucket in 1712.

The build-up of the Nantucket whaling fleets was not a planned sperm whaling venture, though by 1730 there were over 25 ocean-going vessels involved in northern and tropical whaling. New Bedford came late to deep-sea whaling, but by 1775 had nearly a hundred vessels devoted to whale hunting. Almost all the ports along the New England coast became major or minor whaling centres; by 1775 the American whaling fleet totalled about 350 ships. At this time the sperm whalers had penetrated the tropics and were working in the South Atlantic off the coast of Africa.

When the War of Independence came, the whaling fleets were ruthlessly hunted down by the British as part of an attempt to smash the economies of the rebellious colonies. In this they were very nearly successful. Even after the war, when the fleets were rebuilt (over a hundred whaling ships were in service by 1789, most of them new), the Americans still had to face economic warfare by the militarily defeated British which prevented them selling whale oil or other whale products outside the boundaries of the new United States of America. The industry was well on the way to a new peak of good fortune when the naval war of 1812 broke out, and this time the British, and the requirements of the Americans for sea transport for troops and supplies, almost completely destroyed the painfully rebuilt whaling fleets of New England. Yet before this time, the Americans had penetrated round Cape Horn and the Cape of Good Hope, and were active in the South Pacific and the Indian Ocean.

While American fortunes were temporarily on the wane, those of British whaling were improving as American competition was removed from the high seas. The financial returns of companies involved in the industry were good, in contrast to the disastrous history of British whaling in the Arctic. Ships which had been ferrying convicts to the new Australian penal colonies, began to exploit the sperm whale resources of the Tasman Sea and the waters around New Zealand. These vessels were soon joined, in the 1790s, by professional whaling ships in larger and larger numbers, ignoring the edict of the East India Company prohibiting whaling in the region without company permits. Since these

whalers could hardly avoid seeing the large numbers of southern right whales in the coastal areas in the winter, the evolution of bay whaling described in the last chapter was a perfectly natural process, with the ships and men needed to exploit the population already being on the spot. Bay whaling did not begin in New Zealand; the first records are of an enterprise in the Derwent, Tasmania, in 1803 or 1804, and the industry spread from there to Victoria and New South Wales, and then to New Zealand.

The second war between England and the United States came to an end with the Treaty of Ghent late in 1814, and the New England whalers once again set about rebuilding the shattered fleet. This time New Bedford, not Nantucket, became the whaling centre. Over the years the sand bar across the entrance to Nantucket harbour had become progressively worse, and by this time it was not possible for large vessels to get into the harbour.

American whalers had first edged into the eastern Pacific by 1791; by 1814 they reached the vicinity of the Galapagos Islands, and by 1820 the first vessels were working the grounds off the coast of Japan. In 1830 the number of New England ships engaged in this new phase of sperm whaling had risen to more than 400, and in the industry's peak year of 1846, 736 ships were operating, most of them after sperm whales in the Pacific. A recession in the late 1850s caused about 100 vessels to be laid up or put to other uses, and during the Civil War over half the whaling fleet was lost through the action of Confederate raiders or conversion to cargo vessels for the war effort.

The final decline of the New England sperm whale industry is generally taken to date from the production of mineral oil for lighting and heating purposes in 1859. However, a number of other factors were involved as well. In 1866 only 270 ships were whaling, and most of this rapid initial decline can be accounted for by the Civil War. Also, the United States was expanding westward at this time. This, coupled with the unprecedented growth of new industry during the war, provided many safer investment propositions than whaling, always an uncertain venture even at the best of times. As a result fewer new people were tempted into the industry, and when aging vessels were lost at sea or withdrawn from service they tended not to be replaced. A large part of the Pacific fleet was trapped and smashed in the winter ice off Alaska in 1870, reducing the fleet to a total of only 170 ships. At this time much of the fleet had turned its attention from sperm whales to the stocks

Figure 26. (James Johnson, Chester, Nova Scotia.)
Death of a 47ft fin whale *Balaenoptera physalus* on the Sable Island bank off the coast of Nova Scotia in 1967. Photographed from the M.V. *Polar Fish.*

the catastrophe of 1870, the American companies began to review the use of steam vessels, and since transportation by rail was now possible right across the U.S.A., the focus of American whaling shifted to San Francisco, which was used as a wintering port for ships working in the Arctic. The North Pacific fleet in the 1880s still hunted bowheads, at this time pursuing them deeper and deeper into Arctic latitudes. In the 1890s about 20 steam whalers went into the Bering, Wrangel, Siberian and Beaufort Seas after the diminishing bowhead population. This fleet, though dwindling steadily in size, continued to operate throughout the 1890s and 1900s. However, the hunt became progressively more fruitless after the turn of the century. Gradually, the whaling companies, influenced by new techniques developed in Norway for the capture of the fast-swimming rorquals, turned away from the Arctic. Their operations metamorphosed into relatively financially stable, small-scale shore whaling stations (Woollen 1921, Starks 1922), not unlike those of Australia and New Zealand.

The Norwegians without doubt, laid all the major foundations for the last great phase of whaling; the Antarctic rorqual industry (Bettum 1958). In this respect Svend Føyn is often credited with the invention of the first harpoon-firing gun; in fact, primitive versions, frequently as lethal to the firer as the whale, were tried as early as the 1730s. Nevertheless, Føyn, born at Nøtterøy near Husvik, certainly invented the first efficient model, and one that could be used to kill large rorquals as well as slow moving right whales. He was also the first to design a grenade head for the harpoon, capable of killing even the blue whale. The explosion of this device not only fired a charge of shrapnel into the whale's body, but also splayed out a set of long flanges on the head to hold the harpoon firmly in the wound.

Føyn also built the first modern-era whale chaser, the *Spes et Fides,* at Nylands Verksted in 1864, and obtained from the Norwegian government the sole rights to use his inventions on the Norwegian coast for a period of 10 years. The next decade was devoted to very successful whaling on the coast of Finmark, where Føyn recouped the money spent in perfecting his inventions, and made a handsome profit for his company. When his licenced monopoly expired in about 1874 others were quick to open competitive enterprises. By 1886 there were no less than 19 whaling stations on the coast of Norway, operating a total of 35 chasers (Kuekenthal 1890, Isachsen 1927), and catching sei, fin and blue whales.

of bowhead right whales in the Arctic regions of the North Pacific. These had been discovered by vessels exploring into high latitudes in the ever-extending search for new sperm whale concentrations. After the Alaskan disaster the American whaling industry continued to decrease in size, and the sperm whaling phase was, to all intents and purposes, over, even though a handful of ships with motley crews continued as late as 1924 to take sperm whales. The story of the varied fortunes of New England whaling has been related many times, but perhaps the best definitive accounts have been given by Macey (1835), Starbuck (1878), Sanford (1884); Jenkins (1921), and Matthews (1968).

Earlier in this chapter mention was made of the first steam whaler leaving Hull for the Arctic in 1857. After

The Finmark industry began to decline in the 1890s, apparently because the exploitation of the whales in the region was exceeding the recruitment rate of each species. However, the whalers were also having serious troubles with the coastal fishing industry. Fishermen blamed the whalers for ruining coastal fishing grounds with the refuse and filth that are the inevitable offshoot of shore whaling, and for driving fish away from the offshore grounds by the disturbance of chasing and killing. While it is unlikely that there is any truth in the latter argument, the controversy raged for a number of years up to the highest levels of Norwegian Government. In a number of places there were serious civil disturbances as whalers and fishermen fought in the streets. In 1904 the government acted by banning the processing of whales on the coasts of Troms, Finmark and Nordland Provinces, and the catching of whales in coastal waters.

While this ended completely the major phase of Norwegian coastal whaling it had much less impact on the financial fortunes of the companies than might be expected. By this time the Norwegians had opened up whaling enterprises on all the coasts bordering the Atlantic. Coastal whaling began in Iceland in 1888, reaching a peak in 1902 when no less than 30 chasers were operating; in the Faeroes in 1892; in the Hebrides and Shetlands in 1895; on the coast of Newfoundland in 1897; Greenland in the summer of 1900; Spitsbergen in 1904 (Rabot 1919), and northwestern Ireland in 1907. This great coastal phase in the northern Atlantic regions was very short-lived. The climax was reached between 1905 and 1910, when a total of 29 whaling companies operated 63 chasers. It declined very rapidly through 1911 and 1912, and was almost completely ended by 1915. The early years of Norwegian coastal whaling have been documented in great detail by Juel (1888), the middle phase of expansion by Southwell (1905), and the declining years by Fairford (1916).

Probably influenced both by the troubles with the fishing industry and the decline in whale numbers in the coastal regions, Cpt. Chr. Christensen came to believe that the real future of rorqual hunting lay in pelagic whaling. To this end he had the 500 ton *Telegraf* fitted out in the summer of 1903 with on-board rendering equipment and sent her to whale with two chasers in Spitsbergen waters in 1903 and 1904. In the autumn of 1903 the Framnaes Company of Sandefjord fitted out the larger *Admiralen* of 1500 tons and despatched her also to Spitsbergen to whale in 1904. At this time it is true to say that the Norwegian industry

Figure 27. (D. E. Gaskin.)
Leith Harbour, South Georgia in 1961. The last operational British shore whaling station in the Antarctic.

was in a state of great excitement; the more far-sighted executives could see that they were entering a new era. The equipment was now available to take rorquals, and they began to turn their eyes southwards, away from the traditional areas, to completely unexploited regions.

First, sealing crews, and later Sir James Clark Ross, returning from his 1839-43 cruise in Antarctic waters, came back to Europe with stories of vast numbers of whales in the Antarctic regions. However, until the end of the nineteenth century no one could see how this resource could be harvested economically in practice. Following a long lull in activity in these southern waters after the decline of the old sealing industry, four sealers of the Tay Whale Fishing Company of Scotland went to the Antarctic between 1891 and 1893. They were joined by a German vessel from Hamburg in 1892, and then by two more in 1893. None of these vessels took whales themselves, but all confirmed that large numbers of rorquals were seen.

In 1892 the Sandefjord whaler *Jason* was sent on an expedition under the command of Capt. C. A. Larsen to assess the southern whaling potential in the waters east of the Antarctic Peninsula. Here Ross had reported seeing large numbers of right whales during his cruise. The *Jason* found very few rights, but many rorquals,

39

Figure 28.   (D. E. Gaskin.)
Whale chaser *Southern Briar* at Leith Harbour, South Georgia, prior to start of 1961-62 Antarctic whaling season.

and Larsen concluded that Ross had confused their identification. Three steam whalers, each of only about 80 ft overall length and a net tonnage of less than 100 tons, were used to follow up these initial explorations; the *Jason* operated again in the (northern) autumn of 1893, this time in company with the *Hertha* and the *Castor*.

During 1894, Føyn, still active though in his eighty-fourth and last year of life, sent Capt. Christensen southwards with the research vessel *Antarctic;* at this time the companies were still hoping for the discovery of good stocks of right whales. This vessel, carrying an expedition led by Otto Nordenskjoeld, sailed to the South Shetlands and South Georgia between 1901 and 1903 when it was lost in the ice, though the crew were saved. In 1894 (Kristensen 1896), she had explored the Ross Sea.

Christensen, now manager of the Oernen whaling company, despatched two chasers from this company to join the Framnaes *Admiralen* in operations on the Falkland Islands and South Shetland Islands whaling grounds in 1905. This expedition returned to Sandefjord with a full catch.

A very important development in the opening up of the southern whale fisheries came in 1904-05, when Captain C. A. Larsen became manager of the Com-

pania Argentina de Pesca in Buenos Aires, backed by Argentinian capital. This enterprise began to operate on South Georgia for whales and elephant seals using Norwegian personnel and equipment. The remaining years before the First World War saw a tremendous acceleration in the exploitation rate of the southern whale fisheries in the Falkland Islands Dependencies.

Within a few years the British government moved to legislate controls for foreign whaling enterprises in the Falklands. A first attempt to enforce a royalty on each whale taken was a failure. In 1908, seven licences were opened for companies to work at South Georgia, and eight for the South Shetland Islands. All these were taken up by 1911, and in fact two more were offered for the South Shetlands in that year, bringing the total to 17. All permits were issued for an initial 21-year period, to be renewed at the British government's discretion for five-year periods after that. Each allowed a company to operate one floating factory and three whale chasers.

The South Georgia operations consisted of four Norwegian, one Argentinian and two British companies, while those at the South Shetlands included seven Norwegian and three British. In or about 1912 a few small-scale Norwegian operations were opened on the South Orkneys, although they had considerable difficulty with the ice conditions there, and frequent fog. This early phase of factory-ship whaling in the Southern Ocean has been discussed in detail by Salvesen (1912) and Nippgen (1921). The successes in the Falklands tempted the Norwegians eastwards to New Zealand in the abortive enterprises described in the previous chapter.

One Norwegian not satisfied with the licence system operated in the Falklands was Cpt. C. A. Larsen. As a result he initiated the first genuine pelagic factory-ship operation in the south. He assembled a fleet of five whale chasers and the factory ship *Sir James Clark Ross* at Hobart, and after a perilous passage through the sea ice off Victoria Land, moved into the Ross Sea to take whales. It was unfortunate for this expedition, in some ways, that Larsen chose the Ross Sea. The population of rorquals there has probably always been rather small in comparison with other sectors of the Southern Ocean. Economically the expedition was a failure. Not enough whales could be found to keep the factory operating all the time, and flensing was still conducted outboard — in the water beside the factory ship as in the sheltered Falkland operations. Besides being dangerous, this method was time-consuming and often wasteful.

At this time the whole whaling industry was approaching a point of crisis. Companies now found themselves

40

involved with expenditures greatly in excess of those of the earlier northern hemisphere enterprises, and the profit margin was hardly enough to attract a speculator. The humpback whale, which had been a profitable species in the first phase of operations at the Falklands, had declined rapidly in numbers after 1911 because of over-exploitation. Likewise, the number of rorquals within easy reach of the islands was also starting to diminish. One floating factory left the south to work with three chasers after humpbacks on the coast of West Africa, but the real solution for the southern industry appeared to lie in truly pelagic operations, even though the results of the *Ross* expedition had not been very encouraging.

In 1924, whale gunner Petter Sørlle and Naval Architect C. F. Christensen co-operated in the design and manufacture of a stern slipway for the 12,000 ton *Lancing* which would enable whales to be flensed on the deck, and free her from having to seek sheltered bays and ice coves. The invention triggered off the most tremendous upsurge of activity in southern whaling, and all within a very few years of the *Lancing's* first season in 1924-25. Many vessels between 12,000 and 22,000 tons came into the open Southern Ocean, often with as many as seven chasers each. In 1925-26, two pelagic expeditions operated; in 1926-27, three with 15 chasers worked in the Ross Sea alone, although not with conspicuous success for the reason given earlier. In 1928-29 the number of pelagic factories had risen to eight, and in 1930-31 a total of 41 factory ships were criss-crossing the Southern Ocean, a number never equalled or surpassed since. In the 1929-30 season the new *Kosmos* became the first expedition to use light aircraft to search for whale concentrations. While all the factories returned to Europe in the winter, the chasers were wintered at Stewart Island, Hobart, Cape Town, Montevideo and South Georgia. The great phase of expansion of pelagic whaling in the southern hemisphere has been documented by Harmer (1929), Townsend (1930), Bennett (1931), Jenkins (1932), Brooks (1936), Bettum (1958), Johnsen (1960) and Mackintosh (1965).

It is ironic that at the very time the southern industry had mastered its technical problems and had moved to a climax of operation, a sudden external brake was applied in the form of the great depression of the 1930s. The price of whale oil on the world market slumped to a record low, and it became a matter of great urgency to restrict operations to save the financial structure of the industry. As a result almost all the expeditions stayed home in the 1931-32 season, causing a great unemployment problem in the British and Norwegian whaling ports,

Figure 29. (D. E. Gaskin.)
A buoy boat brings more fin whales to the stern of the British factory ship *Southern Venturer* after a heavy day's catching in the Southern Ocean, 1961-62 whaling season. The 'balloon' is a whale's tongue which has become inflated with compressed air.

Figure 30. (D. E. Gaskin.)
The grab goes down the stern ramp of the factory ship *Southern Venturer* to pull a fin whale from the Southern Ocean near Bouvet Island, 1961.

but at least giving the whales a brief respite. In the following year the situation eased, and expeditions began to move to sea again.

Japan entered pelagic whaling in 1934-35, and her industry expanded from a single factory operation in that season to six expeditions by 1938-39. Germany moved her first pelagic expedition into the Antarctic in 1936-37, and by 1938-39 she had five factories, together with two more jointly operated with Norway, in the Southern Ocean.

Until 1937 whale quotas were set on a voluntary basis between Britain and Norway, but in that year the International Whaling Agreement was drawn up, under which operations were to be restricted to a three month season, and the number of ships used was to be limited. This agreement was the forerunner of the International Whaling Convention. In the same agreement the major nations for the first time set a series of minimum lengths for catching whale species, with the aim of conserving stocks by giving young animals time to reach breeding size. Biologists were becoming alarmed about the eventual fate of the blue and humpback whales at least a decade before this agreement (Bryant 1927).

Figure 31. (D. E. Gaskin.)
Fin whale *Balaenoptera physalus* hauled up on deck of FF *Southern Venturer;* this is a relatively small animal, about 64 ft. The grab is being winched down into the slipway to collect another animal.

The outbreak of the Second World War gave southern pelagic whaling a major setback. Some operations were suspended for the duration, others continued, braving the raider and submarine menaces. A German raider attacked some factory ships, but most very soon turned to oil tanker duties, and in the first years of the war they fell to torpedoes one by one. Virtually none, except extremely lucky vessels like the *Pelagos,* survived until the end of hostilities. Meanwhile the whales were given a valuable five-year respite.

Southern whaling regained impetus fairly slowly after the cessation of hostilities. Capital was very short in Europe, and there was, in any case, an inevitable lull while new vessels were built or tankers converted for factory-ship use. In view of her critical position for protein and edible oils Japanese companies applied for permission to resume whaling. The American Occupation Administration agreed, and in the 1946-47 season two Japanese expeditions sailed to the Antarctic. In the same season the Dutch and the Russians entered southern whaling, and the International Whaling Convention was drawn up, eventually to be ratified by 17 nations. This set of articles was really doomed from the start. Proposals for a system of international inspectors were argued over at each meeting for over a decade, but never came to anything, largely because of blatant obstructionism by the Soviet Union, Brazil, Mexico and Panama. The *Olympic Challenger* began operations in 1950 under the Panamanian flag. For five seasons this expedition whaled in violation of almost every article of the Whaling Convention until seized by naval vessels while operating off the coasts of Peru and Chile for sperm whales. The *Olympic Challenger* episode is one of the most unhappy among a series of dismal defeats of attempts at stock conservation in southern hemisphere whaling.

Nevertheless, it is impossible to absolve any of the major whaling nations from blame for what eventually happened to the southern whale stocks. In 1958-59 Antarctic whaling was close to its second and final peak; with nine Norwegian, six Japanese, three British, one Russian and one Netherlands expeditions working in the Southern Ocean. At this stage a reasonable operating schedule still could not be agreed upon, and despite some major biological problems in stock assessment (Ruud 1956, Mackintosh 1959), it was evident that the blue whale at least was in serious trouble, at this time accounting for 3.3 per cent of the total catch, compared with 26 per cent in 1948-49. Nevertheless, the involved politics prevented any really effective action being taken (Elliot 1958), and the persisting operation of a cover

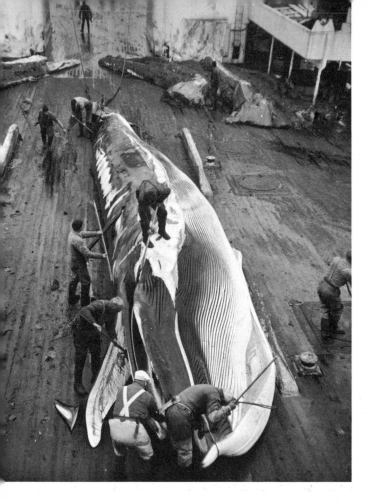

Figure 32. (D. E. Gaskin.)
The preliminary stages of flensing a sei whale *Balaenoptera borealis* on the FF *Southern Venturer* near the South Shetland Islands in March 1962.

unit (Blue Whale Unit: 1 BWU = 1 blue whale, 2 fin, 2½ humpbacks or 6 sei), prevented any one species being given protection.

The Norwegians complained that their abiding by the articles of the Convention sometimes gave great advantage to competitors, even other signatory nations. For example, in 1949 the I.W.C. fixed the maximum humpback catch, after a fairly long period of protection, at 1250 animals per year, and in 1953 the whaling commission resolved that humpback catching should be restricted to a three day period in January. Australian shore stations began to take humpback whales in 1949 (Gates 1963), and by 1958 had taken nearly twice as many humpbacks as the pelagic whaling nations. The Norwegians protested at this state of affairs (Bettum 1958), but since they admitted to having made 1.7 billion kroner from pre-war whaling and 3.6 billion from post-war whaling at this time it is easy to see why the Australians could not bring themselves to feel particularly sympathetic. The Australian government could be blamed, however, for continuing to set unrealistically high quotas in the early 1960s when it was obvious from the biological evidence that the

Group IV and V populations were in grave danger.

In 1958 the Norwegians made a good case for the southern whaling industry suffering from too many ships and too many personnel for the then quota of 14,500 B.W.U.s per annum, and pressed for some limitation of the number of chasers, then 237. The original overall quota had been 16,000 units in 1946, but had been reduced as part of rather weak conservation measures, first to 15,000 and then to 14,500. They were at this time alarmed by reports that new Russian factory ships were to come into operation, with perhaps as many as 40 chasers each. However, the Russian view, even if not particularly laudable from the point of view of conservation was justifiable from that of national politics; the British and Norwegians had made a tremendous amount of money from southern whaling, now it was someone else's turn. There were abortive attempts in both Norway and Japan at this time to get some expeditions to stay home if they were paid a subsidy by other companies, but this proposal was completely unacceptable. The Norwegians were particularly worried at this time by increasing Japanese and Russian competition. These nations had what came to be called a 'meat margin', since whale meat for human and animal consumption was

Figure 33. (D. E. Gaskin.)
Baleen plates of fin whale *Balaenoptera physalus* cut free in a mass and swung over the side on the FF *Southern Venturer* in January 1961. Rorqual baleen is not in demand in the modern world.

Figure 34. (D. E. Gaskin.)
Foreplan (meat deck) of the FF *Southern Venturer.*

almost as important to them as oil. Norwegian whaling companies were not geared to meat production, and the British attempts to sell whale meat just after the war had been dismal failures. In the late 1950s the world oil price began to fall. The Netherlands expedition would have been in the same situation as the Norwegians had it not been backed by a huge government indemnity. Through an effort by whaling companies, the number of chasers in use was limited to 220 in 1959-60. When a new Russian expedition came in as part of the 1956-60 Five Year Plan, the number of chasers operating in the Southern Ocean jumped to an all-time post-war high of 261 in 1961-62.

In the pre-war years the Antarctic industry had been supported largely by catches of blue whales. By 1949, as mentioned earlier, these still made up 26 per cent of the total catch. However, the period 1949-63 could well be called the 'fin whale era', because this species accounted for about 60 per cent to 80 per cent of all catches in these years. By 1963 the Committee of Three (later Four) scientists of the International Whaling Commission were sounding urgent warnings about the three species on which the industry had hitherto been based, the blue, the humpback and the fin whale (Brown 1963). The catches of blue whales had been declining alarmingly

in the early 1950s (Clarke 1955b), and continued to do so until they formed only a negligible fraction of the total catch by 1962. The 1961-62 season could perhaps be called the last of the really good years for the southern pelagic whaling industry, yet still no real agreement could be reached on conservation. Seeing the inevitable, Japanese and Russians alike were turning more and more effort towards the North Pacific, much to the alarm of Canadian and American coastal enterprises.

In 1963 almost complete protection was gained for the blues and humpbacks in the southern hemisphere, far too late to leave their stocks in a condition for rapid recovery to a commercially exploitable level. In the same year the Committee warned that a drastic decline in the size of the fin whale population was taking place, and predicted that the catches were due to fall sharply. Once again no agreement could be reached between the main pelagic whaling nations on reducing the overall quota below the level of the fin whale population's sustainable yield, with the result that while the catch was 26,438 in 1961-62, it dropped to 18,668 in 1962-63 and then to 13,870 in 1963-64. The European industry was by now in desperate straits. The Norwegians cut back their effort, and the British withdrew completely after sending only one expedition in 1962-63. The Dutch followed them out a little later.

The pleas for reductions of effort were now slowly, too slowly, beginning to have some effect. Between 1961 and 1965 the number of expeditions dropped from 21 to 15, though this was largely because it was becoming totally uneconomic to continue operations on western European-style lines. The period from 1963 to the present day is the 'sei whale era', as the industry utilises the last commercially exploitable rorqual species. In the late 1950s the number of sei taken had been creeping up slowly, but in 1965-66 the catch leapt up from 8,286 to 19,874, at once giving the International Whaling Commission another conservation problem. In the same period the fin catch dropped again, from 7,308 to 2,318. Between 1965 and 1967 the number of floating factories fell from 15 to only 9, and in 1966, at the eleventh hour for the fin whale, the pelagic whaling countries at last agreed to a catch slightly lower than the sustainable yield of the remaining populations. By 1967 the Norwegian whaling industry was finally finished in the south, and her last assets were sold to the Japanese, who were left to contest the Southern Ocean with the Russians. Southern pelagic whaling still continues today, but the industry is only a fraction of its former size.

7. Sealing, with Special Reference to the New Zealand Region

The fur seals had the misfortune to evolve a pelage with two layers; an outer one of relatively coarse guard hairs, and an inner one of very dense waterproof fur. When correctly treated the latter is of great commercial value. Before the 'seal skin' can be used the outer hair must be removed by scraping or careful cutting. A successful process for removing this outer layer was developed in China in or about 1750, and in a relatively short time the 'China Trade' for fur seals was so great that a major surge in fur seal exploitation was initiated in the latter half of the eighteenth century.

However, sealing more or less grew up with the various whaling industries, and the animals were hunted for skins and oil long before the development of the China Trade. The grey seal *Halichoerus grypus* and the harbour seal *Phoca vitulina* had populations well within the range of the activities of mediaeval European seamen. In the year 1604 the small vessel *God Speed* of the British Muscovy Company sailed to Bear Island in the Arctic, where the crew hunted walrus (or the 'morse' as it was then called), with some success. The population was only large enough to supply a few seasons of good oil production, but within a few years British, Dutch and German ships had penetrated into Spitsbergen waters, and the same pattern of over-exploitation was repeated. In the late seventeenth and early eighteenth centuries the whalers and sealers scoured the accessible regions of first the European and then the American Arctic for their prey. Fortunately in many areas the walrus population was not accesible to European hunters because of ice conditions, and from these remnants the populations were able to recover substantially again. Unfortunately the walrus population is now shrinking again, apparently because of wasteful over-hunting by northern aborigines and a European-American inspired demand for tusks as souvenirs (Scheffer 1958, King 1964).

Sealions and elephant seals have been somewhat luckier; the former has only a single coarse hair layer which is not much sought after commercially, and the pelage of the latter is useless. However, the elephant seal possesses a thick layer of blubber which produces a respectable amount of oil. Despite this, quite large populations of both animals still survive in the southern hemisphere, and the elephant seals of South Georgia have for a number of years supported a rational, controlled industry worked in association with the shore stations of the Pesca Whaling Company. Since these have recently ceased operations, the sealing has also come to an end. The northern hemisphere populations of elephant seals have been very much reduced by over-exploitation, but are now protected and showing signs of recovery. Large populations of the Californian sealion *Zalophus californianus* and Steller Sealions *Eumetopias jubata*, are still present in the North Pacific.

The development of the 'China Trade' resulted in massive exploitation of the North Pacific Fur Seal *Callorhinus ursinus*, first by the Russians and then by the Americans, with lesser participation by the Japanese and Canadians. At first the Russians killed very large numbers of seals at the Pribilof and Commander Islands without thought for the future, striking at the North Pacific Fur Seal on its breeding grounds. Fortunately the Russian government showed creditable farsightedness concerning the conservation of this resource, and in 1834 forbade the killing of females. Under this regulation the numbers began to increase again. However, the purchase of Alaska and its surrounding islands by the United States saw an end to this protection; in the very first open season the crews of vessels chartered by private companies slaughtered 300,000 seals (Baker, Wilke, and Baltzo 1963).

Fortunately the American government eventually showed as much common sense as the Russians and ordered the killing completely stopped in 1868. It began again under rational management in 1870, the killing of females being prohibited as under the Russian administration. In the 40-year period from 1870 to 1910 some 2,300,000 seals were killed on the islands, and the pelts processed. Unhappily, the good work of this management programme was nearly brought to naught by the activities of pelagic sealers working outside territorial limits and killing seals at sea with lances and guns. More than 1,000,000 were known to have been taken by pelagic operations between the early 1800s and 1910, with no selection against the killing of females. However, the damage caused by pelagic sealing did not prove fatal to the fur seal population. At the present time the Pribilof/Commander Island colonies are under the control of the North Pacific Fur Seal Commission, which in its current form has operated since 1957 and includes the USSR, the USA, Japan and Canada (Roppel and Davey 1965). The herds are now estimated to be back to a level of more than 1,500,000 animals, and there are plans to maintain the killing of a percentage of females each season in an attempt to stabilise the numbers at the population size from which the maximum sustainable yield could be obtained, which is somewhat smaller than the size of the present stock.

On the Atlantic coasts of America, especially around the mouth of the St Lawrence and Newfoundland, profitable sealing for the harp seal *Pagophilus groenlandicus* is still carried out by shore-based Canadian and pelagic Norwegian enterprises. Other populations of this species, which was almost completely ignored by earlier exploiters, are still found at Jan Mayen Island, in the White Sea and at a number of other localities throughout the Arctic (King 1964).

The first real inroads were not made into southern seal populations until 1784, when the Boston vessel *States* plundered the Falkland Island colonies and returned home with 13,000 pelts. However in the years that followed it was the British who were largely respon-sible for the destruction of fur seal stocks in the Falkland Islands and their Dependencies, although sealing vessels from America, Russia, Germany and many other European countries sailed southwards in search of new sealing grounds. Pockets of fur seals, sea lions and elephant seals survived on the coasts of California, the Argentine, Chile, the Galapagos Islands, South Africa, in the Caribbean and on the subantarctic islands, but millions more fell to clubs and muskets. At one time the island of Juan Fernandez supported a population of several hundred thousand *Arctocephalus;* in fact Kellogg (1942) noted that more than 2,000,000 skins were taken from this and the rest of the islands in the group between the time the population was discovered and the end of the year 1824. Now only two skulls are known of this race of *A. philippii;* the animal is completely extinct on the Island of Juan Fernandez.

The New Zealand Fur Seal *Arctocephalus forsteri* has been known to occur around the southern coasts of the archipelago since the latter quarter of the eighteenth century. Captain Cook killed some for food in 1778, and the number of bones found in Maori middens on the Otago Peninsula shows that the animal was extensively hunted for food by the southern tribes.

The focus of sealing activity began to move southwards from the North Pacific at the end of the eighteenth century. Although the *William and Anne* under the command of Captain Eber Bunker visited New Zealand as early as 1792, there is no evidence that her crew took any seals (McNab 1907). The first positive record of sealing in New Zealand waters seems to be of the British vessel *Britannia,* commanded by Captain William Raven, which anchored in Dusky Sound on 6 November, 1792. Raven put a party of sealers ashore on Anchor Island, with sufficient provisions for half a year and tools with which to build a vessel large enough to carry them across the Tasman Sea to the east coast settlements if the need should arise. Raven returned to collect his men at the end of September 1793, and found them in good health and with a partially completed boat, but with only 4,500 seal skins to show for a 10-month stay. Long periods of bad weather, always the curse of the south-west coast of the South Island, even in summer, curtailed their hunting activities.

At about the same time a sealing ground in Bass Strait was being exploited by vessels sailing south from Sydney, the earliest recorded voyages being by two ships in 1791, which returned with little to show for their efforts (McNab 1907). However, this industry developed very rapidly, and by 1804 at least 11 sloops and schooners were working in the waters between Tasmania and the Australian mainland. The Bass Strait sealing proved to be of very short duration; even as early as 1802 both fur seals and elephant seals were disappearing from islands where they had previously been found in large numbers.

In April 1803 the small vessel *Endeavour*, commanded by Captain Oliphant, dropped a party of sealers in Bass Strait and then sailed eastwards to investigate the sealing potential of the islands of New Zealand. This vessel worked down from Dusky Sound to Breaksea and then to the Solander Islands, apparently in such a cloak of secrecy that even the band of sealers left in Bass Strait were not aware of her destination.

McNab (1907) pointed out that this period is a singularly difficult one to research, for two reasons. Firstly, the secretive attitude of the owners and masters of the vessels is understandable, since they were engaged in a highly competitive business. Secondly, the comings and goings of small local trade vessels from the new colony of New South Wales were poorly documented by the port officials. However, there is a positive record of another sealing vessel visiting Dusky Sound, the *Scorpion*, commanded by Captain Dagg. Although primarily engaged in pelagic whaling the *Scorpion* took 4,759 seal skins, though not necessarily all from New Zealand, since she made a further call into Port Jackson before returning to England (McNab 1907). The length of time spent by this vessel on the New Zealand coast is not known, except that it occupied most of the year 1804.

In the following year the 45 ton *Contest* arrived in Sydney with 5,000 seal skins taken from the New Zealand coast. The vessel, owned by Kable and Co., is presumed to have worked around Dusky and Breaksea Sounds, but may have gone further afield. Between 1804 and 1807 the number of vessels working the New Zealand coast for seals increased, and the little colonial fleet suffered quite high casualties. The *Speedwell* was stranded in 1804, the *Governor King* was wrecked in 1806, and the *Contest* was lost in 1807.

The year 1805 was a notable one in the history of southern colonial sealing, since public opinion had been aroused by the sufferings of some of the sealing parties to such an extent that legislation was passed prohibiting any vessel leaving Sydney without entering into a bond to set up food and clothing depots for its sealing gangs. At the same time another most peculiar law came into being, one which prohibited companies from operating sealing vessels south of latitude 43°39′S, effectively put-

Figure 35. (Alexander Turnbull Library.)
A lithograph of a sealers' camp on Byers Island, Falkland Islands.

ting a legal end to the New Zealand sealing, since all the big colonies were south of latitude 44°S. Despite extensive researches, McNab (1907) failed to find any logical reason for this prohibition, except that it is reminiscent of the obstructive attitude of the British East India Company to whaling activities in the same region at this time.

However, the result is most unfortunate for the student wishing to research the period; there is no doubt at all that the companies continued to work the New Zealand coast for seals, since there was no way of enforcing the legislation, but to cover themselves against legal proceedings they naturally enough falsified the catch localities. Many cargoes of pelts ostensibly from 'Bass Strait' and similar southern Australian localities were actually taken from New Zealand. So, just as New Zealand sealing began to attain the status of a major industry, the published records of the colonial New South Wales companies become completely unreliable.

While the Australian vessels were exploiting the seal colonies of the South Island and the adjacent offshore islands, American sealers had begun to operate around the distant islands to the east of the archipelago, especially the Antipodes. In 1803 and 1804 two ships, the *Union*, commanded by Cpt. J. Pendleton, and the *Independence*, under Cpt. F. Smith, went to the Antipodes (Fanning 1924), and also to Kangaroo Island off South Australia.

Both McNab (1907) and Fanning (1924) have given detailed accounts of the subsequent events, of which there are at least two descriptions conflicting on a number of major points; it is impossible to decide at this distance in time exactly what did occur.

However, a number of facts appear to be indisputable. The sealing gangs of this ill-fated pair of ships took 14,000 pelts from Kangaroo Island and 60,000 from the Antipodes. The former consignment was lodged with agents in Sydney, and appears to have found its way safely to Canton and the Chinese market. Pendleton, for some unfathomable reason, never returned to pick up the men he had left for a year on the desolate Antipodes, but took another charter northwards to the tropical islands. In October 1804 he and some of his men were clubbed and speared to death at Tongataboo by hostile natives. The story of the killing was brought back to Sydney by the first officer, Mr D. Wright, who then took over command and again sailed north. They reached Fiji, where the *Union* ran on a reef. While the ship was helpless Fijian warriors sallied out, and in the battle that ensued the whole crew was killed after the vessel was over-run. Thus ended the career of the ill-fated *Union*, and the men abandoned on the Antipodes proved to be the luckiest in the final event.

They were eventually rescued by the American vessel *Favorite,* which left Sydney in company with the *Independence.* Both vessels were headed for the Antipodes, but either during bad weather or at night, the *Independence* became separated from the *Favorite,* and was never seen again. The *Favorite* went on to pick up the sealing gang and their 60,000 skins, of which 32,000 were sent on to Canton (Fanning 1924). What happened to the other 28,000 is not known. Both McNab (1907) and Fanning presented evidence for and against the accusations of fraud which followed the mysterious disappearance of this valuable consignment.

Australian and American sealers met in the latter part of 1804, when the former had begun to explore eastwards towards the Bounty Islands, and the latter were considering taking part in the Bass Strait seal fishery. The surviving documents of the period show that relationships between the rival groups frequently degenerated into open violence.

There is every reason to believe that the Bounty, Chatham and Antipodes Islands were exploited by quite a number of colonial and American vessels during the first decade of the nineteenth century. In addition to the 60,000 skins taken by the gang from the *Union* we know that the *Star,* commanded by Cpt. Wilkinson, took another 14,000 in 1805. This total confirmed figure of 74,000 in only two years at least gives an idea of the magnitude of the rookery sizes on these desolate islets at this time.

In 1806 the whaling vessel *Ocean* discovered the Auckland Islands, and her master, Cpt. A. Bristow, remarked in his report that the group probably abounded with seals. By the end of 1807 no less than three sealing gangs were at work there slaughtering the fur seals and and sea lions, one gang being from the Sydney vessel *Commerce.*

In view of its large size and close proximity to New Zealand it is very surprising that Stewart Island was not confirmed as an island until 1809, when the *Pegasus* reported finding Foveaux Strait, albeit a treacherous passage for sailing vessels, between this new island and the mainland of the South Island (McNab 1907). Fur seals were found to be abundant in the Strait, which is of course scattered with many small rocky islands and islets, and the discovery of the new rookeries helped to prolong the New Zealand sealing boom for a few more years. Despite a drop in the price of seal skins from about six shillings to less than four shillings each, the coasts of New Zealand and the surrounding islands continued to produce thousands of skins for the Sydney merchants to ship to Canton.

After 1820 the pattern of sealing around New Zealand began to change; a direct reflection of the over-exploitation of the stocks. No longer were vessels able to bring back an economic load of skins from the region. Trade with colonists and natives for other commodities, especially native flax, assumed more and more importance.

The same pattern was being repeated all round the southern hemisphere; many of the remotest islands of the southern Pacific, Atlantic and Indian Oceans were discovered during the wild search for fresh sealing grounds. Not all had seal populations; those that did were soon relieved of them by the horde of exploiters which followed any new discovery. However, the hunters from the west at last met those from the east in temperate and subantarctic latitudes, and the industry appeared to be finished.

There was, however, one last big killing to make, even though the insatiable appetite of the sealing companies was self-destructive. The South Shetland Islands were discovered more or less by accident by the British vessel *Williams* early in 1819. One of these islands bears the name of the discoverer, Captain Smith. These islands lie far to the south of Cape Horn, and form a chain along the northwest coast of the Antarctic Peninsula. Smith Island is a forbidding place, long and narrow, with two high mountains, Mt Pisgah and Mt Forster, rearing up to more than 6,000 ft at each end of the island. Not far away is Low Island, an almost featureless, flat, snow-covered hummock. Famous names are remembered in the archipelago; King George, Livingstone, Nelson. Further east beyond the mouth of the Bransfield Strait are two more islands, Clarence and Elephant, in an area famous for violent westerly gales which have caught many a whaler and sealer unawares. In the Strait itself lies Deception Island, used as a base by the early pelagic whaling factory ships which were able to tie up in safety in the harbour, which is in reality the crater of a partially sunken volcano. The main enemy of man in the South Shetland Islands, however, is fog; it creeps up before the dawn, blurring the contours of the land, making ephemeral castles of the mountains, and deadening the siren of the catcher boat seeking its mother ship. There is no record of the number of sealing parties lost in the cold clammy mist of this archipelago, but it was under these circumstances that the last great seal kill was staged.

In the middle and later years of the nineteenth century the fur seal was a rare animal in the New Zealand region (Chapman 1893). The species was not finally given protection until 1894, and in the following 19 years the populations around the mainland began to make a slow recovery. Some restricted licences were granted between 1913 and 1916, but there were still not enough seals to make the ventures economically profitable, and labour became very short when the war began. In 1922 the New Zealand government once again allowed a three-year experimental period of seal exploitation, but no company was able to demonstrate the ability to produce

Figure 36. (D. E. Gaskin.)
Snow-covered Smith Island in the South Shetlands, March 1962.

a sustained economic industry under the prevailing conditions. In 1946 the government once again permitted sealing. There was a short period of chaotic unrestricted killing, which ended on 30 September 1946 when the licences were revoked. Many of the skins taken from the several thousands of animals killed were spoilt in the hands of operators with no experience in their treatment; many that did reach the commercial market were of the poorest quality.

After the 1946 open season the Dominion Museum despatched two scientific expeditions under the leadership of Dr. R. A. Falla to the sealing grounds, the first in November 1947, the second in July 1948. They found that so many pups had been killed that recruitment into the adult population had been severely affected. There was a large surplus of large bulls, fighting over very small harems, in fact every sign of a population with a drastic imbalance that would take years to stabilise again (Sorensen 1969a). Since 1946 there has been no commercial sealing in New Zealand or on her island territories.

Conditions are not the same as in early European times; Fiordland is no longer a desolate unknown wilderness where the seals could be left undisturbed; now all the sounds swarm with crayfishing boats in the summer months. Fur seals are now once again common as far north as Cook Strait, but it has yet to be proved whether the extension of range is brought about by a great increase in numbers or redistribution of a population of stable size.

The results of two decades of census work by staff of the Dominion Museum, Wellington, and the Marine Department, Wellington, have recently been published (Sorensen 1969a, 1969b). Based on these studies and the postulated trends in temporal and spatial movement of population which can be gained from the accumulated data, the government will formulate policy for any future sealing industry in New Zealand. However, the preliminary results are not encouraging. It seems most unlikely that the total size of the population, and its scattered distribution, would allow a sustainable yield that would support a local industry of economic size. Consequently it is not likely that we can look forward to the magnitude of harvest taken by the North Pacific nations from the Pribilof fur seal colonies, by South Africa from the fur seal colonies of western South Africa, or Argentina and Uruguay from the rookeries of the South American fur seal on the South Atlantic coast of that continent.

Certainly one must hope that no New Zealand government will ever allow a repetition of the 1946 open

Figure 37. (Alexander Turnbull Library.)
A lithograph from Fannings' *Voyages* 1833, of the seal rookery on Beauchene Island, Falkland Islands.

season, which was merely another dose of the 'bad old days' (Dr R. A. Falla, in Sorensen, 1969b).

One of the best examples of sealing under rational control is the Pribilof Island/Bering Sea industry, where the North Pacific Fur Seal Commission exercises control. Here expert sealers still kill the animals by the time-honoured and humane method of a sharp blow on the head; but the colonies are carefully worked in rotation through the sealing season in summer. The sealing parties make their way to the hauling-out beaches by truck and work along parallel to the water, always up-wind, and drive groups of bachelor seals into fields behind the high water zone (Baker, Wilke and Baltzo 1963).

Animals in good physical condition above the minimum size limit are killed, others are allowed to pass between the sealers and return to the sea. There is careful super-vision by Bureau of Commercial Fisheries' officers who ensure that only animals of the correct size range are killed, and that they are despatched as humanely as possible. The pelts are stripped from the carcases and taken back to the local village with the blubber layer still attached. The skins are then processed to remove the guard hair layer, and the blubber is used in the manufacture of soap.

Management practice and research go hand in hand on these sealing islands, and the US Bureau of Commercial Fisheries can be justly proud of the fact that as part of the programme to maintain research information on the size and distribution of the population, up to 50,000 animals are tagged or branded each season. Only by the application of such methods in other areas can rational use of the world marine mammal resources be made and sustained indefinitely.

50

8. The Marine Environment of the New Zealand Region

The families and genera of Cetacea and Pinnipedia are discussed in the chapters which follow; particular attention is given to those species which occur in the New Zealand region. Before embarking on the systematic descriptions of these animals and a discussion of their distributions, movements and life histories, it would be wise to spend a little time examining the environment in which they live, and the factors which influence their distribution and movements.

The lay mind has a natural tendency to think of whales as distributed more or less at random throughout oceans of largely homogeneous composition, and seals as scattered along the coasts of land masses abutting on these oceans. However, this is a gross over-simplification. The oceans have structure, although this is less immediately evident than are differences between one terrestrial ecotype and another. There are static features such as deeps and shallows, basins and seamounts; and mobile features such as currents and water masses of varying temperatures and densities. Both, together with the changing seasons which bring alternate cool and warm temperature regimes to the temperate and sub-polar regions, exert important influences on the lives and distributions of marine mammals.

Different species favour different temperature ranges, and their metabolic processes frequently show appropriate adjustments. For example, the dolphins of the genus *Stenella* are restricted to the tropics and subtropics, while the Ross Seal *Ommatophoca* rarely ventures away from the southern pack-ice. Many marine mammals make long migrations each year from the polar regions in summer to the lower temperate or even tropical regions in winter. Such migrations appear to have evolved as a compromise existence for a species, so that it can benefit from the best feeding conditions in high latitudes with great plankton productivity, and in the same year experience water temperatures warm enough for mating and calving. Other species are pelagic for much of the year, but need to come into shallow waters to calve, for example the southern right whale. Seals, of course, are not so completely adapted for aquatic existence as whales and dolphins, and must haul out on ice or beaches for mating and pupping.

Townsend (1935) was one of the first to remark that the distributions of whale species appeared to be determined almost solely by food requirements and suitable conditions for reproduction; only in the last few years have we begun to accumulate evidence to show how rigorous these limits can be. The very non-random distribution of whales soon becomes evident to anyone cruising high latitudes in search of these animals. Vast tracts of sea are in fact empty of cetaceans, and yet suddenly one finds large numbers within a relatively small area.

Learning to find these concentration areas has always been one of the keys to success in commercial whaling; in the last three decades much scientific effort has been bent towards finding out what factors govern the formation and composition of these favoured regions. One way to do this is to note carefully the distribution and movements of marine mammals and then match these data against the best available oceanographic data for the same area. In fact, so little is known of the detailed hydrology of much of the world's ocean surface that this is often impossible. However, the seas of the New Zealand region have been the subject of consistent work by the New Zealand Oceanographic Institute and the Australian Division of Fisheries and Oceanography for a number of years. As a result it is now possible to give a general picture of the marine environment in this area.

Static Features — Bathymetry of the Western South Pacific

The major details of the bathymetry of the New Zealand region were summarised in a very useful bulletin by Brodie (1964). The New Zealand archipelago is in fact the above-surface fraction of a large area raised above the lower ocean floor level in the western South Pacific. Brodie has given this the name 'New Zealand Plateau'. As well as the main island chain and the more isolated offshore islands such as Campbell, Auckland, Bounty, Antipodes and Chatham, this plateau includes a large area of adjacent sub-surface terrain at depths of up to 500 fathoms, including the Norfolk Ridge, the Lord Howe Rise, the Campbell Plateau, the Chatham Rise and the Kermadec Ridge. The New Zealand plateau region has some relatively shallow connections to the northwest with New Caledonia and the New Guinea area at depths of 500 to 1000 fathoms, but is isolated from the Australian landmass by the Tasman Basin (1,000-2,500 fathoms), from the islands to the north by the South Fiji Basin (1,000-2,000 fathoms) and is bounded on the south, east and northeast by the huge South Western Pacific Basin (2,000-3,000 fathoms). Detailed examination of geological and biogeographical evidence suggests that this plateau has been removed from other land or shallow-water connections since at least the Middle Cretaceous (Fleming 1962, Gaskin 1970a), except possibly via the 'Melanesian Arcs' of islands and former islands between New Zealand and New Guinea.

When the coastal bathymetry is examined we find that in comparison with islands such as those of Britain, the North Island continental shelf is narrow and restricted. There are few wide shelf areas; only those of the Hauraki Gulf, Hawke Bay and the South Taranaki Bight off the western south coast of the North Island are of significance. Even these areas include much water which is actually deep (100-250 fathoms) by North Sea standards. The South Island has a coastal shelf of similar depths along the east coast from Kaikoura southwards, and also in the Foveaux Strait region. The last is continuous with the northwestern part of the Campbell Plateau. To the east of Banks Peninsula on the Chatham Rise lie the Mernoo and Veryan Banks, from 100 to 200 miles off the South Island coast (Fleming and Reed 1951). Small areas of continental shelf water less than 250 fathoms in depth also surround the Bounty and Chatham Islands.

Subsurface troughs and trenches, as seen in a relief map, cut deep gashes into the New Zealand Plateau. The 1,000 fathom Caledonia Basin pushes in towards Cape Egmont from the northwest on the south side of the Norfolk Ridge; the White Island Trench (Fleming 1952) forms a deep cleft in the northern Bay of Plenty; and the 2,000-3,000 fathom Hikurangi/Kermadec Trench complex flanks the whole sheer eastern side of the New Zealand Plateau, terminating abruptly in the vicinity of Kaikoura at the junction of the eastern South Island continental shelf with the Chatham Rise. Further south the Bounty Trough runs east-west offshore from the Otago Peninsula, and to the south of the South Island the famous Solander Trough comes close to the southwestern tip of Stewart Island. The coast of Fiordland falls away sheer into a trough of the same name which is contiguous with the Tasman Basin.

In addition to these major features there are numerous small subsidiary canyons and sea mounts. One small canyon juts from the northern side of the Hikurangi Trench into the eastern mouth of Cook Strait, with the result that pelagic deep-water species such as the sperm whale can sometimes be seen only a very few miles from shore in this area. The extremity of the Hikurangi Trench is also very close to land near the Kaikoura Peninsula; this is another area where sperm whales can be seen close to shore. The very presence of Cook Strait is an important factor in influencing the north-south migrations of rorquals and the humpback whale, permitting interchange between Tasman Sea and Pacific, although the configuration of the Strait in relation to the north-east-southwest axis of the east coast of the archipelago makes it more likely that an animal will pass through the Strait when going north than when going south.

Since the bathymetry of any region influences the hydrology it is difficult to point to discrete bathymetric features (other than those discussed in the last para-

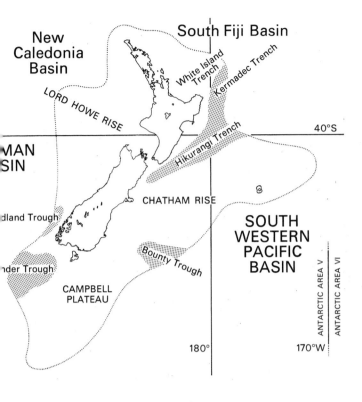

New
Caledonia
Basin

South Fiji Basin

LORD HOWE RISE

White Island Trench

Kermadec Trench

Hikurangi Trench

40°S

MAN
SIN

CHATHAM RISE

dland Trough

SOUTH
WESTERN
PACIFIC
BASIN

nder Trough

Bounty Trough

CAMPBELL
PLATEAU

ANTARCTIC AREA V

ANTARCTIC AREA VI

180°

170°W

Figure 38. (Compiled from maps published by J. W. Brodie.)
The major bathymetric features of the New Zealand region.
Stippled areas indicate trenches and troughs in the margin of
the New Zealand Plateau, which is the area enclosed within
the continuous dotted line.

graph) which directly affect cetaceans. Indeed, apart from
the humpback and the right whale, baleen whales in particular seem to be largely independent of the proximity
of land. However, water movements over the coastal
shelf margin and around isolated islands provide rich
feeding grounds for toothed whales, but although the
topography may be the prime agent causing such water
movements, it is not the factor which acts directly upon
the whales.

However, topography and terrain type are of tremendous importance in determining coastal seal distribution patterns. The flat beaches of the large subantarctic islands are ideal for sea lions, while the sheer
Fiordland coast with its deep sounds and rocky coves is
just as ideal for fur seals. Only at similar localities in
Foveaux Strait and on the Otago and Kaikoura Penin-
sulas are fur seal colonies found which even approach
the Fiordland colonies in size. Once the isolated islets
of the Bounties and Antipodes had huge fur seal
rookeries, but these colonies were totally exterminated
by the early sealers.

Mobile Features — Hydrology
of the Western South Pacific

The New Zealand archipelago does not extend quite
far enough north to be washed by waters which are by
definition tropical. The Tropical Convergence in this
region, which marks the interface between tropical and
subtropical water masses, occurs between latitudes 26°S
and 30°S, running a little north of the Kermadec Islands
and a little south of Norfolk Island (Wyrtki 1960,
Highley 1967). A study of temperature changes between New Zealand and Fiji was made by Garner (1954,
1955), who found a regular cycle of annual temperature
changes in the region, with a summer high and a winter
low. He suggested that the movement of surface waters
in this tropic/subtropic region could be related to the
behaviour of the Southeast Trade Drift, which according to Newell (1966) and Highley (1967) has separate
northern and southern components. Garner found a
relatively small seasonal fluctuation in surface temperature at latitude 20°S, and a somewhat larger fluctuation
at latitude 35°S.

In the northern Tasman Sea one of the most
prominent features is the East Australian Current. This
is a variable component (Hamon 1965), but consistently
dominates the western and southwestern parts of the
Tasman Sea (Starr 1961), and forms part of the water
mass on the northern side of the Subtropical Convergence in the Tasman Sea, which can be traced from
the eastern coast of Tasmania (Highley 1967) to the
coast of Fiordland (Burling 1961, Garner and Ridgway
1965). The East Australian Current follows the east
coast of Australia southward to about the latitude of
the southern New South Wales border, where it is
deflected seawards across the central Tasman Sea as
the less well-defined Tasman Current, a small part of
which passes into the western end of Cook Strait as the
D'Urville Current. However, the East Auckland Current, which flows down the east side of the Northland
Peninsula into the Bay of Plenty, does not appear to
be derived from the Tasman Current turning the North
Cape of the North Island, but from waters originating
northeast of New Zealand which are pushed along the
coast by the Trade Wind Drift mass (Brodie 1960).
A distinct water movement up the west coast of the
South Island from Fiordland, called the Westland
Current, does seem to be part of the Tasman Current
circulation (Brodie 1960). Another component called
the Southland Current (Burling 1961) pushes round the
southern extremity of New Zealand from west to east,

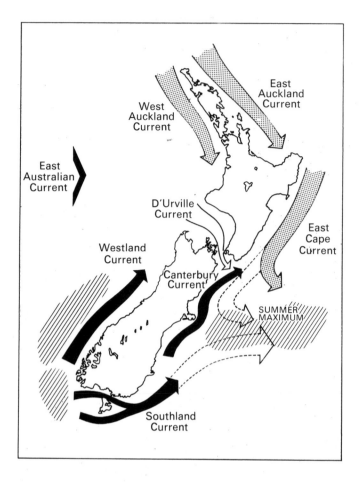

Figure 39. (Compiled from diagrams and maps published by the N.Z. Oceanographic Institute.)

Local currents of the New Zealand region. Black arrows indicate relatively cold water; white arrows, water of intermediate temperature; and stippled arrows, relatively warm subtropical water. The hatched area indicates the approximate summer position of the Subtropical Convergence. In winter it tends to lie a little further north.

forming part of the continuation of the Subtropical Convergence off the east coast of the archipelago. This water is relatively cold; a small tongue called the Canterbury Current (Brodie 1960) forces its way inshore up the east coast of the South Island to at least the vicinity of Cook Strait. The main part of the Southland surface current appears to be deflected eastwards towards the Chatham Islands, where the Convergence region becomes less well-defined in surface waters (Garner 1959). Further offshore in waters to the east of New Zealand there is a south-pushing mass of subtropical water called the East Cape Current (Fleming 1950, Garner 1953), which is deflected eastwards when it meets the cooler water coming from the south. This current originates in the Trade Wind drift. A detailed

study of inshore waters to the east of New Zealand (Garner 1953, Ridgway 1960, Garner 1961) has shown that this current is important in influencing seasonal changes in surface temperature in the eastern Cook Strait region and local circulation in Cook Strait and Hawke Bay, and that the southward prevalence of this current is greatest in summer, when it reaches southward to the latitude of Banks Peninsula, and least in winter, when its southernmost extent is as far as offshore waters east of the Wairarapa coast of the North Island. The Subtropical Convergence is a somewhat mobile yet constant feature off both the east and west coasts of New Zealand, and has important influences also on the climate. It has characteristically steep temperature and salinity gradients at the surface, and is associated in summer with the 15°C isotherm and in winter with the 10°C isotherm (Garner 1959).

South of the Subtropical Convergence the New Zealand subantarctic islands are washed by the waters of the great southern Circumpolar Current, which is turned somewhat to the south in this region by a local 'eddy' over the Campbell Plateau, called the Bounty-Campbell Gyral (Burling 1961). A little to the south of Macquarie Island another water mass interface with

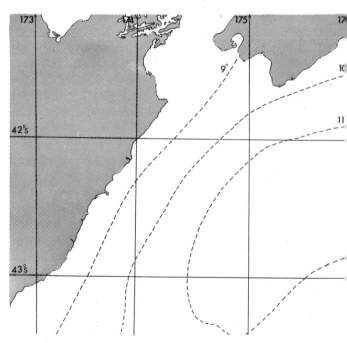

Figure 40.

'Winter stable' surface oceanographic conditions. Relatively cold water dominates the Cook Strait whaling grounds. Isotherms are in degrees centigrade; this and subsequent figures modified largely from Garner (1961) but including some supporting temperature data gathered during the Marine Department whale survey of 1962-64.

steep surface temperature gradients is encountered; this is the Antarctic Convergence (Garner 1958). North of the Ross Sea the position of this interface seems to be influenced by the Pacific Antarctic Ridge. South of New Zealand its position varies from about 54°S in winter to 62°S in summer, and the temperature gradient ranges from about 5°C down to only about 1.5°C.

The importance of these major water mass boundaries in the life of marine mammals is only just being realised. The Subtropical Convergence probably exerts a greater direct and indirect influence on the lives and feeding habits of toothed whales in this region than any other hydrological structure. For example the author (Gaskin 1968c) has been able to demonstrate that four different species of dolphins are associated with different temperature regimes off the east coast of New Zealand; and in Cook Strait and local east coast waters the density of the sperm whale population fluctuates with the seasonal north-south movement of East Cape Current waters (Gaskin, in press[1], 1971).

Upwellings of a relatively local or restricted nature have been demonstrated around New Zealand, so that sharp temperature and salinity gradients can be encountered in relatively small areas. One of these occurs near the Kaikoura Peninsula (Garner 1953) in association with the extremity of the Hikurangi Trench, and may play a substantial role in the consistent occurrence of sperm whales in the area. In other parts of the world upwellings around oceanic islands appear to be important in maintaining sperm whale populations (Clarke 1956a).

While temperature gradients are important factors in the distribution of whales and seals (although there is much less information on the distributions of pelagic seals than cetaceans), the actual temperature is not the cause of the whales congregating in such areas, be they localised like coastal upwellings, or generalised in the form of massive global convergence regions. On these water mass interfaces basic inorganic nutrients are often concentrated by upwelling and stirring, with the result that in the surface waters of these regions there may be rich phyto-plankton blooms, on which feed zooplankton, and other pelagic or benthic animals of small size. These in turn are eaten by small fish and squid, which are then eaten by larger fish and larger squid. Toothed whales subsist on the latter, so they are really only the last large elements in a fairly complex food web.

Some evidence from the Cook Strait area suggests that the distribution of sperm whale feeding grounds may depend on the presence of the right magnitude of

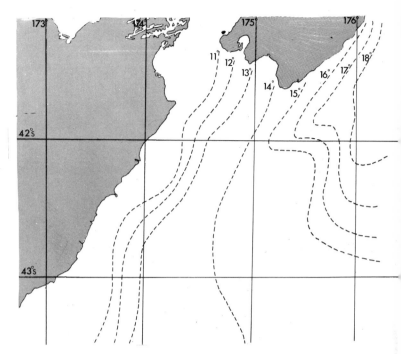

Figure 41.

'Spring transition' surface oceanographic conditions. Warm water from the East Cape Current moving into northeast sector of the whaling grounds, and a relatively steep surface gradient building up against the northeast coast of the South Island. Isotherms centred on November.

Figure 42.

'Summer stable' surface oceanographic conditions. Warm East Cape Current water dominates the whaling grounds. Isotherms centred on February.

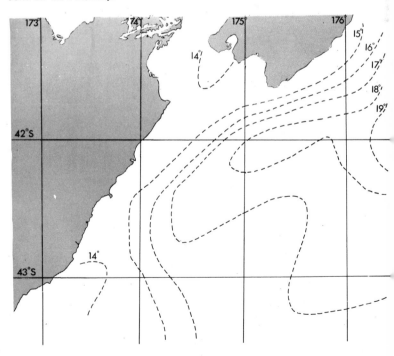

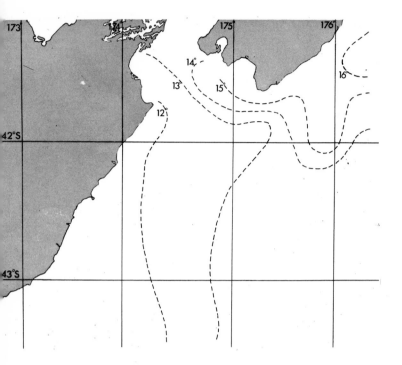

Figure 43.
'Autumn transition' surface oceanographic conditions. Warm surface water retreating northeastwards from the Cook Strait whaling grounds. Isotherms centred on May.

vertical temperature gradients as well as horizontal ones, since squid appear to school abundantly only under rather restricted conditions (Uda 1959). Vertical gradients are confirmed as important in the distribution of baleen whales, and presumably other Euphausid shrimp feeders like the Crabeater seal; baleen whale feeding grounds seem to be more complex hydrologically than toothed whale feeding grounds (Beklemishev 1960; Nasu 1963; Uda 1954, 1962; Uda and Dairokuno 1957; Uda and Nasu 1956; Uda and Suzuki 1958).

I have a large body of unpublished information on sei whales which suggests that this species has a significant association with the disturbed layers of the Subtropical Convergence waters off both the east and west coasts of the South Island. I have also found sei whales feeding on the boundary of the coastal Canterbury Current and the warmer offshore waters southeast of Cook Strait, so even these relatively local interfaces may be important in determining whale distributions. The fur seal colonies of New Zealand are almost all distributed in the southern region which is influenced by the waters of the Subtropical Convergence, as are those of the Chatham Islands. Only the former Bounty and Antipodes colonies were outside this region and one might suppose that local upwellings provided rich feeding.

II

9. Identification of Whales, Dolphins, and Seals

Since knowledge of the anatomy and distribution of many whale and dolphin species is relatively limited, the public can greatly assist scientists by reporting stranded marine mammals. There are few specialists in New Zealand and eastern Australia, so they may not be able to attend a stranding in person through pressure of other work or the sheer distance involved. But when no specialist is available to attend a stranding any member of the public can collect valuable details through making some straightforward observations.

The first and most important point is to try to establish the exact identity of the specimen(s) involved. Newspaper reports abound with rough approximations and local names such as 'blackfish', 'beaked whale', 'toothed whale', 'whalebone whale'; these are useless for identification.

A number of publications, most not easily accessible to the general public, or out of print, or with incomplete coverage, provide guides to identification of species of whales and seals. The book *Guide to the Giant Fishes, Whales and Dolphins* (Norman and Fraser 1937) is still one of the most comprehensive of its kind, but is unfortunately out of print at present. Another guide prepared by Dr Fraser (1966b) is in print and available from the British Museum, but deals with cetaceans only, and then only those of the northern hemisphere, and British coastal waters. A guide to the difficult family Ziphiidae in New Zealand waters (McCann 1962a) and a guide to the seals of New Zealand (McCann 1964c) were published in *Tuatara,* a periodical held by some New Zealand libraries. An earlier publication of mine, the *New Zealand Cetacea,* published by the

Fisheries Research Division (Gaskin 1968a), illustrates all the whales and dolphins known to occur in New Zealand waters, but does not contain a key and does not deal with seals.

However, keys for both cetaceans and pinnipeds have been designed; these are incorporated in this section, together with a worked example. With a little practice any reader will be able to use these to identify most marine mammals occurring in the New Zealand region. In addition each species in Part III is illustrated with line drawings showing the most important diagnostic features, and where possible, photographs are included as well.

When the finder has identified the animal as exactly as possible, any obvious external parasites or growths should be noted, and either drawn or better still cut out and preserved. Any scars should be carefully noted. Sketches of the colour pattern of fresh specimens are very useful, and of course photographs, including the whole animal and a close-up of the head, are the most valuable of all. If the specimen is in an advanced state of decomposition it may be possible to salvage at least the skull and lower jaws for a museum. If the whole skull cannot be saved, a few teeth are useful. These must be pulled from the gum, and not sawn off short at the gum level. If the animal(s) have stranded alive, with no hope of them being returned to the sea, they should be put out of their misery with a rifle bullet through the brain or the heart. The local police will usually do this job if no one else is available. Any obvious cruelty — for example some spectators at strandings have been observed to poke the eyes of living

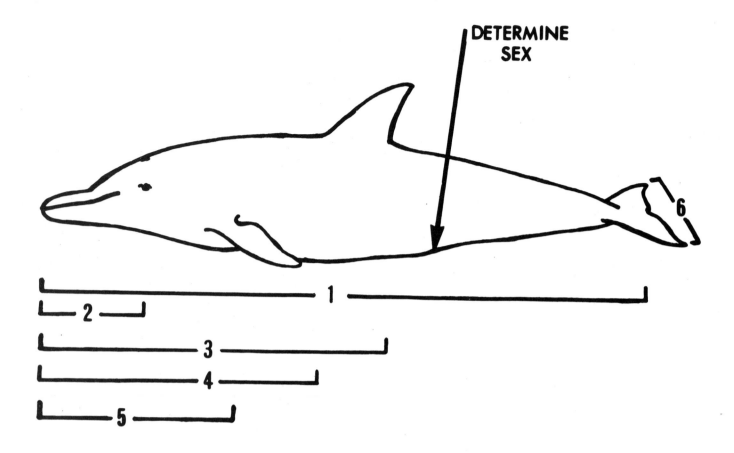

DETERMINE SEX

Figure 44.
Useful measurements to take from a stranded cetacean.

whales with sticks — should be reported at once to the police and appropriate charges laid. Disposal of carcases is sometimes the responsibility of local authorities, sometimes that of the Marine Department. In any case a record of any stranding, together with good photographs or drawings and other details, should be forwarded to the Marine Department Head Office in Wellington in case other records sent elsewhere should go astray.

The next step is to take standard measurements. A guide to the standard measurements of whales and dolphins was prepared by the North American Committee of Marine Mammalogists, and edited and published by Dr K. S. Norris in the *Journal of Mammalogy* (1961). This is not a periodical available to the general public, though it is held in some government department and university libraries. Their format of measurements is very comprehensive, perhaps more than most non-specialists would be prepared to take. However, the most useful of these measurements are those given below.

One of the most useful is the *Standard Length*. This is taken from the tip of the jaw to the crotch of the flukes in a straight line, not round the curves of the body. Other useful measurements that can be made are:
Tip of snout to eye
Eye to anterior flipper base
Centre of flipper base to vent (genital aperture)
Eye to anterior edge of dorsal fin
Length of dorsal fin
Height of dorsal fin
Width of tail flukes
Anterior edge of dorsal fin to crotch of flukes
Vent to anus
Anus to crotch of flukes
Depth of body at eye, dorsal fin centre, vent and tail stem just before flukes.
Girth
Length of flipper
Width of flipper, noting the shape
Depth of notch between flukes.

58

The presence, shape and number of teeth should be recorded, the shape of the head, the presence and number of throat grooves, and if possible the weight of the animal, if small, should also be recorded. If dissec-section is possible, note the thickness of the blubber on the flank posterior and lateral to the dorsal fin, the contents of the stomach (preserve these in alcohol or 10 per cent formalin if possible), and the sex. Preserve any foetuses found if their size permits, and also if preservative is available, the entire ovaries and 1 cm cubes of testis or mammary gland. If such material is ever made available from a large number of animals by the work of a team of people who happen to be available to examine a mass stranding, a very great deal of most valuable information will be saved. The measurements which should be taken from seals are given just before the key to those animals a little further on in this section.

A Key to Stranded Whales and Dolphins

Note: For the benefit of the many people who will have had no experience with this kind of biological key, a worked example is provided at the end as a typical example of the procedure to follow.

1
- Whale having long fringed baleen plates in the mouth 2
- Whale having teeth or bare gums 8

2
- Whale without dorsal fin. Mouth strongly arched with long black plates. Whitish or yellowish lumps on snout and around jaws. Body blackish .... Southern Right Whale (*Eubalaena australis*)
- Whale with dorsal fin 3

3
- Flippers inordinately long in relation to length of body. Small triangular dorsal fin with a series of small humps posterior to it. Animal black above, white beneath, flippers similar. Deep grooves on throat, usually encrusted with barnacles. Posterior margin of flukes scalloped, without central notch. Baleen plates black   Humpback Whale (*Megaptera novaeangliae*)
- Flippers not inordinately long in proportion to body length 4

4
- Baleen plates greyish-blue, except on anterior third of right side of jaw, where they are white. Grooves of throat white. Body grey above, white below. White colour extending on to underside of flukes and flippers .... Fin Whale (*Balaenoptera physalus*)
- Baleen plates yellowish 5
- Baleen plates black 6

5
- Baleen plates and their fringes yellowish. *Line of mouth nearly straight*. Body length 10-25ft, bluish-grey above, white below   Minke Whale (*Balaenoptera acutorostrata*)
- Baleen plates yellowish with dark margins. *Line of mouth strongly arched*. Plates very narrow. Body black above, yellowish or dark grey beneath. Length 20ft or less. Dorsal fin sometimes without posterior concavity .... Pigmy Right Whale (*Caperea marginata*)

6
Baleen plates black, but with *white hair fringe*. Upper surface of snout with single central ridge. Body grey above and white below with strong pink hues. White undersurface not extending to flukes and flippers. Throat grooves end just posterior to junction of flippers .... Sei Whale *(Balaenoptera borealis)*

Baleen plates black, sometimes with patches of grey, white or yellowish hue, but *without white hair fringe*. Throat grooves extending to vicinity of navel. Upper surface of snout with three ridges, one central, two lateral. Baleen plates near tip of snout usually whitish .... Bryde's Whale *(Balaenoptera edeni)*

Body slate blue all over, sometimes with patches of lighter grey dappling and yellow on belly, baleen plates black **7**

7
Baleen plates black. Body surface slate blue dorsally and ventrally. No areas of white on ventral surface other than small flecks. Belly often coated with yellow diatom film .... Blue Whale *(Balaenoptera musculus)*

A smaller subspecies, the pigmy blue whale, matures at about 10ft less than the blue, but is difficult to distinguish from the latter.

8
Toothed whale with dorsal fin **10**

Toothed whale without dorsal fin **9**

9
Animal purplish-brown dorsally, white ventrally, with short beak and no trace of a dorsal fin, 6-8ft long .... Right Whale Dolphin *(Lissodelphis peroni)*

Whale with a dorsal hump instead of a fin, or a series of small corrugations. Gigantic box-like head taking up about one third of body bulk. Long slender lower jaw with powerful teeth. Body length 15-55ft .... Sperm Whale *(Physeter catodon)*

10
No central notch between flukes. Jaws narrow, often forming a prominent beak. Two conspicuous grooves beneath chin form a V shape not fused at apex **11**

Whale or dolphin with distinct central notch between flukes **12**

11
Head with a prominent beak, small teeth clearly visible in both jaws .... Shepherd's Beaked Whale *(Tasmacetus shepherdi)*

Head with prominent beak. A pair of teeth, longer than broad, projecting upwards from the middle of the lower jaw like a pair of tusks .... male Strap-tooth Whale *(Mesoplodon layardi)*

Several other beaked whales, identifiable by the pair of throat grooves, occur in the New Zealand region. These, and the female of *M.layardi,* need to be identified by a specialist.

12
Small whale, less than 13ft, with a small mouth with sharp curved teeth, set back well beneath head. Flippers pointed, dorsal fin present. General appearance rather shark-like .... Pigmy Sperm Whale *(Kogia breviceps)*

Small or medium-sized whale not fitting this description **13**

| 13 | Head distinctly bulbous, very short beak if any visible at all | .... | 14 |
| | Head not prominently bulged, beak present or absent, general body shape porpoise-like | .... | 16 |

| 14 | Teeth absent from upper jaw, or if present, buried in gum. Body grey, lighter beneath. Flippers of moderate length. Prominent dorsal fin .... | Risso's dolphin (*Grampus griseus*) |
| | Easily visible teeth present in both jaws .... | 15 |

| 15 | Peg-like teeth, restricted to the anterior part of the jaws only. Head very bulbous. Dorsal fin very large and deep, flippers very long, narrow and sharply elbowed. Body colour black, sometimes with white chin area, and white belly streak, a small white mark above and behind the eye, and sometimes a white saddle-shaped mark behind dorsal fin .... | Pilot Whale (*Globicephala melaena*) |
| | Teeth not peg-like, not so restricted in distribution. General appearance similar to pilot whale, but head more narrowly bulbous and somewhat elongate. Flippers not unusually long in relation to total body length. Teeth powerfully developed, round in cross-section. Body black, length 8-18ft .... | False Killer Whale (*Pseudorca crassidens*) |

| 16 | Wedge-shaped head with little or no beak. Body colour grey or black and whitish .... | 17 |
| | Porpoise or dolphin with distinct beak. Less than 15ft in total body length .... | 19 |

| 17 | Animal large, 12-30ft. Head with slight beak. Dorsal fin very tall, up to 6ft high in large animals. Body dark purplish-brown with white patches on belly, flanks and back. These patches yellowish in older animals. Flippers paddle-shaped, rounded at apex. Powerfully developed teeth, oval in cross-section .... | Killer Whale (*Orcinus orca*) |
| | Animal small, 3-8ft long .... | 18 |

18	Small dolphin, black and grey above, white below. Dorsal fin very bluntly triangular	Hector's Dolphin (*Cephalorhynchus hectori*)
	Body colour black above, white below, with patches of grey along the flanks. Flippers black, body without black area on flank posterior to flipper. Very short beak. About 30-32 teeth in each row. Dorsal fin sharply triangular	Dusky Dolphin (*Lagenorhynchus obscurus*)
	Body colour and general description more or less as above. Pronounced black area posterior to base of flipper. About 28 teeth in each row	White-sided Dolphin (*Lagenorhynchus cruciger*)

| 19 | Head mobile; very stout neck. Beak only 3-4 inches long, but well defined. Colour grey or grey-green dorsally, whitish ventrally. About 20-22 teeth in each row. Animal 8-15ft in length ..... | Bottlenosed Dolphin (*Tursiops truncatus*) |
| | Prominent beak, streamlined head. Dark stripe running through the eye region. Body dark purplish-brown, above, yellowish-white beneath, with greyish shading on flanks. About 40-50 teeth in each row. Body length up to 8ft .... | Common Dolphin (*Delphinus delphis*) |

Guide to the Use of the Key

Assume a small whale has come ashore which we want to identify. To start into the key, read couplet 1, look at the animal.

Q. Does it have fringed plates, or true teeth?

A. True teeth, very prominent ones. *Go to 8.*

Q. Is there a dorsal fin present?

A. Yes. *Go to 10.*

Q. Is there a notch between the flukes?

A. Yes. *Go to 12*

Q. Does it fit the description of the pigmy sperm?

A. No. *Go to 13.*

   Q. Is the head bulbous, that is, with a very prominent bulging forehead?

A. Yes. *Go to 14.*

   Q. Does it have teeth in both jaws?

A. Yes. *Go to 15.*

   Q. Are the teeth peg-like and the body coloured with a number of white patches?

A. No.

   Q. Are the teeth prominent, the head elongately bulbous, and the body colour black?

A. Yes. Identification, *Pseudorca crassidens.*

Figure 45. (George E. Gale, University of Guelph.)
Fin whale *Balaenoptera physalus,* showing characteristic view of the back in a shallow dive. Spruce Island, off the southeastern coast of New Brunswick, visible in background.

Notes: If after carefully following the key through, you are not able to determine the species, take all the careful notes and illustrations you can, and refer the problem to a specialist. You may have one of the beaked whales, which are almost impossible to key for the non-specialist, or even a species not normally found in the region.

The Identification of Whales at Sea

Correct identification of whales at sea is very difficult without long experience, but a number of pointers can be provided for the layman.

The shape and form of the spout is generally characteristic for the species of large whales. The sperm whale has the blowhole set at the left tip of the snout, and the blow rises obliquely forward. The spout of the right whale is doubled into a V-shape after a deep or relatively long dive, but not always. In the rorquals the spout is single, high and narrow. When whales are chased the shape of the spout changes, it becomes somewhat shorter and broader as the whale begins to pant. The spout of the humpback, high and bushy, is characteristic even at quite long ranges.

When a ship is relatively close to a whale the external features are of help. All the rorquals have prominent dorsal fins. The dorsal fin of the humpback is small and triangular, usually with a series of corrugations behind it. The right whale lacks a dorsal fin, and has a large yellowish encrustation on the snout. The sperm whale also lacks a dorsal fin, having a hump or series of blunt corrugations where other species have a fin.

The rorquals do not show their flukes when diving, except under rather unusual circumstances. The humpback, right and sperm usually show the flukes when making dives, especially when alarmed. The posterior margin of the humpback's flukes lacks a central notch, and has instead a line of corrugations. Size is also a guide, especially among the rorquals, but estimation of whale lengths at sea requires experience.

The killer whale is recognisable by the high dorsal fin with the straight posterior margin in the male, and the black and white markings. The pilot whale is recognisable by its very broad dorsal fin and bulbous head. Hector's dolphin has a quite small, blunt dorsal fin. The right whale dolphin completely lacks a dorsal fin and has striking black and white colouring. The wedge-shaped heads and stark black and white colouration of the dol-

Figure 46. (D. E. Gaskin.)
Head and back of large male sperm whale *Physeter catodon*, photographed just as 'blow' commences. At latitude 42°29'S 173°50'E on 21 March 1961 from *Chiyoda Maru No. 5*.

phins of the genus *Lagenorhynchus* are very characteristic, and the prominent beak and sharply divided markings of the Common dolphin are also easy to recognise.

Some whales, such as the pilot whale and the false killer, move in large schools, acres in extent. Other species, such as the pigmy right whale, have never been recorded at sea.

The Stranded Seal

In the southern parts of Australia, and in Tasmania and New Zealand, seals, especially fur seals, are a common element of the coastal fauna. They have well established colonies, which, under the protection imposed by the regions' governments in the last decades have prospered. The animals are reasonably numerous in suitable areas, and rarely excite the interest of coastal residents unless they appear in very unusual numbers, pup in unexpected areas or out of season, or are blamed for taking commercial fish or for wrecking nets.

Of more interest are the fleeting visitors; the Southern Elephant Seal, the Leopard Seal and Weddell Seal. These may turn up inexplicably in places where they have never been seen before, and perhaps after a few days vanish again, just as inexplicably. Frequently such seals strand in an exhausted and near-starving condition; sometimes they are badly injured and require shooting.

A seal from Macquarie Island, Heard Island or Campbell Island may bear a brand or have a small alloy tag through its flipper. The shape and size of such a tag or brand, and any lettering should be carefully noted and the information reported to the nearest wildlife organisation, Marine Department office, university or museum. If the seal is dead, any tag present should be cut free and sent to the Dominion Museum, Wellington, together with as much information as possible.

The Committee of Marine Mammalogists in the United States has prepared a schedule for taking standard measurements from seals (Scheffer 1967) and suggest the following measurements be taken as a minimum requirement.

The standard length, taken as the straight line distance from snout to tail tip along the ventral surface.

The curvilinear length along the side from snout to tail tip, when the animal is too heavy to roll over, or is wedged between rocks.

The length of the front flipper along the anterior margin with the flipper held out at right angles to the body.

The girth just posterior to the fore flippers.

The weight.

All these should be metric measures if possible.

The number of seals occurring in this region is small compared with the cetaceans, but any attempt to key them runs into some practical difficulties. Only experience suffices to separate immature fur seals and sea lions unless the animals are dead or unusually cooperative, and examination of the teeth is likewise somewhat difficult unless the animals are dead. If any doubt is held concerning the identification, take notes and drawings, and preferably a photograph, and consult a specialist biologist.

Key to the Seals of the New Zealand Region

1
- Ears small, but clearly visible externally; hind limbs turned forward in locomotion on land .... Otariidae **2**
- External ears absent; hind limbs not turned forward for locomotion on land .... Phocidae **3**

2
- Pelt having only a single layer of coarse hair. Genus *Neophoca*, sea lions. *N.cinerea,* the Australian sea lion, found on the coasts of South Australia and *N.hookeri,* the New Zealand sea lion, found on the islands of the New Zealand subantarctic, especially Auckland Island, only straggling to the New Zealand mainland.

- Pelt with an outer layer of coarse hair and an inner layer of thick water-resistant fur. Genus *Arctocephalus,* southern fur seals. *A.doriferus,* the Australian fur seal, found in southwestern and southern Australia; *A.tasmanicus* from Bass Strait and Tasmania; *A.tropicalis gazella* the Kerguelen fur seal, and *A.forsteri,* the New Zealand fur seal, ranging to Macquarie Island.

3
- Nostrils of males inflatable to form proboscis (males are the sex almost always encountered in temperate latitudes). Animal may be very large, weighing more than one ton. Pelt dark brown or greyish brown .... Southern Elephant Seal *(Mirounga leonina)*
- Nostrils of male not inflatable. Pelt black or dark grey above, whitish beneath, or grey with rings or blotches, or white with a few dark markings, or otherwise variegated **4**

4
- Eye sockets and eyes very large. Outer and inner incisors of nearly equal length. Found only in Antarctic regions, usually among pack ice .... Ross Seal *(Ommatophoca rossi)*
- Eye sockets and eyes not abnormally large **5**

5
- Seal dark grey above, shading to pale grey, with irregular splashes of paler or dark pigment in some areas, but not with dark blotches or rings in striking variegated pattern, head not semi-reptilian in appearance .... Weddell Seal *(Leptonychotes weddelli)*
- Pelt blotched or ringed, especially around shoulders **6**

6
- Body relatively long and sinuous. Head with long incisors and semi-reptilian in appearance. Pelt blotched and spotted with two or more shades of grey .... Leopard Seal *(Hydrurga leptonyx)*
- Body relatively short. Head rounded, not elongate. Pelt dark grey dorsally, shading to light grey on the flanks and white on the ventral surface. The backs and flanks are also marked with reddish-brown circles and irregular-shaped patches, especially around the shoulders .... Crabeater Seal *(Lobodon carcinophagus)*

III

10. The Whalebone or Baleen Whales — Mystacoceti

This suborder contains some of the largest animals ever to exist on the face of the earth, larger even than the biggest dinosaurs. All are characterised by a double row of whalebone or baleen plates hanging from the roof of the mouth. These plates are used to strain the food from seawater, usually euphausid shrimps, copepods or fish. There is a thick layer of oil-rich blubber beneath the skin.

These animals have been the most harried and hunted of all the whale species, and some are now close to extinction. Originally they were hunted for their baleen alone, later for their oil and eventually for their meat as well.

There are two families in the suborder; the Balaenopteridae or rorquals, and the Balaenidae or right whales.

The Rorquals — Family Balaenopteridae

All the rorquals possess a dorsal fin. They have the neck vertebrae free, not fused as in right whales. Each species has a number of characteristic grooves running along the ventral surface from the tip of the lower jaw to about the base of the thorax. Relative to the length of the body the whalebone plates are much shorter in the rorquals than in the right whales. All species except the humpback are slim-bodied, and even out of the water give an immediate impression of excellent adaptation to high speed swimming. The flippers are relatively small and tapered, except in the case of the humpback, and the large flukes have a distinct notch between them.

Authorities are not in agreement on the exact number of species in this family, but only six have general acceptance. These are split between two genera: *Balaenoptera* with five species, *musculus, physalus, borealis, edeni* and *acutorostrata,* and *Megaptera* with a single species *novaeangliae*. The humpback, with its short squat body and very long flippers, is very divergent from the usual rorqual form.

Another whalebone whale, the grey whale *Eschrichtius gibbosus,* differs from both the rorquals and the humpback in a number of important basic characteristics, including the possession of only two throat grooves instead of the hundred or so found in the other species. In some ways the grey whale is intermediate between the rorquals and the right whales. It is placed in a family of its own, the Eschrichtiidae. Two distinct populations are known, one on each side of the North Pacific. Its range does not extend into the western South Pacific, and it is not discussed in detail in this book.

Except for the humpback, all the rorquals appear very similar in the water, and the shape of the spout is rarely constant for a given species, varying according to the depth of dive from which the whale has surfaced, and the wind and other weather conditions. However, experienced gunners have little difficulty separating blue, fin and sei at reasonably close ranges. It is difficult to distinguish Bryde's whales from sei except from a very close view of the dorsal surface of the snout, although Bryde's whales are not found in the Southern Ocean.

A century ago there was confusion among both whalers and naturalists concerning the exact identities of rorqual species, for until the invention of the grenade harpoon and the speedy steam chaser these fast-swimming animals were hardly ever taken. Even in the late nineteenth century many scientists recognised several species of blue and fin whale. On general morphological grounds the situation is now reasonably clear, but recent detailed population studies (Fujino 1964, Chittleborough 1965) indicate that fairly distinct races and breeding units exist, as in human populations. The concept of whale populations as wide-ranging homogeneous units is probably very far from the truth. Some discrete subspecies have now been shown to be justified; for example, the pigmy blue whale *Balaenoptera musculus brevicauda*. The population of this subspecies is distributed in the Southern Ocean south of the Indian Ocean. The main locus of the population appears to be to the southeast of South Africa (Ichihara 1966b), with one specimen being recorded as far north as Durban (Gambell 1964).

There is also some evidence that northern and southern hemisphere minke whales may need to be considered as separate species, and some Japanese workers (Kasuya and Ichihara 1965) believe that two distinct species exist in the southern hemisphere, one concentrated around New Zealand and Tasmania and the other around Cape Horn and the South Shetland Islands. Minke whales seen around New Zealand do not always have the white patch on the flipper which is characteristic of the North Atlantic form (Cowan 1939, Sergeant 1963), though some certainly do have this patch.

THE BLUE WHALE
Balaenoptera musculus (Linnaeus 1758)
Southern sub-species: *B.m.intermedia*
(Burmeister 1866)

DESCRIPTION: The skin of the blue whale is bluish-grey, liberally dappled with pale grey on the throat grooves and often on the flanks as well. The dorsal fin is small relative to the size of the animal. The baleen plates are black. After blue whales have been in high latitudes for some time a yellowish scum of diatom film accumulates on the ventral surface. At one time these 'sulphur-bottom whales' were thought to belong to a distinct species.

Within the limits of the fossil record, it is almost certain that the blue whale is the bulkiest animal that has ever existed. The longest authentic measurement is of a female 102 ft in length, recorded in the International Whaling Statistics. Accurate recordkeeping began some time after the birth of the Southern Ocean whaling industry, so the existence of 110 ft animals in the original unexploited population is not impossible. Estimates of 120 ft, which are often heard in nostalgic whaling circles, were probably based on measurements made round the curves of the stomachs of very large females. International statistics demand that a standard straight-line

Figure 47.

Balaenoptera musculus, blue whale: 1. slate-blue belly, 2. black baleen, 3. dorsal fin.

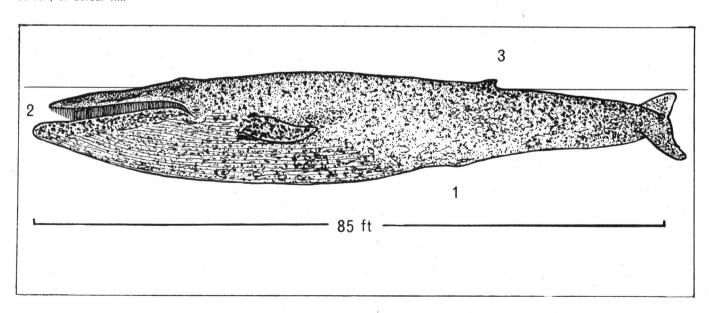

85 ft

body length be recorded, from the tip of the jaw to the crotch of the flukes.

DISTRIBUTION: Although its numbers are now greatly reduced compared to former times, the blue whale occurs in all the major oceans of the world. In the North Atlantic there appears to be coastwise segregation into two populations (Jonsgaard 1966). The one in the west ranges from the coast of the Carolinas in winter, but not as far south as Florida (Moore 1953), and up to the pack ice of Davis Strait and southern Greenland in summer (Allen 1916). Blue whales taken around Spitsbergen by Norwegian whalers in the 1930s (Jonsgaard 1966) might have come from either the eastern or western North Atlantic populations. On the eastern side of the Atlantic blue whales appear to winter around the Cape Verde Islands (Kirpichnikov 1950) and move in summer perhaps as far north as Spitsbergen and the Barents Sea. Christensen (1955) and Jonsgaard (1955) discussed the low level to which North Atlantic stocks had been reduced by over-exploitation.

Despite heavy pelagic whaling activity there in recent years, little has been published on the distribution and movements of blue whales in the North Pacific. The population size there, even before catching, was almost certainly much smaller than in the Southern Ocean (Mackintosh 1965, Nishiwaki 1966a). A map of catches of blue whales shown by Dr Nishiwaki suggests that there are east and west populations of blue whales in the North Pacific, the western population migrating in summer to the vicinity of the northern coast of Japan, and also off Kamchatka, while the eastern population centre is close to the central and eastern Aleutian Islands and the coast of Alaska.

Virtually nothing is known concerning blue whales which inhabit the northern Indian Ocean (Daniel 1963, Mackintosh 1965), but blue whales found in the southern Indian Ocean are probably best regarded as part of the series of populations which exist in the circumpolar Southern Ocean. A map comparing the distributions of blue whales and pigmy blue whales published by Dr T. Ichihara (1966b) indicates that in the southern Indian Ocean the pigmy blue whale predominates, particularly in lower and temperate latitudes. Above 55°S, especially between longitudes 80°E and 120°E, great blue whales were more often taken. Blue whales have been seen and taken near the Western Australian coast (Chittleborough 1953), but only in small numbers.

The greatest concentration of both pelagic and shore whaling effort has been exerted against the blue whales migrating to and fro from the South Atlantic and the

Figure 48. (D. E. Gaskin.)
British and Norwegian whalers pose before a 94ft blue whale *Balaenoptera musculus*, on the FF *Southern Venturer* in the Weddell Sea in February 1962.

coast of South Africa to the vicinity of South Georgia, the Weddell Sea and the coast of Queen Maud Land. Study of the population dynamics by the scientists of the International Whaling Commission confirm that the numbers of the population in this region have been decreased to a dangerously low level. The breeding grounds of blue whales in this area are not known, but may be in offshore zones not too far from the coasts of Southwest Africa and Brazil (Mackintosh 1966). Other breeding loci may be out in mid-South Atlantic. However, blue whales begin to move into this sector of the Antarctic whaling grounds from the South Atlantic in October and November, feeding on *Thysanoessa macrura* and *Euphausia vallentini* on the outer edge of the Antarctic whaling grounds (Nemoto 1962), and *Euphausia superba* when they finally reach their peak numbers on the southern ground close to the Antarctic mainland in February and March. After the end of March the number of whales in this region decreases steadily as they move away north again. A few will still be found in Antarctic latitudes even in May and June, and correspondingly some individuals appear not to make a complete migration, but remain in temperate latitudes all the year.

Seasonal movements of blue whales in the New Zealand region follow an almost identical pattern. Migration streams pass both the eastern and western coasts of

67

the archipelago, usually several score miles from shore. Tory Channel whalers have taken a number of specimens in Cook Strait during their operations (Gaskin 1968a), and blue whales appear to move from the east coast of the South Island to the west coast of the North Island by passing through the strait. There are a number of authenticated stranding records from both coasts of both main islands of New Zealand (von Haast 1883, Archey 1926, Gaskin 1968a).

The population of blue whales in the New Zealand region is now very small. Catches by pelagic factory ships to the south, southeast and southwest of the archipelago reached a peak of 2,200 in the 1950-51 season, declining drastically to only 74 in the 1961-62 season (International Whaling Statistics 1967). Catches continued at a very low level until complete protection was introduced by the International Whaling Commission in 1963. It is agreed that the blue whale is close to extinction near New Zealand. During the four month whale-spotting cruise of *Chiyoda Maru No. 5* in 1966-67, only 5 specimens were seen (Gaskin 1967a), four of these in the Tasman Sea in temperate latitudes

BIOLOGY: Because of the former commercial importance of the blue whale in the Antarctic, its biology has been studied in considerable detail (Mackintosh 1942, 1965; Mackintosh and Wheeler 1929; Ruud, Jonsgaard and Ottestad 1950; Nishiwaki 1950, 1952; Nishiwaki and Hayashi 1950; Nishiwaki and Oye 1951; Omura 1950).

If the number of laminations in the earplugs and the number of ripples laid down on the baleen plates each year have been correctly interpreted, then it appears that blue whales may live to thirty years of age unless taken by whale chasers or dying from disease or accident.

The pairing season for blue whales in the waters around Australia and New Zealand reaches a peak in May and June, judging from studies on foetal length distributions. The breeding grounds are not known, but as in the South Atlantic, are probably in temperate or even subtropical latitudes. The gestation period is between 10 and 12 months, and most calves are born in April and May, at the very end or even after the end of the southern whaling season. In the years before complete protection was given to the blue whale in the Southern Ocean, catching was prohibited before 1 February, so that by the time biologists on factory ships or at antarctic land stations were able to examine specimens, the foetuses were already very large. Blue whales are about 24 ft long at birth. The calves are suckled for about 7 months, and at the beginning of the antarctic whaling season following birth, mothers are still often accompanied by very large adolescent young, between 58 and 68 ft in length. These animals reach sexual maturity at about 5 years of age (Nishiwaki 1952), and at body lengths of about 74 ft in males and 78 ft in females (Mackintosh 1965).

In the southern latitudes of the western South Pacific, blue whales feed almost entirely on euphausid shrimps, and their feeding is restricted to the upper few fathoms of the sea where the whale krill accumulates in large quantities. The blue whale is, however, capable of quite extensive deep dives (Slijper 1962), and although the normal time taken for a single dive or 'sound' is 5 to 10 minutes (Mackintosh 1965), there are records of blue whales staying submerged for up to about 20 minutes.

We know little of the composition and stability of school structure for this species. Nemoto (1964) reported that schools of up to a dozen individuals are still seen very occasionally in the North Pacific, and that mother and calf combinations often swim apart from small schools of males and pregnant and non-pregnant females. In the Antarctic blue whales used to be seen much closer to the Antarctic mainland than fin whales and actually in leads and pools in the pack ice. Traditionally the British and Norwegian factory ships used to move south to the

Figure 49. (D. E. Gaskin.)
Whalers peeling back the blubber from the head of a 91ft blue whale on the deck of the FF *Southern Venturer* in the Weddell Sea, 1962.

ice edge as the blue whale season began. The reason for this high latitude-biassed distribution is not known, and it does not seem to occur in the North Pacific (Nishiwaki 1966a). Blue whales have been known to maintain speeds of about 10 knots for long periods, and over relatively short distances can stay ahead of a chaser doing speeds of 15 to 17 knots. The species can undoubtedly sprint at 20 knots for very short distances.

The normal whale ectoparasites have been recorded from blue whale skin from time to time. These include *Penella* and *Cyamus* and a few others which may be more or less accidental in their attachment. Blue whales have not been reported with the large encrustations of barnacles which occur on the skin of humpbacks and right whales. Tapeworm cysts are frequent in the blubber, and the alimentary canals are usually infested with cestodes and nematodes. There is, however, little evidence (Mackintosh 1965) that these parasites do the whales any real harm. Oval scars, often raw, are frequently found on the skin of blue whales, especially on the flanks and ventral surface. These marks are probably made by lampreys in temperate waters. The wounds are completely unlike the rows of small circular scars found around the mouths and across the heads of sperm whales, which are apparently caused by squid on which sperm whales feed.

The drastic decline in the size of the blue whale population in the Southern Ocean can be at least in part blamed on the blue whale unit system, as considered earlier. All the time that baleen whales were considered as part of a unit catch rather than as individual species, the blue whale could not be efficiently protected. As long as it was economic to base the hunt on fin whales and blue whales were not protected, they were taken whenever a chaser encountered them. Population studies by the International Whaling Commission's 'Committee of Three Scientists', Mr K. R. Allen, formerly Director of Fisheries Research in Wellington, Dr D. G. Chapman of the U.S.A., and Mr S. J. Holt of the Food and Agriculture Organisation, indicate that the original stock size of the population of blue whales in the Southern Ocean was about 150,000 before extensive exploitation began. The decline in population size was already serious before the pelagic whaling industry began to recover from the effects of the Second World War, but pleas for restrictions on the taking of blue whales only served to slow, rather than halt or reverse the trend towards extinction. Moves to limit the blue whale season's length were brought in during the early 1950s and implemented in 1954 (season to start from 21 January) and 1955 (season to start on

Figure 50. (D. E. Gaskin.)
A whaler demonstrates the gape of the lower jaw of the large female blue whale shown in Figure 49.

1 February). However, by 1962 the stock was obviously in a critical condition. The Committee estimated that only one or two thousand blue whales survived in the whole Antarctic Ocean. With the exception of an area south of the Indian Ocean where the pigmy blue whale predominated (Ichihara and Doi 1964), final protection for the blue whale was agreed at the 1963 Whaling Commission meeting, far too late to save the stock.

THE FIN WHALE
Balaenoptera physalus (Linnaeus 1758)
Southern sub-species: *B.p.quoyi* (Fischer 1829)
DESCRIPTION: In the southern hemisphere the average male fin whale is about 67 feet and the female about 73 feet. The largest fin whale recorded in official whaling statistics was a female 88 ft in length which was taken during the 1931 Antarctic whaling season. In post-war years fin whales larger than 80 ft have become rare owing to the heavy catches. Few animals manage to avoid the attentions of pelagic whaling fleets for enough seasons to reach full physical maturity. The author measured an 85 ft female during the 1961 season in the Southern Ocean, and several others over 80 ft were also recorded in the same season, but these were only a small handful of the 27,099 fin whales taken in the Southern Ocean by the land stations and pelagic fleets that season.

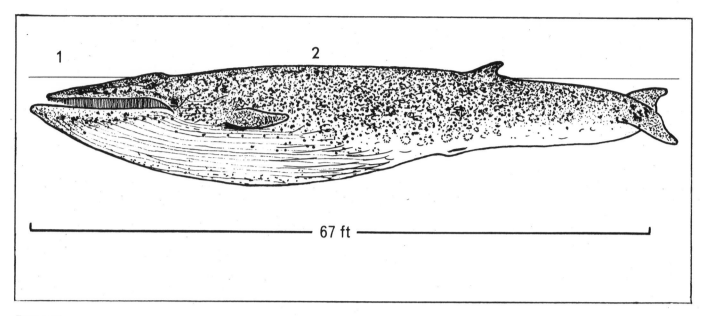

Figure 51.
Balaenoptera physalus, fin whale: 1. yellowish or bluish grey baleen, 2. white belly.

The fin whale, also known as the finner, finback or common rorqual by whalers, is somewhat sleeker in build than the blue whale, with a less blunt snout and a more streamlined head. The dorsal surface and flanks are dark grey dappled with pale grey, while the ventral surface, including the undersurface of the flukes, is white. The lower jaw is asymmetrically pigmented; the right side is white and the left side grey. This is associated with the habit this species has of swimming tilted to the right while feeding, with the grey above and the white below the water. Because of this habit the fin has also been called the side-whale by some whalers, but this name is not generally favoured because of possible confusion with the sei whale. There is little variation in the relative amounts of grey and white in this species, but in a few cases there is less white on the ventral surface than usual. The whalebone plates of the fin are bluish-grey except for a white section on the right side towards the tip of the jaw. One Japanese worker has recently found that there is a significant difference in the shape of the tip of the snout in male and female fin whales. Northern hemisphere races or subspecies of the fin whale appear to become sexually mature at slightly shorter lengths than their southern counterparts (Slijper 1962; Mackintosh 1965), and are not as long at physical maturity, but are otherwise very similar.

DISTRIBUTION: The distribution of the fin whale is similar to that of the blue whale, although there are important differences within given areas. In the North Atlantic fin whales range from the vicinity of Spitsbergen at about 80°N in summer (Jonsgaard 1966) down to Florida in the west (Moore 1953) and the Canary Islands in the east (Kirpichnikov 1950). Ruud (1949) and Postel (1956) recorded strandings on the coast of Tunis, and Tamino (1953c) recorded one on the coast of Italy.

In the North Pacific two distinct populations occur (Nishiwaki 1966a), one in the eastern and the other in the western part of the Pacific. A small eastern sub-population has been recorded from the East China Sea. The distinctness of these populations has been demonstrated by analysis of whale marks recovered in the North Pacific marking programme (Nishiwaki 1966a) and by blood serum classification (Fujino 1953, 1960).

Relatively little is known of fin whale distribution in the Indian Ocean, although those in the southern portion appear to form part of what is generally regarded as the Southern Ocean circumpolar population. Generally whale biologists recognise a large stock of fin whales in Antarctic Area II south of the South Atlantic, another in Area III south of South Africa, another in Area IV to the south of Western Australia, smaller stocks in Area V south of New Zealand and in Area VI in the southern South Pacific, and finally a small stock in the Bellingshausen Sea to the southwest of Cape Horn (Ottestad 1956). The breeding grounds of this species are assumed to lie in temperate or subtropical waters, perhaps 150 or more miles from the coasts of major land masses (Clarke 1962, Mackintosh 1966). Near South Africa the northern limit of distribution on migration appears to be about 35°S (Bannister and Gambell 1965), unless the breeding grounds lie far offshore out of range of land-based chasers. South African catches generally show a predominance of immature animals.

The distribution of fin whale populations to the south of and around the coasts of Australia and New Zealand is not known in detail. Only two specimens have been taken at Australian whaling stations in the last 15 years (International Whaling Statistics, 1968), and aerial surveys off the coast of Western Australia indicated that fin whales generally stay many miles offshore (Chittleborough 1953). Migration streams appear to pass both coasts of New Zealand, and whalers in Cook Strait have observed fin whales passing through the narrows while going north, crossing from the east side of New Zealand to the west. All known fin whale strandings in New Zealand have occurred either on the east coast or Cook Strait coast of the South Island (Parker 1885, Gaskin 1968a).

Large regular concentrations of fin whales are rarely recorded near the New Zealand mainland. Only one sighting was made in 1963, for example. This can be regarded as typical for coastal waters up to 10 miles from shore. However, in January 1964 Royal New Zealand Air Force and civil aircraft reported larger than usual concentrations of whale food along the east coast of the South Island from Christchurch to Cook Strait, especially in the Pegasus Bay area. In February and March 1964 the whale chaser *Orca* began to report fin whales, mingled with a greater number of sei, moving up the coast about 15 miles from shore. These animals were first seen in numbers off the Kaikoura Peninsula, and later off Cape Palliser. In the latter area the stream of whales appeared to be dividing, some turning through Cook Strait, others going north-east along the east coast of the North Island. In mid-March the Russian factory ship *Slava* arrived in Pegasus Bay, and aerial photographs revealed fin and sei whales tied at the stern and being brought in by her accompanying whale chasers. At the same time Mr Allan Wright, the lighthouse keeper at Akaroa, reported boats from the expedition chasing rorquals about five-seven miles offshore. After the Russian fleet moved away on 19 March rorqual sightings were much more sporadic, but were still made up to 3 April, all animals being seen moving north. Two animals were seen by fishermen only one mile from shore near Akaroa.

BIOLOGY: In the southern hemisphere the majority of calves are born between June and July (Mackin-

tosh and Wheeler 1929; Laws 1959, 1961). Sextuplets have been recorded (Jonsgaard 1953). The gestation period is about 11 months. The calves are suckled for about 6 months and are about 20-22 ft in length at birth (Symons 1955b). They grow about another 17 ft by the time weaning is complete and they then begin to feed on euphausids and other pelagic crustaceans like the adults (Pike 1953a). Sexual maturity is attained at body lengths of about 63 ft in males and 65 ft in females. Northern hemisphere fin whales are mature at shorter lengths (Jonsgaard 1952, Ohsumi 1960, Ohsumi, Nishiwaki and Hibiya 1958, Mackintosh 1965). Relative growth of various parts of the body was studied by Kramer (1954), and Fujino (1954).

Figure 52. (D. E. Gaskin.)
Rare dorsal view of a fin whale *Balaenoptera physalus* in January 1962 on FF *Southern Venturer,* demonstrating the fit of the upper jaw into the lower, and the greatly elongated rostral region of the head extending from the centrally placed nostrils.

Intensive research on this commercially important species indicates that sexual maturity is reached at about five years (Laws and Purves 1956; Laws 1961). Age in fin whales has been estimated by three methods; earplug lamination counts (Roe 1967), baleen plate ripple counts (Ruud 1945, van Utrecht 1966, van Utrecht-Cock 1966), and counts of the number of corpora albicantia in the ovaries resulting from pregnancies (Mackintosh and Wheeler 1929). A life span of at least 26 years has been proven by direct means. A stainless steel *Discovery* whale mark fired into a fin whale in the Southern Ocean prior to World War II was recovered 26 years later, and the target animal, according to the records in London was listed as a large animal at the time of marking. Indirect ageing methods described above are still not completely reliable and results are contested by a few authorities, but it appears that under natural conditions a fin whale might live to an age of 30 or even 40 years. At present there seems to be nearly a consensus of opinion that two laminations per earplug are laid down each year (Mackintosh 1965).

The diet of fin whales in the Southern Ocean consists almost entirely of the shrimp *Euphausia superba,* also the predominant food item of southern blue whales. However, at the coastal whaling stations of Norway the fin whale is known as the Sildhval, or Herring whale, since in the temperate waters of the North Atlantic and the northern North Sea the species feeds mainly on surface schooling fish, and herring forms a major fraction of the diet.

School structure has been studied by Andrews (1909), and in more detail by Nemoto (1964), but only on the feeding grounds of the North Pacific and Southern Ocean. They conclude that in general fin whales only congregate into large schools when actually on feeding grounds or on migration, that otherwise the usual school size is not more than about five animals, and that cows with calves tend to swim apart from other whales. These conclusions are drawn with the warning that the heavy exploitation of the species may have brought about considerable changes in the behaviour patterns.

The size of the fin whale population in the seas south of New Zealand and eastern Australia was probably never as large as in other sectors of the Antarctic whaling grounds. Catches were good near the Balleny Islands and in the Ross Sea when pelagic whaling first began again after World War II, but steadily declined after the 1959-60 season, when 4,730 whales were caught, to 1,444 in 1963-64, and to only 304 in the 1966-67 season. In that season, only 22 fin whales were seen by the Japanese

Figure 53. (D. E. Gaskin.)
Flensing a 78ft. fin whale *Balaenoptera physalus,* near the South Sandwich Islands in December 1961, on the FF *Southern Venturer.* Note the large dark tongue adjacent to the baleen plates in the mouth, and the loose tissue underlying the throat grooves.

whaling research vessel *Chiyoda Maru No. 5* in three and a half months steaming (Gaskin 1967a), and only two of these were seen to the south-east of New Zealand. There is no doubt that the fin whales of Antarctic Areas IV, V and VI are in desperate need of protection.

The condition of the blue and fin whale stocks is such that one can only paint a most sombre picture; under these circumstances it might be well to leave the fin whale on a slightly lighter note.

There is a stiff penalty for taking a lactating fin whale with a calf, although the gunner who would deliberately shoot a mother with young is rare, i.e. not from any sentimentality, probably, but because a calf grows into a big whale worth money the following season. However, lactating whales are occasionally taken by accident. While serving as biologist/whaling inspector on the *Southern Venturer* in 1961, I received a cable from the Dairy Research Institute of Reading, England, requesting a quart of whale milk for experimental analysis. A fortnight later one of our chasers did take a lactating whale; the streams of milk could be seen while the animal lay belly up in the water at the stern of the factory. Unfortunately by the time it was hauled up several hours later the flow had ceased. I tugged at the teats to the great amusement of the deck crew, but succeeded only in wringing out about three tablespoonsful of thick creamy fluid. Finally, the bosun came to the rescue. Half a dozen heavyweight flensers were detailed off to stand on top of the whale above the teats, the whale being on its side, and at the shouted command began to jump up and down in unison above the glands. Despite the near-hysterical state of both myself and the watching crowd, the Dairy Research Institute obtained its quart of whale milk.

THE SEI WHALE
Balaenoptera borealis (Lesson 1828)
Southern sub-species: *B.b.schlegeli* (Flower 1865)
DESCRIPTION: This is the smallest rorqual normally exploited in the Southern Ocean. Because of its relatively small oil yield it has not been taken in large numbers by pelagic factory ships until recent years, when schools of blue and fin have become hard to find; and in addition the Japanese pelagic industry has shifted emphasis from whale oil to whale meat.

In general outline it is much like a smaller replica of the fin whale. The average body length at maturity is only about 46 to 50 ft. The dorsal surface is bluish grey, and the ventral surface white, mottled with pink. The white zone does not extend on to the undersurface of the flukes and flippers as in the fin whale. In the water the sei whale is recognised by experienced whalers by the shape of its back, which is very narrow compared to the larger rorquals, the sharply recurved dorsal fin, and the shape of the spout under conditions of good visibility. Except by a close view of the snout it is not possible to distinguish the sei in the water from the next species, *Balaenoptera edeni,* Bryde's whale. The latter has a central ridge down the dorsal surface of the snout from nares to tip, together with a pair of lateral ridges which arise from grooves towards the rear of the snout (Omura 1962a, 1962b). The sei whale has only the central ridge. The whalebone plates of the sei are black with a fine white hair fringe. The throat grooves extend to the base of the thorax, about in a line with the flipper, while the grooves in *B.edeni* extend back as far as the navel (Omura 1959, 1962a, 1962b).
DISTRIBUTION: Although it was once believed to

Figure 54.
Balaenoptera borealis, sei whale: 1. black baleen, with white hair fringe, 2. white belly, tinged strongly with pink, and with throat grooves extending only to vicinity of sternum.

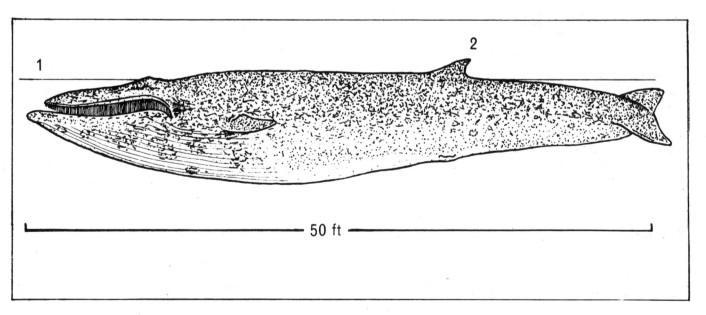

50 ft

Figure 55.   (D. E. Gaskin.)
Sei whale on deck of FF *Southern Venturer* during 1962, near
South Shetland Islands.

range throughout all latitudes in the major oceans of the
world, relatively recent studies suggest that many 'sei'
whales of tropical and subtropical latitudes were in reality
records of Bryde's whales (see distribution section under
B.edeni). However, Bryde's whale does not appear to
extend very far into temperate latitudes (Omura and
Nemoto 1955, Nishiwaki 1966a), unless water temperatures reach nearly 20°C, and based on New Zealand
experience, it seems probable that all animals are sei
where surface temperatures fall below 15°C.

In the North Atlantic the normal high latitude limit of
sei whale migrations appears to be about 72°N (Ingebrigtsen 1929), though the same author did record one
small school at latitude 79°N, and Collett (1912) gave
records from the Barents Sea and even the coast of
Novaya Zemlya. On the western side of the Atlantic the
coast of Labrador appears to be close to the northward
limit of movement. Sei whales figure in catches at whaling stations on the coasts of Newfoundland and Nova
Scotia at the present time. On both sides of the Atlantic in

subtropical latitudes 'sei' whale strandings have been
recorded, but Jonsgaard (1966) suggested that some or all
of these could be Bryde's whales.

Sei whales are caught in large numbers in the North
Pacific by pelagic expeditions, and land stations in Japan
and Russia (Omura and Fujino 1954, Omura and Nemoto
1955). In 1967 the total North Pacific catch was a little
over 6,000 animals, including catches from stations then
operating on the coasts of British Columbia and California.
Sei whales are the most important part of the Japanese
coastal whaling catch (Nishiwaki 1966a), and are taken
in June-July and again in September-October, during the
northward and southward legs of migrations which take
them to the vicinity of the Aleutian Islands, where large
numbers are seen in summer.

Sei whales have been intensively studied in the Southern
Ocean in the last few years, as the catches have risen at
an alarming rate since the southern fin whale stocks
declined sharply after the 1961-62 season. While only
5,503 sei were taken in the 1962-63 season, the catch
was 20,380 in 1964-65. Since then it has declined to
12,368 in 1966-67. Although the number of pelagic
factories operating in the Antarctic has dropped from 15
in 1964-65 to only nine in 1966-67, the size of the sei
whale stock appears to be declining faster than the
decrease in effort. Fortunately the pelagic whaling
nations, now only Japan and the Soviet Union, have been
able to agree at a catch quota of only 3,500 blue whale
units to try to ensure that catches would be less than the
combined fin and sei whale sustainable yields. This
should enable stocks of both to recover slowly from the
over-exploitation to which they have been subjected.
North Pacific stocks are still protected only by mutual
agreement on catch levels between the major whaling
countries in the area; this is most unsatisfactory, and is no
substitute for scientifically backed international control.

According to the International Whaling Statistics, in
1965-66 the major sei whaling area was Antarctic Area
II, while in 1966-67 effort had shifted largely to Area
III. In the same period exploitation of sei in Area V
south of New Zealand declined from 1,008 in 1965-66 to
717 in 1966-67. It is probably true to say that sei are
still fairly common in the New Zealand region; sightings
by *Chiyoda Maru No. 5* in 1966-67 suggested more so
than to the south of Australia (Gaskin 1967a). There
are few stranding or catch records for this species on the
New Zealand coast; a single sei was taken at Tory Channel in Cook Strait during the 1956 whaling season, and
four more off the Marlborough coast in December 1964.
Whalebone plates were collected by one of my assistants,

Mr Colin Humphrey, and from these it was possible to confirm the catches as sei and not Bryde's whales. Most New Zealand strandings of 'sei' probably were Bryde's whales, judging from the localities (Gaskin 1968a). However, a sei was stranded at Lyall Bay in August 1948, one at Porirua in 1922 (Oliver 1922a), and most recently, at Hokitika in late 1968. Sei whales have been observed in migration off the east coast of the South Island on a number of occasions, and Tory Channel whalers confirmed that sei whales moved through Cook Strait from east to west very early in the local humpback season. Their opinion was that sei were not very common, but this appears to have been because observations from the Tory Channel did not start until late April, as pointed out by Dr W. H. Dawbin in one of his papers on New Zealand humpbacks (1956a). In fact a number of sei were sighted by the whale chaser *Orca* in December 1964, and four animals taken from these. More animals were seen in March of the same year at the time when fin were recorded in some numbers, and the Soviet factory ship *Slava* came into Pegasus Bay.

Sightings of sei whales off the east coast of the North

Figure 56.   (D. E. Gaskin.)
Gunner of the *Southern Lotus* preparing to kill a sei whale *Balaenoptera borealis* near the South Shetland Islands in early April 1962.

Island in June 1963 (Gaskin 1963) suggest that the animal winters in temperate latitudes in this region. However, no southward movement in Spring has been noted in coastal waters. Observations by the *Chiyoda Maru No. 5* expedition in the summer of 1966-67 indicated that concentrations of sei whales occurred both to the east and west of New Zealand, in the southern Tasman Sea and between New Zealand and the Chatham Islands. A great number of any given sei whale population appear to stay in temperate and subantarctic waters, rather than migrating down close to the pack ice like blue and humpback and many fin whales. In the 1966-67 season no sei whales were taken below latitude 70°S, although factory ships operated between 70°S and the edge of the Antarctic mainland taking fin and sperm whales (International Whaling Statistics, 1968). In fact the data shows that more than twice the number of sei were taken between latitudes 40 and 50°S than between 50 and 60°S, and 60 and 70°S combined. The general conclusion is that in this part of the southern hemisphere sei whale migrations almost certainly occur, but are not very distinct on the basis of present information. However, work by Bannister and Gamble on sei and other baleen whale migrations in South African waters (1965) showed a clearer picture. Sei whales are the last of the baleen whales to reach southern latitudes on migration, often not arriving until February and March. As a result, although fin whales reach their maximum numbers in July on the northward phase of the migration in South African waters, the number of sei whales does not reach a peak until September.

BIOLOGY: The blubber of sei whales is very thin in comparison to that of larger rorquals, and is usually no more than about two inches thick on the flanks (Andrews 1916). Some whalers contemptuously refer to it as 'oil cloth'. The oil yield is correspondingly low, and some small sei produce no more than two tons of oil, worth less than $360. If hunted for oil, they need to be taken in very large numbers to make the venture even marginally profitable. As mentioned earlier, the Japanese pelagic industry in the Southern Ocean is now largely geared to hunting sei for meat.

Sei whale diet varies with latitude. In the Southern Ocean the species feeds almost exclusively on *Euphausia superba* (Matthews 1938c), but in the northern Atlantic Ocean sei feed on the euphausid *Thysanoessa inermis* and the common copepod *Calanus finmarchicus,* as do other baleen whales (Mackintosh 1965). Matthews (1932) noted that in temperate and subantarctic waters off the shelf of Patagonia sei whales feed on swarms of *Munida gregaria.*

Figure 57. (D. E. Gaskin.)
Lemming (opening the body cavity and separating meat) a sei whale *Balaenoptera borealis*, on the FF *Southern Venturer* in March 1962.

The breeding season and reproductive cycle of sei whales is slightly different from that of the other large rorquals; this correlates with its retarded migratory cycle in the southern hemisphere discussed in the last section. Maximum pairing occurs about July (Matthews 1938c), and calves are born at about the same time of year, after a gestation period of 12 months. Mating is presumed to take place about a week after the calves are born, although catches of mature resting females suggest that calves may not be carried every year. Calves are about 12 ft in length at birth, and grow rapidly during a five-month weaning period. Sei in the southern hemisphere are about 28 ft long when weaning is complete. Sexual maturity is attained in both sexes at body lengths of about 45 ft.

As in other species of rorqual, females are consistently longer at a given age and state of maturity than males. Since exploited populations of whales generally soon show a reduction in mean body lengths of the animals caught, we need to look back a number of years for large sei. In the 1961-62 Antarctic whaling season the mean length of a catch of 1,977 males was 48½ ft, and that of

a catch of 3,181 females was little over 51 ft. Lengths recorded of the longest male and female were 56 ft and 64 ft respectively. In the 1966-67 season the mean length in a male catch of 6,464 was 47 ft, and that in a female catch of 5,896 was just under 50 ft. While a 57 ft male was taken in this season, the longest female was also only 57 ft.

Sei whales often form very large loose aggregations on feeding grounds; for example an aggregation of 140 sei was seen from *Chiyoda Maru No. 5* in 1966-67 (Gaskin 1967a), and altogether nine groups of more than 30 animale were sighted. It might be wrong to call these large associations schools. However, tight schools of six to eight animals are not uncommon unless catching becomes heavy in an area, although Nemoto (1964) considered three or four a normal upper limit for school size. In fact a study of sei whale sightings by *Chiyoda Maru No. 5* (Gaskin 1967a) verifies Nemoto's view; of 209 sei whale contacts 165 were of 1, 2, 3 or 4 animals. Segregation by sexes appears to be common in sei whale schools; Bannister and Gamble (1965) found that of 32 sei taken in one period of a few days, 30. were females. Although gunner selection towards large animals may have partly accounted for this, it is unlikely to be the only factor.

While large external parasites such as *Penella* are only rarely observed on the skin of sei (Matthews 1938c) the small *Balaenophilus* frequently covers the baleen plates (Bannister and Grindley 1966). Lamprey scars are also common; sei are attacked in temperate waters. The same scars are normally almost healed by the time the animals move into relatively high latitudes (Matthews 1938c). A knife-cut into a sei whale intestine almost invariably reveals large numbers of endoparasites. The most common species are the cestode *Tetrabothrius affinis* and the acanthocephalan *Bolbosoma turbinella* (Matthews 1938c).

In conclusion: a sad story associated with sei whales in New Zealand waters — although some may think it questionable taste to move from a discussion of intestinal parasites to reporters. One of that venturesome fraternity came on board *HMNZS Paea* while she was engaged in whale-marking off the Palliser coast in February 1963, with the intention of making a tape recording of the procedure for the New Zealand Broadcasting Commission. Both whales and saturated biologists put on a fine display for the public performance. However, the reporter's satisfaction at his eloquent commentary was somewhat marred; first, by discovering on playback that nothing had gone on tape, secondly, by being offered red, half-cooked saveloys to eat while feeling desperately seasick on what was a rough and blustering day, and thirdly, by having to suffer the indignity of being hauled out by a boat hook after falling in Napier harbour with all his equipment just at the moment the voyage ended.

BRYDE'S WHALE
Balaenoptera edeni (Anderson 1878)

DESCRIPTION: Bryde's whale can be distinguished from its close relative the sei whale by four characteristics (Omura 1959, 1962a, 1962b). Firstly, the upper surface of the snout of *B.edeni* has a pair of lateral ridges as well as a ridge down the centre from the vicinity of the blowhole, while the sei whale has only a single median ridge. The general build of the body is much slimmer than that of the sei while the throat grooves extend to the navel, much further posteriorly than in *B.borealis*. The baleen plates are more robust and coarse than those of the sei, and are anteriorly white and grey, not black as in the sei whale.

Variation in Bryde's whales from one locality to another is considerable. Their real status may eventually come to rest as one or more subspecies of the sei, and specimens intermediate between Bryde's and sei have been encountered on occasions. However, there is no doubt that in most places the populations of Bryde's whales and sei whales are distinct and do not interbreed.

DISTRIBUTION: *Balaenoptera edeni* is a tropical and subtropical species, its range not extending further south than about 40°S nor further north than about latitude 40°N. The species has been identified off the coasts of Japan (Omura 1959, 1962b), South Africa (Olsen 1913), Equatorial Africa (Ruud 1952), both the Brazilian and South African coasts of the South Atlantic (Omura 1962a, Best 1960), in the Caribbean (Soot-Ryen 1961), in the Straits of Malacca (Junge 1950), in the Bay of Bengal (Slijper 1962), in the South-east Pacific (Clarke and Aguayo 1965), off the coast of southern West Australia (Chittleborough 1959b), and around the north east coast of the North Island of New Zealand (Gaskin 1968a).

In New Zealand waters the animals are particularly common in the vicinity of volcanic White Island in the Bay of Plenty, and around both Little Barrier and Great Barrier Islands in the Hauraki Gulf. Schools are also encountered near Motiti Island not far from Tauranga harbour. Movements of quite large schools have been noted; in June 1963 the Royal New Zealand Airforce reported a sighting of a school of 15 medium sized rorquals, which were almost certainly Bryde's whales, moving south from White Island towards the mainland.

The whaling station at Great Barrier Island exploited the Bryde's whales of the Hauraki Gulf, in addition to its main catches of humpback whales, between 1956 and 1961, when the station finally closed down permanently. After these very small catches, which totalled only six males and 13 females, local fishermen reported that sightings of whales became scarce near Little Barrier Island until about 1963, although now they have been recorded again quite commonly all along the Thames and Coromandel coasts, and even on the edge of Auckland

Figure 58.

Balaenoptera edeni, Bryde's whale: 1. anterior baleen plates grey and whitish, 2. grooves extending back as far as navel.

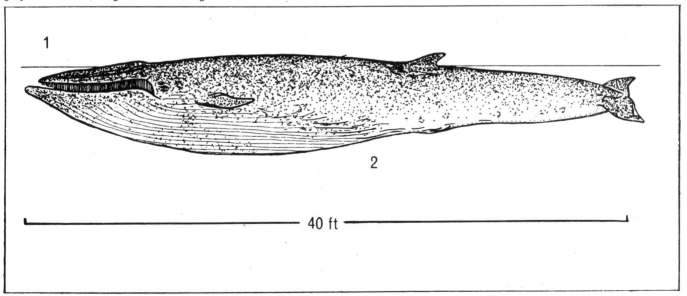

1

2

40 ft

harbour. It is not likely that the population around the Barrier Islands is large enough to stand commercial exploitation at even a very low level for any length of time. There is no evidence of migratory movements on any scale near New Zealand. My own records are scanty, but they suggest that Bryde's whales occur in the Bay of Plenty and the Hauraki Gulf all the year round, feeding on surface school fish, and that any movements they make are purely local and coincide with seasonal movements of their food fish. Dr P. B. Best (1960) found that the species' movements in South African waters were restricted to a very limited north-south seasonal migration following the movements of fish schools, although Symons (1955a) considered the possibility of migration to the Antarctic. Best concluded that these movements might be better interpreted as inshore-offshore shift of population centre. However, in the western North Pacific the species makes seasonal migrations which cover a span of about a thousand miles, wintering around the Bonin Islands and migrating in spring and summer into the waters adjacent to the northeast coast of Japan (Omura and Nemoto 1955, Nishiwaki 1966a). The whole migration range of the species however, is greatly affected by the movements of the warm north-flowing Kuroshio current, and Bryde's whale in this part of the Pacific appears to have a northward range generally restricted by the 20°C surface temperature isotherm.

I do not think that the surface temperature can possibly be so restrictive for the distribution of the species in northeastern New Zealand waters. A study of the movements by surface isotherms recorded by Dr D. M. Garner and Dr N. M. Ridgway (1965) shows that in June, July and August the 20°C isotherm lies far to the north of the North Island of New Zealand, while the Hauraki Gulf and Bay of Plenty are dominated by surface waters of between 16° and 17°C. In late September and early October 1963 three Bryde's whales were marked (Gaskin 1964a) in north-eastern New Zealand coastal waters, two of them within a few miles of Little Barrier Island. We later learned that these apparently belonged to a small school which had been in the area, seen almost every day by local fishermen, throughout most of August.

On the other hand, the lack of records of Bryde's whales further south than the East Cape of the North Island is closely consistent with the southward extent of temperatures of 19-20°C in mid-summer off that part of New Zealand. It is a pity that the cessation of whaling activity in New Zealand, although affording protection to this species, also prevents us from obtaining more information on the distribution and movements of the animal.

BIOLOGY: The coarse whalebone plates of Bryde's whale, relative to those of the sei whale, are associated with its diet and feeding habits. While the sei whale feeds habitually on *Euphausia superba,* the whale 'krill' in the southern hemisphere (Matthews 1938c), Bryde's whale is generally a fish feeder, taking pilchards and other school fish of similar size (Chittleborough 1959b, Best 1960). However, Omura (1962b) noted that the stomachs of Bryde's whales taken on the coast of Japan often contained large quantities of pelagic crustaceans. Larger fish have been recorded from time to time, but these were probably harrying the school of smaller fish taken by the whales, being swallowed by accident rather than design. Dr W. H. Dawbin examined a number of Bryde's whales at the Great Barrier Island whaling station and found the species to be eating pilchards, herring and mackerel. Quantities of the crustacean *Munida gregaria* were also found in some stomachs.

While conducting whale-marking in the Hauraki Gulf in September-October 1963, I had an opportunity to watch a small school of Bryde's whales feeding among a number of dense patches of surface school fish, which we could not see clearly enough to identify. Two whales in particular drew our attention, racing along side by side among the fish and suddenly leaping vertically out of the water so that the flippers were visible. They would then slide back into the sea but immediately begin to move rapidly forward again with their snouts well clear of the surface. Only when the marking launches came within about 50 yards did the behaviour change. The process of catching fish, for this they were clearly doing, with some fish jumping clear of the closing mouths at the last second, was conducted at speeds of from 8 to 12 knots.

Neither while feeding nor being chased during actual marking operations did the whales dive for any length of time. The longest dive timed was one of three and a half minutes. When travelling with some apparent purpose the animals spouted at intervals of two to two and a quarter minutes. The spout was quite thin and hard to see in cloudy conditions, and at no time did the animals show their flukes when diving.

We know very little about the reproduction of *B.edeni.* It has been suggested (Slijper 1962) that there is no regular pairing or calving season in either hemisphere, and Dr Best found that this appeared to be true at least in South African waters, after a study of foetal length distributions (Best 1960). However, Dr Omura (1962b) found strong evidence for seasonal mating and calving in the north-west Pacific population, associated with the migratory habit of this particular population.

The six males taken at the Great Barrier Island whaling station were from 36 to 48 ft in length, and the 13 females ranged from 39 to 49 ft in length. These New Zealand specimens appear to have been of comparable size to those taken in Japanese waters, where sexual maturity is attained by males at a body length of about 36 ft and by females at about 37 ft (Omura 1966), but somewhat larger than specimens taken at the whaling station in western Australia (Chittleborough 1959b). Japanese studies suggest that both sexes reach sexual maturity when about 5 years of age, and that calves are about 11 ft in length at birth.

THE MINKE WHALE
Balaenoptera acutorostrata (Lacépède 1804)

DESCRIPTION: This rarely exceeds a length of 25 ft in the southern hemisphere, although in the North Atlantic the mean length of males is about 23 ft (within a range of up to 27 ft) and the mean length of females is about 25 ft (within a range of up to 29 ft) (Sergeant 1963). The dorsal surface of the minke whale is bluish-grey and the ventral surface is pure white. It is often stated in general literature that the minke can be easily recognised and separated from the young of larger rorquals by the white patch on the outside of the flipper. However, this character is only a constant diagnostic feature in the northern hemisphere populations. Several different subspecies have been described, but the taxonomic status of

these is still not clear. There may be two species of minke whale in the southern hemisphere, *Balaenoptera acutorostrata* Lacépède and *B.bonaerensis* Burmeister. The latter species was first described by Burmeister in 1867 but has long been considered a synonym of *acutorostrata* until Williamson (1961) opened the question again. The problem of these two forms has been considered yet again by Kasuya and Ichihara (1965), who presented much stronger evidence for two different colour phases, if not separate species, of minke. It appears that *B.acutorostrata* possesses a white patch on the outside of the flippers, while *B.bonaerensis* does not. Both forms have been stranded on the New Zealand coast. The whales described as *B.bonaerensis* by the two Japanese workers were caught south of Australia. While a specimen of minke stranded at Haast in August 1958 clearly lacks a white patch on the flipper (Gaskin 1968c, fig. 18), another specimen stranded in 1968 near Timaru just as clearly did have such a patch. In both colour phases the baleen plates are short, flexible and yellow or pale ochre in colour. The dorsal fin is relatively small, but very prominent.

DISTRIBUTION: The minke is now the commonest rorqual in the Southern Ocean, mainly because it has the lowest oil yield of the commercially exploited baleen whales, and it has been ignored while the industry concentrated on blue, fin, humpback and sei. In the northern hemisphere the species has not been quite so lucky,

Figure 59.
Balaenoptera acutorostrata, minke whale: 1. short yellowish baleen plates, 2. white belly.

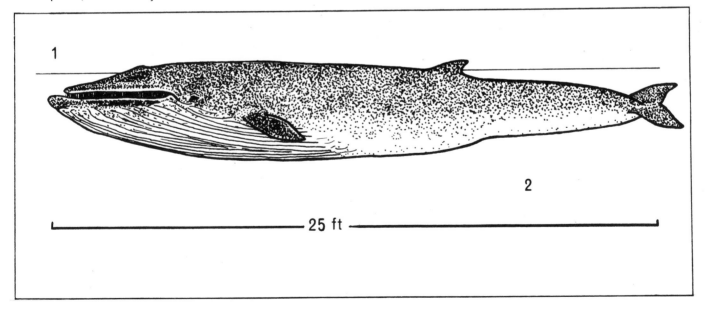

though there is no indication that it is in any danger. The Japanese, Newfoundland, Norwegian and Icelandic coastal operations, with their relatively low costs, have found it profitable to hunt the minke (Sergeant 1963).

In the western North Atlantic the species seems to be most common in cold waters, ranging from the latitude of Long Island (Allen 1916) northwards in summer to Hudson Strait north of Quebec. However, there are also records of minke from Florida (Moore 1953) and the Gulf of Mexico (Moore and Palmer 1955). On the eastern side of the North Atlantic, Turner (1891), Ellison (1951, 1954b), Stephenson (1951) and Jonsgaard (1955) found that minke occurred widely through temperate and arctic waters, and that although there was evidence of seasonal migratory behaviour, immature whales did not make regular movements. This view was supported by Fraser (1953). Norwegian catches of minke whales have declined steadily since 1963 (International Whaling Statistics 1968), giving some suspicion of overexploitation.

In the North Pacific catches are much smaller, two to three hundred compared with two to three thousand in the Atlantic, and have risen steadily since 1963. The species has been studied in the North Pacific by Scattergood (1949), Omura and Sakiura (1956), Omura (1957), and earlier by Matsuura (1936b). Fairly regular seasonal migrations in the coastal waters of Japan were observed, together with some segregation of schools into mature and immature animals while passing the coast.

In the Southern Ocean the species has been taken in very small numbers by Japanese and Russian pelagic expeditions (Arseniev 1961) on a tentative basis, to see if it might be possible to take the species economically, and to provide specimens for scientific examination. Only in the tropical South Atlantic has the species been exploited to any degree in the southern hemisphere; Brazil began a minke whale fishery in 1963 with a catch of two animals, and this increased to 488 in 1967.

The species is known from Western Australian waters (Chittleborough 1953) although it has never been taken commercially on that coast. The minke occurs around Tasmania (Davies and Guiler 1958) and on the coast of Victoria (Wakefield 1968), but records are sparse. There are definite records of minke whales on each side of the New Zealand archipelago (von Haast 1881, Gaskin 1968a), but there is nothing to suggest that the species is common in coastal waters. None were sighted during a survey conducted by the Marine Department along the east coast of both main islands in 1963 (Gaskin 1963, 1964a), and only two were sighted by the whale chaser

Figure 60. (Dominion Museum, Wellington.)
Head of a minke whale stranded at Haast, New Zealand, showing the arrangement of the baleen plates.

Orca during two seasons' operation in Cook Strait and off the Marlborough coast. Most strandings have occurred on the east coast of the South Island (Gaskin 1968a), but the coastwise flow of a northward cold current (Brodie 1960) may account for these coincidental strandings, which were at localities where other marine animals have stranded (McCann 1964b).

Minke were not seen in numbers anywhere near the New Zealand archipelago by the Japanese whale research vessel *Chiyoda Maru No. 5* during its cruise in the 1966-67 summer, although single animals were seen in both the Tasman Sea and off the east coast of New Zealand about 100 miles from shore. More were seen in colder waters close to the Antarctic Convergence at about latitude 60°S to the southeast of the country (Gaskin 1967a). Minke do in fact appear to be distributed throughout the Southern Ocean in higher latitudes than most of the larger rorquals except perhaps blue whales (Mackintosh 1965). The species is often seen in leads into the Antarctic continental shelf ice pack, and Taylor (1957), found minke whales trapped in gradually closing pools in the ice. Sergeant (1963) recorded a similar occurrence on the coast of Labrador.

BIOLOGY: While minke whales in the Southern Ocean appear to feed on euphausids like larger rorquals, Sergeant (1963) found that the dominant food in Newfoundland waters was the small fish, the capelin, *(Mallotus villosus* Mueller), which swarms in coastal waters at the time the minke whales are most abundant, i.e. between late May and early August.

The minke whale occupies a unique place in the annals of whale behaviour studies, since to date it is the only baleen whale to have been kept in an aquarium (Kimura and Nemoto 1956). This specimen was kept in a large tidal pool with a sea gate. The animal finally managed to escape while the gate was open, but a number of interesting observations were made before this. The whale seemed to be most active at night, and did not in fact seem to sleep at all in the sense which we know it. Schools of small fish were let in through the gate on the flood tide, and appeared to be eaten at night.

Most of the biological information we have on the anatomy and life-cycle has come from work in the north (Carte and MacAlister 1868; Omura and Sakiura 1956; Jonsgaard 1951, 1962; Sergeant 1963). Most pairing takes place in the North Atlantic between January and May, and most calves are born between November and March after a 10 month gestation period. Weaning appears to take only $4\frac{1}{4}$ months, a shorter time than in any other rorqual. Calves are about nine feet in length at birth and become sexually mature at about two years of age. There appears to be close agreement between results obtained in the North Pacific and the North Atlantic for these biological data (Sergeant 1963). Benham (1902b) described the external features of a calf stranded on the coast of Otago.

THE HUMPBACK WHALE
Megaptera novaeangliae (Borowski 1781)

DESCRIPTION: The humpback is the best known large cetacean in the western South Pacific, although unfortunately no longer one of the most common. At physical maturity the species reaches lengths of 45-50 ft. The dorsal fin is not very prominent or well developed; it is set well back on the crest of the dorsal surface, and is usually in the form of a triangular projection followed by an irregular series of smaller projections. In some specimens a true curved dorsal fin is present, similar to that of rorquals. There is much variation in colour pattern. The typical humpback whale is black dorsally and white ventrally, the two zones of pigmentation meeting low down on the flanks. Animals nearly all white or all black have been recorded. The flippers are very long in relation to the total body length, and their leading edges are covered with a number of lumps and protrusions, and frequently with large barnacles of the genus *Coronula*, which are also found under the chin, among the throat grooves, on the surface of the snout, and around the genitalia of both sexes. The flippers, as well as the flanks, are liberally marked with white splotches of pigment in generally black areas, and black marks in other-

Figure 61.
Megaptera novaeangliae, humpback whale: 1. black baleen, 2. small dorsal fin, often reduced to a ridge, 3. very long flippers.

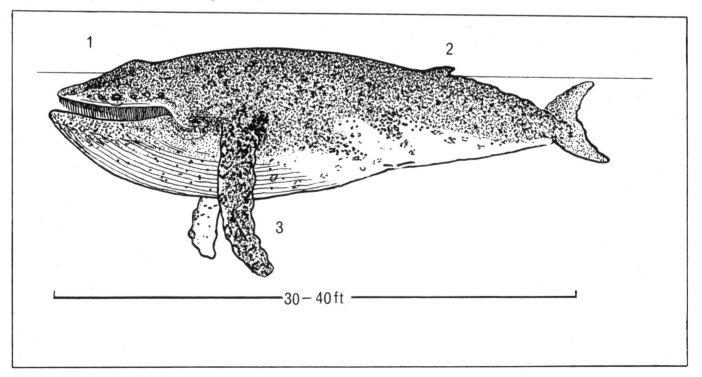

30 − 40 ft

wise white areas. The baleen plates of the humpback are black (sometimes grey) and occasionally marked with white. There are about 250 plates on each side of the upper jaw (Hinton 1925).

Dr D. G. Lillie studied humpback whales caught in New Zealand coastal waters before the First World War, and made detailed observations on the colour patterns (1915). He recognised four types of pattern, in which the black of the dorsal surface encroached in three bands into the white zone of the ventral surface, the final stage being an animal almost completely black. Intermediate coloured animals were most common in New Zealand waters, although humpbacks from South Africa and the South Georgia region are most commonly of the dark type (Matthews 1937). As a rule in both areas, females are darker than males. In the dark animals, blotches of white pigmented areas, or possibly small patches without the dark pigment, are usually lacking.

DISTRIBUTION: Although the humpback whale is much less common now than in earlier times, it is still found in all the major oceans of the world. In the North Atlantic, eastern and western populations make seasonal migrations from the tropical and subtropical Atlantic and the Caribbean (Kellogg 1929) to the coasts of Greenland, northern Norway and Labrador (Jonsgaard 1966). Kellogg (1929) noted that on the eastern side they used to penetrate as far as Spitsbergen and the Barents Sea, but according to Jonsgaard (1966) there are practically no humpbacks left in the eastern stream now, even though the species is at present completely protected in the North Atlantic.

In the South Atlantic humpback whales migrate to and from the coasts of Brazil and the Congo coast of Africa. At the Saldanha Bay whaling station on the coast of South Africa whales were migrating north when taken in July and south when taken in September (Olsen 1915). Most of the population seemed to strike the coast (at that time of writing) north of Saldanha Bay. Heavy catches used to be made on the coast of Angola (Harmer 1931). In former years large numbers of humpback whales were taken by British and Norwegians whaling at South Georgia. The first of the southbound whales arrived off the island in October, and the population reached a peak between November and February (Hinton 1925). However, Matthews (1937) pointed out that a number of immature animals did not take part in this migration, and were taken at South Georgia during winter whaling. Catches of humpbacks taken by American whalers (Townsend 1935) indicate that the humpback whales from both hemispheres winter in the vicinity of the

equator, but that most of the populations stay in the subtropics. Mathew (1948) recorded a stranding on the coast of India.

The observations made by Japanese whale biologists in the North Pacific (Nishiwaki 1966a) suggest two populations as with other large baleen whales, one in the east (Pike 1953b) and the other in the west. However, whales marked near the Aleutian Islands have been recovered at the Ryukyu Islands, where considerable numbers of humpbacks used to be taken (Nishiwaki 1959, 1960). Whaling for humpbacks in the North Pacific is now carried out at a very low level. There seems little doubt that the very large catches of humpbacks taken by Russian factory ships operating from Kamtchatka in 1962 and 1963, when nearly 3,500 animals were taken and processed (International Whaling Statistics 1967), severely reduced the size of the western North Pacific population. Nishiwaki (1959, 1960) noted that a distinct sequence could be seen in the migration of animals reaching the Ryukyu Islands. Immature males arrived first, mature males and females next, and nursing females with calves last of all, the latter group not leaving until the autumn.

The most intensive studies in the world on the biology of humpback whales have been carried out in Western Australian, eastern Australian and New Zealand waters. In Western Australia the species has been studied in great detail by Dr R. G. Chittleborough (1953, 1954, 1958a, 1958b, 1959a, 1959b, 1959c, 1962, 1963, 1965), and pre-Second World War information has been summarised by Dakin (1934). Eastern Australian humpbacks have been studied by Dr H. Omura (1953), Dr W. H. Dawbin (1956b, 1959a, 1964, 1966), Stump, Robins and Garde (1960), and Robins (1954), and New Zealand humpbacks by Dawbin (1954, 1956a, 1956b, 1959a, 1959b, 1960, 1964, 1966). A little further information has been added, together with a bibliographic summary of Dawbin's work, by Gaskin (1968a). Ivashin (1962) summarised Russian humpback marking in the Antarctic.

The Western Australian humpbacks are referred to as the Group IV Antarctic Area stock, while those of the Antarctic Area to the south of eastern Australia and New Zealand together form the Group V stock. These two populations are, by and large, fairly distinct units. There is some interchange between the two south of southern Australia during the species' main feeding period in the Southern Ocean, but whale-marking recoveries indicate that this mixing is very small, only about 5 per cent per year. The two populations are isolated by the whole width of Australia during the breeding season in tropical waters. Although the two populations are thus more or

Figure 62. (Government Printer, Western Australia.)
A humpback whale being dragged up the slipway of the Norwest Whaling Company Station near Carnarvon, Western Australia.

less reproductively isolated, the small degree of mixing is apparently just enough to prevent the two populations becoming genetically isolated (Chittleborough 1965).

Off the coast of Western Australia humpback whales are first sighted going north at the end of May or the beginning of June after the pelagic whaling season closes south of Australia on 7 April. Some humpbacks are still in waters near the pack ice edge at that time. They continue to move northwards until August, when a return migration begins, which is completed by perhaps as late as December, although by then the bulk of the population will be moving back into Antarctic waters again. The species breeds in waters adjacent to the coast between latitudes 25° and 30°S, and in the years between the two World Wars, was heavily exploited in these latitudes (Dakin, 1934). After the Second World War the humpback sustained whaling stations at Albany in southern Western Australia and Carnarvon on the west coast. However, catches declined slowly between 1958 and 1961, and very drastically from 1961, when the total Australian catch was 1,481 for all land stations, to only 87

in 1963. By this time only Albany was left operating, since this was the only station with substantial stocks of sperm whales within reach.

Even before 1949 the humpback Group IV population was still showing signs of the serious over-exploitation in the 1936-39 period (Chittleborough 1965), despite a 10-year period of protection in the southern hemisphere from 1939 to 1949. It has now been demonstrated by the whale biologists in this region (Chittleborough 1965) that the Group IV population has been reduced to about only 800 animals, and that the stock will have to be left to recover for at least 30 years before culling in commercially profitable numbers can begin again.

The Group V population, although not showing signs of being seriously depleted before the 10-year protection period in the Southern Ocean, held its own against pelagic, eastern Australian and New Zealand shore whaling exploitation from 1949 to 1960. After this the same drastic decline in numbers experienced at Western Australian whaling stations also occurred in the east. New Zealand catches dropped from 361 in 1960, to 80 in 1961, 32 in 1962 and only 9 in 1963. The whaling stations at Byron Bay and Moreton Island on the east Australian coast failed, as did the New Zealand station at Great Barrier Island, leaving only Tory Channel station in Cook Strait to struggle on for two more seasons on sperm whales and a few sei. A study of the population dynamics of the Group V population (Chittleborough 1965) showed that the total stock did not exceed 500 animals in 1963-64. The humpback whale is now very belatedly protected completely throughout the whole southern hemisphere.

Dr W. H. Dawbin studied the migration of the Group V population in great detail. During the summer months these animals feed in the Southern Ocean between latitudes 59°S and 68°S and longitudes 150°E and 180°E. During the summer the bulk of the population performs migrations similar to those observed off Western Australia, moving northward into tropical and subtropical waters to breed. Some reach latitude 10°S. The whales pass the coast of New Zealand going north between early May and early October, maintaining average speeds of between two and five knots. Some individuals linger in one place for several weeks (Dawbin 1956a), though this is most frequently observed on the southward leg. Whales pass both sides of the New Zealand archipelago, and a substantial fraction used to pass through Cook Strait going northwest. Dr Dawbin believes (Chittleborough 1965) that the Group V population has within it a number of fairly distinct breeding sub-units, as whales generally seemed to repeat migrations along the

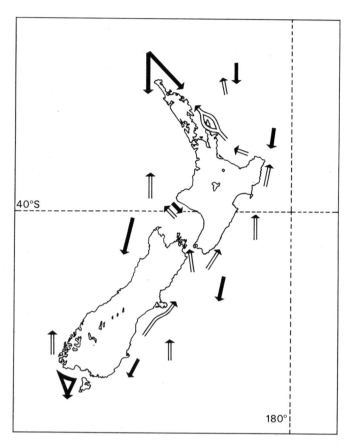

Figure 63. (Compiled from drawings and diagrams published by W. H. Dawbin.)

The seasonal movements of humpback whales around the New Zealand coast. Black arrows indicate southward migration in spring, white arrows northward migration in the autumn. Relatively little southward movement occurs through Cook Strait.

same route year after year. In the tropics the Group V animals became widely dispersed into coastal waters. They have been recorded in the following localities; off the east coast of Queensland, at Norfolk Island, around the Chesterfields, off eastern Papua, in the Solomons, New Hebrides, Fiji Islands, Samoa, the Cook Islands and at Tonga, where a small number are taken for meat by the natives in most years. These breeding units appear to mingle on the feeding grounds in the Antarctic but segregate out on the migrations to the breeding grounds. During migration the whales appear to maintain fairly steady speeds (Dawbin 1966), and examination of their stomach contents shows that humpbacks very rarely feed on the northward migration through New Zealand coastal waters. The whales spend only about six weeks at the most in the tropics, and once breeding is complete they turn southwards, passing New Zealand between late August and early December.

Foveaux Strait is an area where humpback whales gather before leaving New Zealand on the remainder of their southward migration back to the Southern Ocean feeding grounds. Some catches were made at Great Barrier Island from the southbound stream, but the Tory Channel station operated almost entirely during the northbound migration.

Dr Dawbin (1966) found a definite social segregation within the migrating animals. On the northward leg sexually immature males and females pass the coast of New Zealand first, together with mature females which are just at the end of lactation. The middle part of the stream is composed of mature males and non-pregnant females. Pregnant females come at the end of the migration stream. The whales which reach the breeding grounds first also leave first; the southbound sequence is very much the same, with immature whales coming south first. The pattern appears to be the same throughout the southern hemisphere; Dawbin carried out a detailed analysis of all available humpback records from land stations operational in the southern hemisphere since the Second World War, using a computer to test for statistical significance of the periods at which the different social categories predominated.

BIOLOGY: In the Southern Ocean the diet of humpback whales consists almost entirely of the krill *Euphausia superba,* though the Japanese recorded *Thysanoessa macrura* in the stomachs of animals taken between 130°W and 100°W (Nemoto and Nasu 1958). During the summer each humpback whale eats about 1 to 1½ tons of krill per day (Slijper 1962). A few fish are occasionally found, but these are probably swallowed accidentally. Whales during the early phase of the southbound migration past the New Zealand coast have been known to

Figure 64. (D. E. Gaskin.)
A foetal humpback with its membranes. On FF *Southern Venturer* during 1961-62 Southern Ocean whaling season.

feed on the lobster krill *Munida gregaria*. However, in general, whales on migration appear to live almost entirely on their fat reserves. This is dramatically demonstrated by the differences in oil yields from northbound and southbound animals (Chittleborough 1965). While up to 55 barrels of oil might be expected from the northbound humpback, the yield from a southbound one will generally be as much as 10 barrels less.

The life-cycle of the humpback is relatively well known. The males become sexually mature at an average length of about 36 ft, and the females at an average of about 38 ft. Both sexes are about four or five years of age at puberty. The males undergo a definite sexual cycle (Chittleborough 1965) with a recognisable seasonal change in the weight and activity of the testes.

The complete female cycle lasts about two years, although there is frequently a resting year between each full cycle in which a calf is conceived, born and suckled (Chittleborough 1954, 1958a). The incidence of twins in the species is very low, only about 0.3 per cent (Chittleborough 1965). Most calves are conceived in August, and the gestation period is about 11½ months, so that the majority of calves are born in the following July and early August. The mean length at birth is about 14 ft. Lactation lasts about 10 to 11 months (Chittleborough 1958a) although Dawbin believes (1966) that the calves begin to take food themselves when only a few months old, based on observations made in Foveaux Strait. It is assumed that the main causes of calf mortality are attacks by sharks or killer whales, but this is something on which we have very little information.

As in all baleen whales, growth of the calves is very rapid. Growth in the humpback slows to almost nil at body lengths of 45 to 46 ft. There are seasonal changes in girth however, associated with the feeding cycle. The maximum body lengths recorded for male and female humpbacks are 47 ft and 51 ft respectively. The maximum age attained by a humpback is about 48 years, if earplug laminations have been correctly interpreted.

The Right Whales — Family Balaenidae

Two genera are recognised in this family; *Eubalaena* and *Caperea*. In *Eubalaena* there are at least two species or groups of species, *E.mysticetus*, the Greenland or bowhead whale, and *E. 'glacialis'*, the black or right whales. It is very difficult to distinguish the three species recognised within the *'glacialis'* group; but the populations do not mix, and are presumably genetically distinct. The southern right whale is usually known as *E.australis*, and the North Pacific right whale as *E.sieboldii*. The name *E.glacialis* is reserved for the Biscayan or North Atlantic right whale (Andrews 1908a). The genus *Caperea* contains one species, *C.marginata*, the pigmy right whale, which appears to have a circumpolar but discontinuous distribution in the temperate waters of the southern hemisphere.

The mouth of right whales is very strongly arched, and the whalebone plates are long relative to the body size in comparison with the whalebone of rorquals. These plates reach a length of 14 ft in *E.mysticetus*. Throat grooves, which are so characteristic of rorquals, are absent in right whales. *Eubalaena* lacks a dorsal fin, but a small one is present in *Caperea*. Unlike the rorquals, right whales have all the cervical or neck vertebrae fused together.

THE SOUTHERN RIGHT WHALE
Eubalaena australis (Desmoulins 1822)

DESCRIPTION: This whale can be recognised by its lack of a dorsal fin, the high arched jaw, the peculiar eruption on the anterior end of the snout called the 'bonnet' by whalers, and the V-shaped spout, although the latter is not a completely reliable character. The 'bonnet' is actually a heavily cornified skin structure infested with a number of species of parasite, especially crustaceans. Its function is not known. The length of the right whale at maturity varies from 45 to 55 ft, although specimens of up to 60 ft are not unknown.

In the days of bay whaling in the southern hemisphere, and particularly New Zealand, this species was called the 'common whale' or 'black whale'. To the pelagic whalers it was the 'right' whale to catch, since it floated after being killed, and had a very high oil yield compared with rorquals or the humpback. The whalebone of this animal was inferior in length and quality to that of the Greenland right whale, but was nevertheless widely used for a time.

DISTRIBUTION: *Eubalaena australis* has a circumpolar distribution in the southern hemisphere, although the migration streams of different populations appear to be quite widely separated (Townsend 1935), with no evidence of interchange between the groups.

Most of the main right whaling grounds were adjacent to major landmasses; these whales were caught off the coasts of Brazil, Peru, Cape of Good Hope, Western Australia, Tasmania, New South Wales and New Zealand. Subsidiary streams appear to have terminated around temperate and subantarctic islands such as Tristan

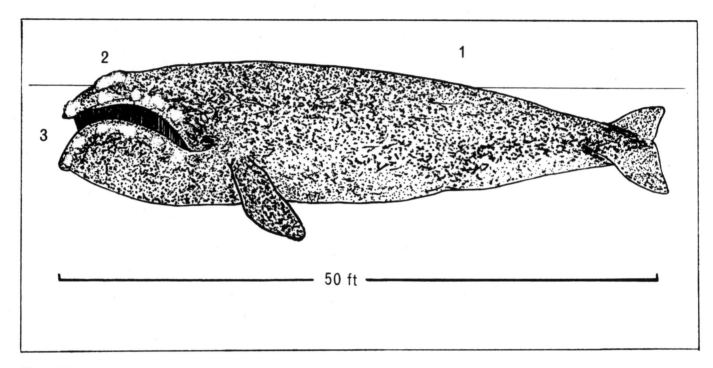

Figure 65.
Eubalaena australis, southern right whale: 1. No dorsal fin. 2. 'bonnet'. 3. arched mouth with long baleen.

da Cunha, South Georgia, Kerguelen, Campbell, Auckland and the Chatham Islands. There were also a number of offshore right whaling grounds, for example, one on the Chatham Rise between New Zealand and the Chatham Islands. Since the right whales were brought close to the verge of total extinction before good records were kept, it is not now clear if these were all separate termination areas of migration routes. Possibly there was considerable lateral movement of individuals, and even year to year variation in the centres of population.

In the New Zealand region only Campbell Island is still regularly visited by right whales (Bailey and Sorensen 1962). Analysis of sightings from Campbell Island (Gaskin 1968a) suggests that the migratory pattern has changed, perhaps through the intense over-exploitation prior to protection. The whales now seem to arrive in the vicinity of Campbell Island about two months later in the year than they did 50 or more years ago. In 1911 and 1912 the peak of sighting occurred between May and July, while weather station records show that there is now a peak between July and September. A number of recent sightings on the coasts of New Zealand (Gaskin 1964c), and also Australia (Chittleborough 1956) were made between the same months.

With the possible exception of Tristan da Cunha, more right whales are seen at Campbell Island than at any other single locality in the southern hemisphere. Most of the Campbell Island sightings have been restricted to North West Bay and Perseverance Harbour, although a few whales have also been sighted from time to time off the south coast. The greatest number of sightings in one year to date was 116 in 1963, a figure approached only by 106 in 1942. It is difficult to estimate the efficiency of the weather station observations from one year to the next, and the type of behaviour exhibited by right whales makes it seem very probable that the same animals were sighted over and over again. Consequently it is not possible to make a reliable estimate of the total population size near the island. However, the largest number of whales seen by the same group of observers on the same day at one time was only between 45 and 60. This probably represents a sizeable fraction of the total population in the whole western South Pacific. Only 14 right whales were sighted by the Russian whaling research vessel *Kooperatsiya* in 1957-58 at various localities in the Ross Sea and the Southern Ocean between the longitudes of New Zealand and Easter Island (Zenkovich 1962)

In this region the right whales spend the summer months feeding in the Southern Ocean and the northern part of the Ross Sea, often very close to the pack-ice, and begin to move northwards at the end of the summer, reaching the subantarctic islands and the New Zealand mainland in July, August and September. Observers at Campbell Island have reported right whales rolling around in pairs in relatively shallow water, and giving every appearance of mating (Gaskin 1968a). Some calves are also born there, and pairs and small family groups of whales seem to spend several weeks at Campbell Island moving in and out of the bays. After September the whales disappear rapidly out to sea, and with the coming of the southern spring migrate back into high latitudes again.

Right whaling on the New Zealand mainland was very intensive. Probably only the fact that a few individuals habitually mated and calved at remote Campbell Island, rather than going further north, saved the species from complete extinction in the western South Pacific. Although small whaling stations were set up on Campbell Island, the adverse weather conditions and the isolation prevented them from ever becoming going concerns. Men could not be persuaded to stay there for the small returns that resulted. One by one the stations were abandoned after only one or two seasons whaling, and in this way the species survived in the area.

Unfortunately the population is undoubtedly still quite small, and the recovery has been very slow. Thirty-six years elapsed between the last sighting of a right whale by the Cook Strait whalers in 1927 and the first recorded sighting on the New Zealand coast in recent times, at Tory Channel in July 1963 (Gaskin 1964c). Altogether five right whales were seen round the New Zealand coast between July and September 1963; the first one, ironically enough, within a few hundred yards of Tory Channel whaling station. The animal moved slowly into the channel, following the edge of the kelp beds, and showed no fear of the boat which closely followed its passage for some miles into the Marlborough Sounds. At the end of August Mr Cornelius de Ryk of Bluff, Southland, was fortunate enough to be able to take about 40 ft of 8 mm colour cinefilm of a right whale which came to the entrance of Bluff harbour. On 1 September another right whale, this time with a calf, was reported from Whangarei Heads, and on 4 September a single rather small specimen came close inshore at Maraetai, near Auckland.

In 1964 there were only two sightings recorded, both from the Marlborough Sounds, but in 1965 at least five were reported from the southern side of Cook Strait. One mother and calf came into Whekenui Bay and stayed for

Figure 66.   (J. Moreland, Dominion Museum.)
Young right whale in Wellington harbour. In this photograph the edge of the arched lower jaw is visible. The white patch of the 'bonnet' can be seen.

several days. In November 1965 a weaned right whale calf came very close to the Seatoun wharf in Wellington harbour, allowing Mr John Moreland of the Dominion Museum, Wellington, to take several magnificent photographs, one of which is used in this book.

Three sightings were made in the New Zealand region in the spring of 1966. A single animal was sighted by the Japanese oceanographic vessel *Umitaka Maru* at 40°21'S 171°33'E on November 13th, and two together at 55°33'S 165°10'W on 24th November (Brown 1967).

The right whales that reached the New Zealand mainland supported a lucrative industry for only about a decade. New Zealand bay whaling began about 1830, before the numbers declined rapidly and drastically. In the years that followed right whales were taken by pelagic whalers who were hunting sperm as their primary quarry; all the time (until about 1870) that it was profitable to take sperm whales, the right whales remained vulnerable, and were given no respite. Only a few right whales were taken at later dates on the New Zealand coast. Seven were killed between 1915 and 1917, four between early 1921 and late 1922, and two between 1924 and 1927, according to official records of the New Zealand Marine Department (Gaskin 1968a). Nine out of the thirteen were killed in Cook Strait, two on the Marlborough coast, and the other two near Auckland. There are no records even of sightings in the period 1927-62. The upsurge

87

in sightings since 1963 cannot be attributed to increased interest in cetaceans alone. Optimistically we hope this means that the numbers have outgrown the handful that the subantarctic islands can support, and that the species will continue to return to its old mating and calving grounds around the New Zealand coast in increasing numbers. In the South Atlantic (Elliott 1953) and the eastern South Pacific (Clarke 1965) there are also signs that the right whale populations are making a slow but steady recovery in numbers.

BIOLOGY: Unfortunately, catches of southern right whales, and Biscayan or North Pacific right whales, had long passed their peak before current methods of biological study applied to whale populations could be used. Dr L. Harrison Matthews (1938b) was able to study a few specimens taken at South Georgia whaling stations before the Second World War, and this paper still summarises most of our current knowledge of the species' biology. Ivanova (1962) described the morphology of *E.sieboldi,* and Omura (1958a) examined two specimens at Japanese whaling stations. However, more recently, Dr S. Klumov (1962) was able to study 10 specimens of the North Pacific right whale *E.sieboldi,* taken under a special licence issued by the Soviet Government. His findings are of great interest, and since this species is so closely allied to *E.australis,* they are worth outlining here.

The peak of pairing occurs in December and January; as one might expect this is roughly opposite to the apparent peak for *E.australis* in the southern hemisphere. Calves are born after a gestation period of between 11 and 12 months, and appear to be weaned when about 5 to 6 months old. Females reach sexual maturity at body lengths in excess of 42 ft, and ovary weights of two kilograms. Klumov found that three distinct populations exist in the northern Pacific region; a local stock is found in the Sea of Okhotsk and another in the western North Pacific. Neither of these Asiatic coastal populations seem to mix with the third, referred to as the American coastal stock.

Figure 67.   (Dr Hideo Omura, Director, Whales Research
                     Institute, Tokyo.)
North Pacific right whale at Ayukawa whaling station, **coast of** Japan, in 1963.

Figure 68. (Dr Hideo Omura.)
Amphipod crustaceans, whale lice *Cyamus ceti*, on the Ayukawa right whale, 1963.

Examination of the stomach contents showed that pelagic crustaceans, especially *Calanus* sp. and *Euphausia* sp. were the major items of diet. The skin and baleen plates were covered with diatom film, and the parasites *Tetrabothrium* and *Bolbosoma* were recorded.

THE PIGMY RIGHT WHALE
Caperea marginata (Gray 1846)

DESCRIPTION: The pigmy right whale reaches a length of about 20 ft, and is in many ways a miniature replica of its larger and better known relative, the southern right whale. However, there are some important differences in internal anatomy.

The skin is black dorsally, shading into dark grey on the flanks and becoming pale yellowish-grey ventrally. The lining of the mouth is white. The mouth is quite strongly arched in profile, and the whalebone plates are long relative to the total length of the animal. They are white, edged with black on their outer margins. There are about 230 plates on each side of the mouth, and the length of individual plates varies from a few inches to about 27 inches (Slijper 1962). There is usually a dorsal fin present, but it varies in shape from a blunt conical projection to a distinct, slightly curved triangular structure.

DISTRIBUTION: The pigmy right whales' movements are confined to the temperate regions of the southern hemisphere. Only about 40 stranded specimens have been recorded, most from Tasmania and New Zealand on the edges of the southern Tasman Sea, and all within a circumpolar belt between latitudes 30° and 60° south (Davies and Guiler 1957). There are also good records from South Africa (Norman and Fraser 1937) and Australia (Hale 1932b, 1939), and two unconfirmed records from southern South America (Davies and Guiler 1957). The species was originally described from western Australian material (Gray 1846). It has not been taken at sea, or even reliably seen, and there is no evidence for migratory movements.

BIOLOGY: We know very little of the biology of this species. Cave and Aumonier (1961) described the histology of the liver and kidney. To date only Dr J. L. Davies and Dr E. R. Guiler of the University of Tasmania have had the opportunity to examine a pregnant female specimen. This animal stranded in late June (Guiler 1961) and was carrying a foetus about 2 ft in length. Dr Guiler also had a reliable record, though unfortunately the specimen had been lost, of another female which stranded in late October or early November, carrying twin late-term foetuses. He suggested on the basis of these two records that the pigmy right whale mated in late summer or autumn and calved in the spring near Tasmania.

Nothing is known of the food or feeding habits of the species, and no parasites have been recorded from stranded specimens.

Figure 69. (C. Gill, South Australia Museum, Adelaide.)
Pigmy right whale stranded on the coast of South Australia; a young male at Port Lincoln Bay, December 1955.

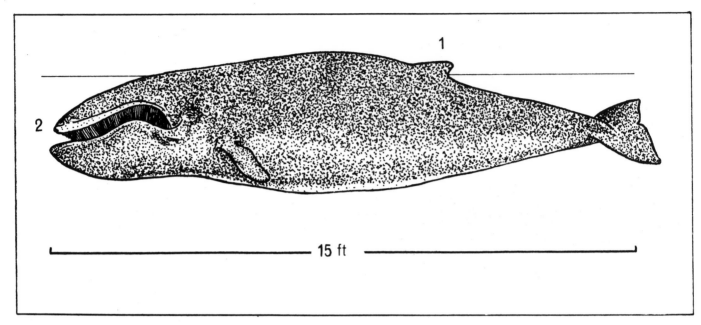

Figure 70.
Caperea marginata, pigmy right whale: 1. small dorsal fin, 2. arched mouth with yellowish baleen.

Davies and Guiler (1957) discussed some of the skeletal peculiarities of this animal. The rib-cage is flattened ventrally, a situation unique to this species among cetaceans. Though we have no observations which might explain what adaptive significance this has, the two Tasmanian workers have speculated that *marginata* lives in relatively shallow waters mainly near the bottom, coming to the surface only to breathe or perhaps to calve and mate. This, they argue, might explain why there are no records of the species being seen at sea. However, it should be mentioned that except at very close quarters it would be fairly easy to confuse the pigmy right whale with the minke whale, which is a common species in the Southern Ocean. Minke whales are frequently sighted by pelagic whaling vessels, but are virtually ignored because they have no commercial value to such operations.

However, even if the pigmy right whale proves eventually not to be particularly rare, merely difficult to find, it still merits special attention as the most unusual whalebone whale. Fossil whalebone whales, such as the Cetotheres, were smaller animals than most of the species living today, and had a number of primitive anatomical features, some of which are possessed by the pigmy right whale. The evolutionary lineage of the right whales is nearly as long as that of the rorquals, and to some extent we might regard the pigmy right whale as a relict species presenting an interesting link to some extinct groups (Davies and Guiler 1957).

Since this is such an unusual and little-known species, I would like in conclusion to entreat any reader who comes across a stranded specimen, to contact the nearest university or museum at once and to make every effort to preserve the body from damage until a biologist can examine it.

11. The Sperm Whales Family Physeteridae

There are two genera in this family: *Physeter* with one species, the giant sperm whale or cachalot; and *Kogia* with two species, the pigmy sperm whale and the dwarf sperm whale. While the pigmy sperm whales only attain lengths of about 12 to 13 ft at maturity, the cachalot male grows to a length of about 57 to 58 ft at physical maturity, and the female to about 40 ft. Despite the vast size difference between the two genera there are sufficient anatomical similarities to consider them as part of the same family. However, some authorities prefer to separate them below this level into the subfamilies Physeterinae and Kogiinae.

Functional teeth are confined to the lower jaw, although a close examination of the gum of the upper jaw in both genera usually reveals small buried maxillary teeth. The nasal passages are so twisted in the cachalot as to bring the nares out on the left side of the tip of the snout. In both genera the head capsule anterior to the brain contains a spermaceti organ, which is of great relative size in the sperm whale. This organ is made up of a complex connective tissue compartment containing a white glistening waxy compound called spermaceti, which at body heat is a clear viscous liquid. The function of this organ is not known, although at present it is assumed to play a part in deep diving.

THE SPERM WHALE OR CACHALOT
Physeter catodon (Linnaeus 1758)

DESCRIPTION: Few males of more than 60 ft are recorded in the International Whaling Statistics. Males 80 ft in length or more have been mentioned by Bullen (1928), McGinitie and McGinitie (1949) and Sanderson (1956), but such can only be attributed to over-active imagination. The sperm whale is easily distinguished from other species at sea; the dorsal surface is grey or purplish-brown, and there is no dorsal fin; instead the species has a hump or series of humps along the posterior crest of the back. The blow is single, rising obliquely about 8 to 10 ft. Except when viewed directly from front or behind it is very distinctly fan-shaped.

The head containing the spermaceti organ, accounts for about one third of the total length of the body. The elongate lower jaw fits snugly into a maxillary depression, and the tip does not reach the anterior end of the blunt head capsule. The eyes are relatively inconspicuous, and the back part of the eyeball has very heavily thickened tissue; this is apparently an adaptation to deep diving.

The flippers are relatively small, and not very flexible. The species does not use them in propulsion, and even animals which have lost a flipper do not appear to be impeded. In Japan sperm whales have been recorded with the rudimentary hindlimbs protruding through the blubber (Nemoto 1963). The flukes of the sperm whale are conspicuously notched in the midline. The blubber is very dense, and composed of millions of white fibres running at all angles to form a dense mat. At the shoulder the blubber may be more than 18 in. thick, and on the flanks 4 to 6 in. The skin is not smooth, but corrugated into many series of longitudinal ripples.

DISTRIBUTION: Sightings of the sperm whale have been recorded in all the major seas of the world, although only very rarely does it now venture into semi-landlocked regions such as the Mediterranean and the Black Sea. However, a few specimens are still recorded in the Mediterranean from time to time (Norman and Fraser 1937), and there is evidence that the Phoenicians hunted sperm whales in the eastern part of that sea many hundreds of years before Christ (Sanderson 1956).

Townsend (1935) plotted the major warm-water concentration areas, but omitted most of the high latitude sperm whaling grounds frequented by modern whaling ships, since his records were based on the logbooks of old American whaleships. There are several important areas of sperm whale aggregation in the North Atlantic. At present the only one of any great commercial importance is that around Madeira and the Azores, which has been studied in detail by Dr R. Clarke (1953, 1954, 1955a, 1956a, 1956b). International Whaling Statistics show that a few sperm whales are taken on the coast of Spain each year, and a small industry on the coast of Iceland also catches these animals. Sperm whales of the western North Atlantic are observed by coastal whalers of Nova Scotia, but are rarely taken. However, an active marking programme has been conducted in recent years (Mitchell 1968c), both off Nova Scotia, and the coast of Newfoundland. Small numbers of sperm whales have also been taken for many years by chasers operating from the coast of Norway and the Faroe Islands, and until 1956 some were also taken by a station operating on the West Greenland coast. The species has stranded fairly frequently on the coasts of the United Kingdom (Norman and Fraser 1937), and also at very widely spaced intervals of time on the Netherlands coast.

Sperm whales are not uncommon in the Gulf of Mexico and the Caribbean, and the old whaleships used to exploit areas of the South Atlantic for them, notably on the edge of the Benguela Current and around Tristan da Cunha. The species is common off the coast of Brazil, small numbers being taken each year by the single Brazilian land station. The International Whaling Statistics have shown a significant rise in the number of sperm whales taken by pelagic factory ships returning home to Europe after whaling in the Antarctic during the summer.

The whaling industries of Cape Province and Natal rely heavily on sperm whale catches, which usually average from 600 to 900 annually, and make up about half the total catch in each area. In the last few years catches at Natal have risen to nearly 3,000; possibly such a level of exploitation is greater than the stock will be able to stand. However, the biology and population dynamics of the

Figure 71.

Physeter catodon, sperm whale or cachalot: 1. narrow lower jaw with large teeth under large box-like head, 2. hump instead of true dorsal fin.

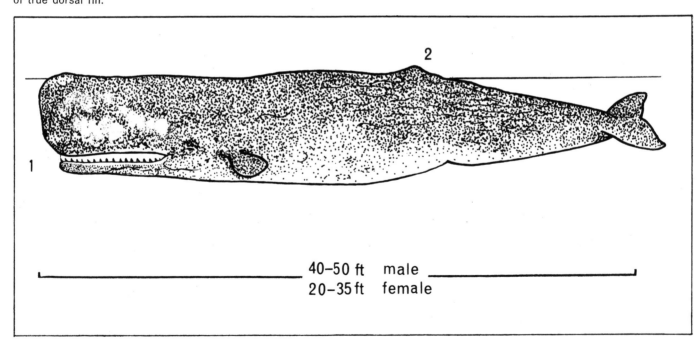

40–50 ft male
20–35 ft female

Figure 72. (D. E. Gaskin.)
A 52ft male sperm whale on the deck of the FF *Southern Venturer* in the 1961-62 Southern Ocean whaling season. This animal had an unusual amount of white on the head and flanks.

species in South African waters are under careful study, and some excellent papers on the reproductive cycle have been published by Mr P. B. Best of the South African Division of Sea Fisheries (Best 1967, 1968, 1969), while an interim study of the population size and the potential sustainable yield has already been published by Gambell (1969). The seasonal movements of sperm whales through the South African whaling stations have been studied by Bannister and Gambell (1965) and Gambell (1967).

As many as 2,000 sperm whales per year have been taken in the Indian Ocean during the 1960s by pelagic whaling vessels, but the only shore stations are at Durban, Natal, and at Albany in Western Australia. Catches of sperm whales by chasers from Albany have ranged from practically nothing in the late 1950s, when the station operated almost exclusively for humpbacks, to 710 in 1964. Since that year the catch has progressively deteriorated; even when the number of whales caught was still as large, the effort expended to take them was much greater than in 1964 (Bannister 1969). The distribution of the population around the southern coasts of Western Australia was examined by a protracted aerial survey

(Bannister 1968), while earlier reports on the same population were published by Bannister (1964), Godfrey (1964), and Gates (1964).

Catches of sperm whales by pelagic factory ships south of Australia and New Zealand and in the South Pacific were significant only between 1962 and 1966; since that time they have been very small. Small sperm whale catches were made by the Tory Channel Whaling Company operating in the eastern mouth of Cook Strait in 1963 and 1964. This station ceased to operate in December 1964. Some biological observations on the sperm whale population in this region have been published (Gaskin 1964b, 1965, 1967b, 1968b; Gaskin and Cawthorn

Figure 73.

Distribution of sperm whales in New Zealand waters. Based on information collected in the period 1962-67. The stippled areas indicate shallow water in which sperm whales are rarely seen, except as sick or stranded specimens. The hatched areas are those, usually coinciding with regions of upwelling or current boundaries, where sperm whales are most abundant. The arrows indicate summer movements, based on several thousands of sightings. Winter movements are less well known, but no marked northward tendency has yet been observed.

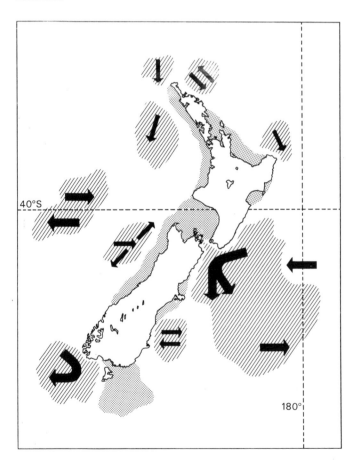

1967a, 1967b). Sperm whales are found all around New Zealand, but are most abundant in the central Tasman Sea, near North Cape, on the shelf around the Chatham Islands, off the Marlborough coast, and over the Solander Trench southwest of Stewart Island. Further north large seasonal concentrations occur round the Kermadec Islands.

Sperm whaling has been carried out in the North Pacific for many years. Pelagic whaling vessels from Japan and the Soviet Union take from 1,000 to 4,000 animals each year, while the total catch of whaling stations on the Kuril Islands and the coasts of Kamtchatka and Japan varies from 4,000 to 6,000 per year. The most productive sperm whaling grounds of the North Pacific-Bering Sea region lie around the islands of the Bering Arc which stretches from Hokkaido to Alaska (Omura 1955). Small catches have also been made in the last two decades by whaling stations at the Ryukyu and Bonin Islands, and on the coasts of Oregon and California. Somewhat larger numbers were taken by a whaling station on Vancouver Island. This station, which killed 320 sperm whales in 1955, is no longer operational.

The most productive localised sperm whaling grounds in the world lie on the Humboldt Current off the coasts of Peru and Chile (Clarke 1962, Clarke, Aguayo and Paliza 1964, 1968, Saetersdal, Mejia and Ramirez 1963). Work by Saetersdal and his co-authors suggests very strongly that the population is being over-exploited, although at present the land station catches still total from 5,000 to 6,000 sperm whales per year.

Another 3,000 to 5,000 sperm whales are taken each year by pelagic factory ships in the Southern Ocean between latitude 40°S and the coast of Antarctica. Although the magnitude of this catch rose steadily during the late 1950s and early 1960s, the number of factory ships operating in the Antarctic in the latter part of this decade has been decreasing rapidly, and the sperm whale catch can be expected to decline also.

Sperm whale migratory movements are complex and less regular than those of some baleen whales such as the humpback. The problem of obtaining an overall picture is further complicated by a behaviour pattern of the males which results in a degree of sexual segregation within any given population. About 40 to 60 per cent of the males in any population disperse into high latitudes away from the breeding herds (Ohsumi 1966). The pattern of world distribution of sperm whales is discontinuous (Townsend 1935). Concentrations of whales coincide with areas of current-mixing or upwelling, where zooplankton and larger animals such as fish and squid

are abundant. Not all sperm whale concentration areas are suitable for breeding. Large aggregations of male sperm whales can be found over deep oceanic banks in the Antarctic; the concentration and dispersal of these whales may be governed indirectly by the phases of the moon influencing their food species (Holm and Jonsgaard 1959).

Breeding herds are confined almost exclusively to relatively warm waters. Gilmore (1959) claimed that the distribution of female sperm whales was limited by the 20°C surface isotherm, while Schubert (1951, 1955) quoted a lower value of 17°C for the coasts of Chile and Peru. In addition Matthews (1938a) quoted 40°N and 40°S as bounding the distribution of female sperm whales; however, the Vancouver whaling station took female sperm whales at latitude 50°N (Mr L. Hume, in letter), and Omura (1955) gave 51°N as the most northerly limit for females off the coast of Japan. I have a record of a harem, from which females were taken by the Tory Channel whalers, which entered Cook Strait in July 1947. Surface temperatures there at that time of year average 10° to 12°C. Dr Keiji Nasu and I also recorded nursery schools with very small calves between 44°S and 45°S to the east of New Zealand in the summer of 1967, moving in surface temperatures of from 14° to 16°C. No calves were present in the 1947 school, according to Marine Department records; 12° to 14°C is probably a more realistic lower limit for the temperature range of female sperm whales than 17°C or 20°C.

As yet we do not know for certain if the mature males which exist in high latitudes ever return to the breeding herds. Clarke (1956b) and Gaskin and Cawthorn (1967b) have suggested that since sperm whales tend to congregate in areas suitable for feeding and breeding in tropical and temperate waters, as shown by Townsend's charts (1935), the drift of males into high latitudes and subsequent return to breeding herds other than those from which they came, could be a means of preventing the localised populations from becoming genetically isolated. Japanese whale marking in the North Pacific (Nishiwaki 1966a) indicates that such mingling does take place.

Since the distribution of sperm whales is not continuous it has been natural for scientists to pay attention to body proportions and colour patterns in order to see if whales from different geographical regions could be readily distinguished. Such studies have been undertaken by Matthews (1938a), Ohno and Fujino (1952), Omura (1955), Ivanova (1955), Clarke (1956b), Gudkov (1963) and Clarke, Aguayo and Paliza (1968). No such differences have been demonstrated; in fact the last-named authors concluded that this was probably a sterile line of research.

Omura (1955) and Nishiwaki (1966a) discussed the movements of sperm whales in the North Pacific-Bering Sea region. They found that only large males moved into the Sea of Okhotsk, and that very few ever move into the Inland Sea, the East China Sea or the Sea of Japan. Most animals migrated along the Pacific coast of the main Japanese archipelago. Young but sexually mature males appear off the coast of Japan in spring, but the main herds, including the harems, do not arrive until July and August. Notable sexual segregation begins at about latitude 45°N; most harems do not move any further north than this, but considerable numbers of males move on to the vicinity of the Aleutian Islands, where the main catches are taken between May and August. Virtually no male sperms pass through Bering Strait, which clearly marks the northern limit of distribution of the species in this region. In the autumn the whales move south, and a winter sperm fishery used to be operated at the Bonin Islands for the same stock.

Matthews (1938a) examined catch and sighting data on sperm whales passing the island of South Georgia, and concluded that there was a rise in sperm whale numbers around the island in summer, and a decline in the autumn. The catch at South Georgia, at latitude 54°S was almost entirely composed of male animals.

The movements of sperm whales in South African waters have been studied by Bannister and Gambell (1965), and Gambell (1967). They discovered that the population density was least in the winter months of June to August, and that more sperm whales began to move into the area in September. Density was greatest in April and May, and female sperm whales were most numerous between November and April.

Working on the population off Western Australia, Bannister (1968) found that sperm whales concentrated over the continental slopes off the southern coast, but had a more widespread distribution in the offshore regions to the west of the continent. Most movement was parallel to the shore, and off the west coast constant southward movement was observed. In the same region a marked

Figure 74. (D. E. Gaskin.)
Ten large sperm whales *Physeter catodon* tied at the stern of the FF *Southern Venturer* in November 1961 near the South Sandwich Islands.

decrease in population density was noted in winter, with high density in spring and autumn and relatively reduced density in summer, as if the population centre had temporarily moved south beyond the catch radius of the shore station at Albany.

Analogous but not identical patterns of density change and movement were observed in the Cook Strait region in 1963 and 1964. In the eastern mouth of Cook Strait and off the Marlborough coast of the South Island population density was least between June and September. A very rapid increase in density was noted in November, and numbers remained high until the autumn, decreasing steadily after April (Gaskin 1968b). Since the above paper was published the catch and sighting data have been analysed in more detail (Gaskin, in press[1], 1971).

The increase in sperm whale numbers in the spring is associated with the southward movement of the relatively warm surface waters of the East Cape Current. About four fifths of the influx is composed of mature or sub-mature males in 'bachelor' schools. Most whales enter the whaling grounds from the west and leave to the south or southeast. South and southeast movement is at a maximum in the spring, accounting for 88 per cent of all movement. In the summer such movement is only 59 per cent of the total, in the autumn only 47 per cent, and in the winter 43 per cent. A retreat of the population centre from the catching radius of the whaling station appears to be reflected in the amount of eastward or offshore movement seen in the autumn (34 per cent), compared with only 16 per cent in the summer and 3 per cent in the spring.

A decrease in population density in the autumn is associated with the retreat of relatively warm surface water from the whaling grounds. Some sub-mature males winter in the Cook Strait region in company with a number of much larger solitary animals. The presence of the young animals is reflected in a sharp decrease in the mean body length of whales taken from bachelor male schools in the winter months.

The Cook Strait region whaling grounds are a rich feeding area for sperm whales, but vertical temperature profiles suggest that conditions favourable for squid schooling occur in the eastern mouth of the Strait in spring and summer but not in the autumn; this correlates with the observed change in sperm whale population density.

Further study may show that movements of the sperm whale population centre in this region are correlated with seasonal population shifts of several squid species. Too little is known about marine food webs in the Cook Strait area to speculate about factors controlling squid movements.

In the subtropical and tropical regions hydrological conditions probably dictate the distribution patterns of sperm whale populations. For example, comparison of the distributions of sperm whales shown by Townsend (1935) in winter and summer in the northern and central Tasman Sea and around the Kermadec Islands with the oceanographic work of CSIRO scientists such as Highley (1967) reveals that the animals appear to be associated with Tropical Convergence waters.

Another area of the world in which sperm whale movements have been studied in detail is the eastern North Atlantic, especially around the Azores and Madeira (Clarke 1956a). The centre of population shifts from the Azores in summer to the vicinity of the Cape Verde Islands in winter. During the summer the mature male population ranges much further afield than the main herds; some males go as far north as Iceland, the west coast of Ireland and even southern Greenland. Clarke suspected some interchange between the Azores and Madeira populations, though he was not able to confirm this by whale-marking.

Figure 75.
Results of aerial survey for sperm whales in the Cook Strait whaling grounds between November 1962 and December 1964. Each search area was approximately 20 nautical miles per side. Key shows density range in terms of numbers of sperm whales sighted per search hour in good weather. Two concentration areas appear to exist, one south of Cape Palliser (1), the other southeast of Kaikoura (2).

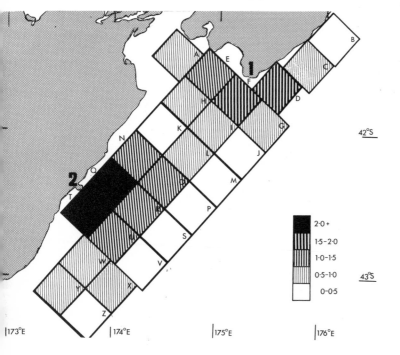

BIOLOGY: Whale population movements are generally explicable in terms of breeding and feeding requirements (Townsend 1935). In baleen whale species an annual cycle is often recognisable (Mackintosh 1965, Dawbin 1966). However, work on sperm whale populations in the Southern Ocean off the coast of South Africa (Best 1968) has shown that the gestation period in this species is 14.6 months; for this reason an annual cycle of migration associated with breeding alone seems unlikely.

Lack of material has precluded work on the breeding cycle of the sperm whale in the vicinity of New Zealand. However, Ohsumi stressed the probable uniformity of events throughout the southern hemisphere, and there is no reason to expect the cycle in this region to differ in any great degree from that found in other southern areas.

Best (1968) provided evidence from very large samples collected off South Africa that sperm whale pairing is at a maximum in November and December, and that the peak of parturition in the southern hemisphere is in February and March. While Clarke (1956b) estimated that lactation lasted for 13 months, Chuzhakina (1961) 10-11 months, Matthews (1938a) 7 months, and Matsuura (1936a) only 6 months, the most recent and reliable work by Ohsumi (1965) and Best (1968) suggests that it is longer than formerly believed. These authors obtained estimates for the lactation period of between 24 and 25 months. This would agree in order of magnitude with the 22 month lactation period found by Sergeant (1962a) in another odontocete, *Globicephala melaena* Traill.

The few available observations from the Cook Strait region support the cycle of events outlined by Ohsumi and Best. Three pregnant females taken from the school hunted in Cook Strait in July 1947 carried foetuses 4 to 5 ft in length. Schools containing calves have been seen off eastern New Zealand only in the summer months. Very young calves of about 15 ft in length were seen from *Chiyoda Maru No. 5* in March 1967, and judging by growth curves drawn by Ohsumi, these animals were probably born in the February immediately preceding. The presence of schools with such small calves strongly suggests that calving does occur in the offshore part of Cook Strait waters in the summer months. Some pairing may also take place, but on Ohsumi's timetable of events in the reproductive cycle this would mostly be complete before the fresh influx of animals into the Cook Strait whaling grounds in November. With the exception of the single 1947 school, mixed-sex schools were never sighted in the Cook Strait region during the winter months. The operational radius of the whale chaser *Orca* was, however, quite limited, and a shift of the breeding centre

northwards and offshore through only about 1½° to 2° of latitude would be sufficient to put it out of range of the whaling station during the winter. In fact, the annual north-south movement of the warm surface waters of the East Cape Current is just about 2° of latitude (Garner and Ridgeway 1965).

Body length of calves at birth is estimated to be between 12 ft and 13 ft 3ins (Clarke 1956b, Ohsumi 1965), with a slight suspicion that southern hemisphere animals are fractionally larger than those of the North Atlantic. Best (1968) found that the mean weight of calves at birth was 774 kg, and that there was a sex ratio of 1:1. He also noted that weaning was protracted and that the calf was still less than 25 ft in length when finally weaned. This is in close accord with Clarke's North Atlantic figure of 23 ft. The annual ovulation rate for females in South African waters was 0.59, thus reflecting the long reproductive cycle. Males start to mature sexually at lengths of about 39 ft (Nishiwaki 1955, Best 1969) in the southern hemisphere, but at only 31 ft in the North Pacific (Nishiwaki and Hibiya 1951, 1952). The corresponding lengths for females appear to be 30 ft and 29 ft (Mackintosh 1965). More work is required to find if this summary is meaningful in reproductive terms, especially where males are concerned. Best (1969) found that in males of body lengths between 45 and 46 ft equal proportions of mature and immature tissue were present in the testes, suggesting that these animals were only just

Figure 76. (Dr S. Ohsumi, Whales Research Institute, Tokyo.) Foetus of sperm whale, *Physeter catodon*. An 86 cm female removed from a whale at Ayukawa station on 2 September, 1963.

Figure 77.   (D. E. Gaskin.)
Flensing a male sperm whale *Physeter catodon* on the *Southern Venturer* in November 1961 near the South Sandwich Islands. The white tissue below the blubber is tough connective tissue covering the muscles.

Figure 78.   (D. E. Gaskin.)
Cutting up the head of a 46ft male sperm whale *Physeter catodon* on the FF *Southern Venturer* in November 1961 near the South Sandwich Islands.

coming into full reproductive condition. Best could not find histological changes in the size or condition of the seminiferous tubules to suggest a male sexual cycle which correlated with the female one, but did discover that the androgen-secreting interstitial cells were enlarged during the breeding season. Increased hormone production could well initiate marked changes in male behaviour, both in a sexual and a general social context.

The maximum life span of this species could be as much as 50 years; at present sperm whale research workers accept a single tooth lamination per year, although some have argued in the past that two laminations were deposited each year. Each lamination consists of a clear zone and an opaque zone, and these zones are themselves divisible microscopically into thinner layers. Only very recently has agreement been reached on the correct method for interpreting these laminations (International Whaling Commission Report No. 19, 1969).

A number of studies have been made on the social structure of sperm whale populations, the earliest by Beale (1839) and Beddard (1900). More recently Clarke (1956a) described the various social aggregations observed near the Azores, and Gambell (1967, 1969) carried out a similar study in South African waters.

There is still some argument as to whether this species can really be called polygamous; Matthews (1938a) and Clarke adhered to this theory; Ohsumi (1966) considered the idea to be only partly true, pointing out that harem schools usually contained more than one large male. The whole discussion has been reviewed by Caldwell, Caldwell and Rice (1966). However, Clarke and Gambell recognised a number of basic categories; females are invariably gregarious, they may or may not be accompanied by mature or immature males, or calves. In temperate waters all-male schools are encountered; these are generally called 'bachelor' schools. In all waters, but mainly in the subantarctic and antarctic, large solitary or semi-solitary males occur. The distance between individuals is generally several miles, though numbers will be found together on rich localised feeding grounds.

Observations made in New Zealand waters (Gaskin 1970b) can be summarised as follows. Six social categories were recognised, some of them obviously of a very temporary nature. These groupings can be recognised by the sex and size of the animals in the schools. 'Solitary' whales are invariably male, are from 47 to about 56 ft in length, and are distributed thinly over any given whaling ground. Other all-male groupings are pairs and bachelor schools. Pairs seen in the Cook Strait region usually consisted of two large males (47 - 56 ft), or one large and one small (46ft or less), or, very rarely, two small animals. Bachelor schools were almost invariably composed of pubertal and sub-mature males (33 - 46 ft), and the smallest individuals were usually found in the largest bachelor schools (10 - 18 animals). It was suspected that some pairs were formed by a solitary male joining a young male survivor from a school broken up by the chaser.

Immature mixed-sex schools were observed in the summer months, the animals in these were estimated to be between 25 and 35 ft in length. Harem schools were predominantly female, with from one to four large males (47 - 56 ft) present, and frequently one or more younger males 25 - 35 ft). These were presumed to be adolescents from previous seasons. Nursery schools were noted on several occasions, these, of course, being characterised by the presence of calves (14 - 25 ft). The number of calves was always considerably less than the number of females in the school. Males were not always observed in close association with nursery schools, but large specimens were sometimes seen lying perhaps a mile or so away from the main school clusters.

Female sperm whales differ morphometrically from males in several important respects (a full discussion with tabulated values was given by Matthews (1938a)). The most easily recognisable feature is that the length from snout to flipper in the female accounts for only 27 per cent of the total body length, compared with 49 per cent in males. This difference is slight in calves, but detectable in larger animals at sea, with experience. The girth of the head is proportionately less in the female, the dorsal surface slopes more steeply down to the anterior nares, and the crest of the back is often smoother than in the male, with only one small hump instead of a series. Kasuya and Ohsumi (1966) have described a callus on the hump of sperm whales which appears to be a secondary sexual character not present in large males, but present in immature males and mature females. This character has also been observed on whales in the New Zealand region, and was noted on five immature males and two mature females brought into the Tory Channel Whaling Station.

In broad outline, the social structure of the sperm whale population to the east of Cook Strait showed many similarities to those described by Clarke and Gambell for Azores and South African waters, but differed in several important respects from that noted by Caldwell and his co-workers off the coast of California. The solitary and bachelor school conditions seemed to be more sharply marked in Azores, South African and New Zealand waters than off California.

The social status of an individual sperm whale of either sex obviously alters with age, and it is very likely that harems and nursery schools are the same entities seen at different times of year. It seems unlikely, on the other hand, that immature mixed-sex schools persist for more than a few months. Bachelor schools are presumed to be the next stage in male social structure after these, while the females are taken into harem schools by males, or, being gregarious, join them without coercion.

Squid are the main item of diet of sperm whales in every part of the world in which these animals have been studied (Clarke 1956a, at the Azores; Pike 1950, off Vancouver Island; Rice 1963b, off California; Okutani and

Figure 79. (D. E. Gaskin.)
Mandibles ('beaks') of squid taken from stomachs of sperm whales in the Cook Strait region. Above: upper (left) and lower (right) beaks of *Moroteuthis* sp., taken from squid which were about 18in. long. Below: upper mandible of *Architeuthis* sp. This beak was alone, not with recognisable remains, but the squid from which it came was probably about 8ft in length.

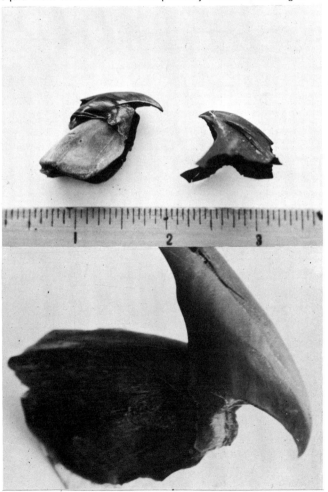

Nemoto 1964, off Japan; Akimushkin 1955, Betesheva and Akimushkin 1955, Tarasevich 1963, around the Kurile Islands; Matthews 1938a, off South Georgia; Gaskin and Cawthorn 1968a, 1967b, in the Cook Strait region).

Elasmobranch (sharks, skates and rays) have been found in sperm whale stomachs examined at Durban (Matthews 1938a); Californian whaling stations (Rice 1963b); Vancouver Island (Pike 1950); the Azores (Clarke 1956b), where a common skate *Raja rhina* accounted for about a quarter of the total contents; and in the Cook Strait region, where the remains of several sharks were found.

Quantitative analysis showed that onychoteuthid squid were the most common food item in the Cook Strait region both by number and weight, although ommastrephid squid were also quite abundant as stomach remains. Very small squid mandibles were never found in the stomachs examined, supporting to some extent the statement by M. R. Clarke (1962) that sperm whales may have a preference for squid of a particular size range or may not be able to catch small specimens.

Very large mandibles, attributable to squid of the giant genus *Architeuthis,* were occasionally found in stomachs opened at Tory Channel. Thus sperm whales in this area appear to take squid of considerable size, larger than any recovered intact in the stomachs. Clarke (1956a) reported seeing a harpooned sperm whale near the Azores vomit up the trunkless head of a very large squid, and also measured a specimen of squid in the same locality (Clarke 1955a) which was 38 ft in length including tentacles, and weighed 184 kg.

Matsushita (1955) thought that sperm whales feed mainly at night when some squid rise closer to the surface. *Histioteuthis* is one of these; however, *Moroteuthis* is not (Clyde F. E. Roper, in letter), so we cannot postulate that Cook Strait region sperm whales feed almost entirely in surface waters.

During the day, dives of up to 75 minutes have been recorded by New Zealand whalers for large male sperm whales (Gaskin 1964b). The occurrence of sperm whales tangled in deep-sea telephone cables has been cited as evidence that sperm whales are bottom feeders (Yablokov 1962), but it has also been suggested that cables snared in this fashion might have been hanging across submarine canyons or gullies (Gaskin 1964b).

The regular occurrence in the stomach of sperm whales of typical New Zealand bottom fish such as groper and ling is much clearer evidence for bottom feeding. The long narrow jaw may be used to agitate the bottom or probe among rocks, making fish swim upwards so that they can be taken. Damage to anterior mandibular teeth was suggested as indicative of feeding by this method (Clarke 1956a).

Sperm whales from the Cook Strait region were also frequently recorded as feeding on southern kingfish *Jordanidia solandri;* these were almost certainly taken away from the sea-bed. Some squid are also probably taken in mid-water regions.

Coconuts and wooden slats have been recorded in stomachs by Nemoto and Nasu (1963) and Gaskin and Cawthorn (1967b), suggesting that sperm whales will snap at interesting objects on the surface. Considering the total evidence, it would seem that there is little point to arguing about the depths at which sperm whales feed; they are undoubtedly catholic feeders with a wide range of diet, a general preference for squid, and the ability to take food from surface waters, mid-water regions, or off the ocean-bed to depths of 1,100 fathoms (Yablokov 1962).

The actual feeding method used by sperm whales is still a controversial matter. In a recent paper (Gaskin 1967b) I suggested that the sperm whale was probably a passive feeder, and that by actively feeding on a number of luminescent organisms may be able to convert its mouth into a deep-water lure. Others (Slijper 1962) have considered that the contrast between the white mouth lining and the purple tongue may act as a lure. In any case it is hard to imagine that the sperm whale is a totally active feeder; most of the food animals are quite small, and the long, narrow jaw has virtually no lateral movement, so that the sperm whale is capable of little more than a straight up and down biting movement; nor does it give the impression of being a particularly agile swimmer.

When swimming unalarmed the sperm whale moves at the surface with a slow, shallow, porpoising movement, at about three to four knots. The spouting rhythm under these conditions is very regular, although the interval between blows varies from one individual to another; it may be as short as twenty seconds, or as long as four minutes. The sperm whale executes two different movements when diving. When it intends to make a relatively short dive the back humps, rises and sinks, without the flukes coming clear of the surface. Before a long dive the flukes almost always rise completely clear of the surface (Gaskin 1964b). Sperm whales may surface in porpoise fashion from both long and short dives, but sometimes they breach right out of the water in spectacular fashion. After deep feeding dives the whale is most vulnerable to whale chasers. It lies at the surface spouting very rapidly, almost appearing to pant. The blood and muscle respiratory pigments have to be recharged with oxygen before the animal is capable of a second prolonged dive. If chased at such a time the whale will try to escape by swimming away from the pursuing vessel; it is capable of short sprints at 8 to 10 knots, but rarely attempts to dive again.

Large, predominantly female schools have been noted as showing remarkably little concern at the ravages of a whale chaser in their midst (Caldwell, Caldwell and Rice 1966). A whale may be shot dead, and yet another only a few dozen yards away will swim on as if nothing had happened.

On the other hand, large male sperm whales have earned a (somewhat exaggerated) reputation for ferocity when attacked. Of course, times have changed, and the sperm whale which could demolish a wooden whale boat is no match for the cruising whale factory. Even so, steel whale chasers have had plates buckled and propeller blades broken by wounded sperm whales. Another common type of damage caused to such vessels is a shattered asdic dome, since this projects below the hull when in use.

During 1964 a small ocean-going yacht bound across the Pacific was attacked by a sperm whale north of the Marquesas Islands and holed in several places. The boat reached safety, but not without some hasty repairs at sea (*Wellington Evening News* report). The crew claimed that no provocation had been given. This might not have been an intentional attack as such; sperm whales have been observed to play with logs and other debris at sea.

While the New Zealand whaling research team was undertaking marking operations off Kaikoura in May 1963,

Figure 80.   (M. V. Brewington.)
Cachalot Fishery (Peche du Cachalot), Garneray. A French version of the second 'Moby Dick' print.

one animal lifted its head from the water and moved towards the boat at a speed of about three knots. Since the boat was only 38 ft in length and the whale about 50 ft, we did not follow this interesting behaviour through to any conclusion, but departed rapidly for other waters.

Despite the inevitable additions to stories of the old whaling days, there are a number of very well authenticated cases of sperm whales destroying whaleboats, and three whaleships are known to have been lost through attacks or accidental rammings. These include the *Ann Alexander* and the *Essex*.

The *Essex* saga is one of the most famous of the old whaling days, and is worth retelling briefly. This vessel was working off the coast of Peru in November 1820, many hundreds of miles out to sea. While hunting a large bull sperm whale one boat was destroyed, and a short time after, the *Essex* was rammed amidships by the same animal. The planking was ruptured in two places and the ship sank rapidly. The surviving boats were launched and these tried to reach Easter Island, the nearest landfall with favourable currents. Of the original crew of 20, only 8 were rescued by the whaler *Dauphin* in late February 1821, admitting to the men who had saved them that they had only been able to survive by resorting to cannibalism.

It might be appropriate to end this section with a brief note on ambergris, that prize for which so many amateur beachcombers search each year. It is used by the cosmetics industry as a base to hold perfumes, and although some synthetic substitutes have been invented, the original product still has value if of good quality. It is formed in the rectum of sperm whales, apparently as a secretion associated with the irritating effect of squid mandibles, which are frequently found embedded in it. The obstruction continues to grow in size unless the animal is able to void it. Fresh ambergris usually comes in lumps about the size of a tennis ball; it is sticky, blackish brown and very malodorous. Weathered ambergris is preferred by the cosmetics manufacturers; this is light in colour, fibrous, with the smell of a good cigar. Many things on the beach can be confused with ambergris, especially a number of small deep-water sponges which are occasionally washed up on New Zealand beaches after bad weather. Real ambergris will burn with a clear blue flame.

THE PIGMY SPERM WHALE
Kogia breviceps (Blainville 1838)
DESCRIPTION: 'Pigmy sperm whale' covers two species: *Kogia breviceps*, which is sexually mature at about 8 to 10 ft in total body length, and *Kogia simius,* maturing at lengths of 6 to 8 ft (Handley 1966). Dr Handley has proposed the name 'dwarf sperm whale' for *simius* However, most general texts on cetaceans still recognise only a single species in the genus *Kogia*. To date, all the specimens recorded from New Zealand and Tasmania appear to have been *breviceps*.

The head of the pigmy sperm whale is bluntly pointed and the lower jaw is small and set back beneath the head, so that to the layman it has a strong superficial resemblance to a shark. The whale has well developed slender pointed teeth in the lower jaw; there are usually from 12 to 14 pairs, though the rear ones are sometimes buried in the gum. Teeth are also present in the upper jaw, but usually buried and hard to find without thorough examination. The blowhole is set slightly to the left of the middle of the head immediately above the angle of the lower jaw. The eyes are placed laterally, slightly further back than this line.

The flippers are quite small for the size of the animal. They are placed in line with the jaw angle, roughly twice as far from the tip of the snout as the eyes. They are narrowly triangular and seem to be held pointing downwards and outwards while the animal is swimming. There is a small but prominent dorsal fin set about half way along the crest of the back, strongly recurved in most specimens. As in all whales the tail flukes are horizontal, so that even a layman should instantly be able to separate *Kogia* from any shark, and in this genus there is a distinct notch between the flukes.

Kogia has a number of skeletal characteristics not shared by any living cetacean. The rostrum of the very asymmetrical skull is very short, the nasal bones are virtually absent, the maxillae are very large in proportion to the size of the skull, and the dorsal surface of the braincase is divided by a distinct body septum (Handley, 1966). All the cervical or neck vertebrae are fused into one unit and the sternum or breastbone consists of only three pieces.

DISTRIBUTION: The type specimen of *Kogia breviceps* is from the Cape of Good Hope. The distribution of the species seems to be throughout temperate, subtropical and tropical waters (Hubbs 1951b), though it has not yet been recorded from the South Atlantic or the Indian Ocean other than the coast of South Africa. Only one specimen is known from the eastern South Pacific (Handley 1966). Specimens from the coast of South Australia (Hale 1962a), the eastern United States (Allen 1941, Caldwell 1960, Manville and Shanahan 1961) and Japan (Yamada 1954) should be referred to as *Kogia*

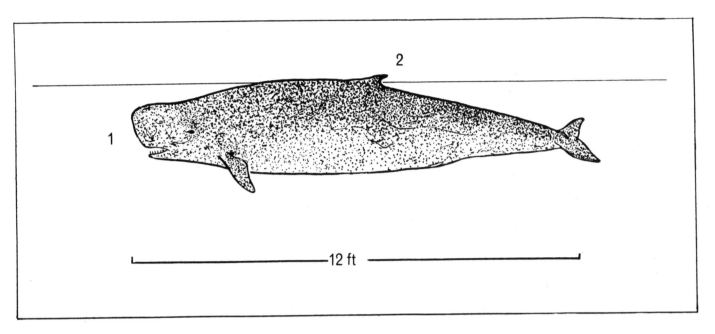

Figure 81.
Kogia breviceps, pigmy sperm whale: 1. short pointed jaw with sharp teeth, set back beneath head, 2. dorsal fin.

simius according to Dr Handley. This species of *Kogia* has also been recorded from Hawaii (Edmondson 1948), India (Owen 1866), Ceylon (Deraiyagala 1962), South Africa (Sclater 1901), and Cambodia (Serene 1934).

New Zealand *breviceps* records have been published by Benham (1901, 1902a, 1902c), Dell (1960) and Gaskin (1966), and all records to June 1965 summarised by Gaskin (1968a). The distribution of strandings must be interpreted with some care, since Mr Charles McCann, formerly of the Dominion Museum, Wellington, pointed out in a recent paper (1964b) that the coastal currents around the New Zealand archipelago result in coincidental strandings in a number of specific areas. For example, a tongue of the East Australian current is responsible for the large number of strandings of many species of whales on the west coast of the North Island north of Wellington. However, the map of stranding localities which I gave in my 1968 paper (Gaskin 1968a, fig. 32) suggests that the species is probably fairly common off both the Tasman Sea and South Pacific coasts of New Zealand. Dr C. M. Hale (1947, 1962a) suggested that the *Kogia* found off the coasts of South Australia have migratory movements, but we have no real evidence for definite migrations of either species. Unless the pigmy sperm whale attracts the interest of commercial whaling enterprises, and this seems most unlikely, we are not going to obtain large enough samples on a sufficiently regular basis to make a real study of the movements and population structure. At this time we do not even know if the populations of *Kogia breviceps* and *Kogia simius* across the world are continuous or discontinuous.

Dr G. M. Allen suggested that records from North Atlantic coasts indicated a northern biased distribution of *Kogia* in the autumn and a southern biased distribution in the spring. Similarly, South Australian records reveal that most strandings occurred in the colder months. Dr Hale interpreted this as indicating a seasonal inshore-offshore movement. If these observations are valid there appears to be a shift of population by *Kogia simius* from high latitudes to low latitudes in the summer in the North Atlantic, and the reverse movement from high latitudes to low latitudes in the winter off the coast of South Australia. This is not at all what one might expect. In other species of toothed whale such as the sperm, seasonal movements in both hemispheres appeared to be synchronised.

BIOLOGY: Pigmy sperm whales are blackish-blue dorsally and pinkish-white ventrally. The skin of both males and females is frequently marked by parallel rows of scars which are hard to explain except as tooth marks. This leads us to suppose that *Kogia* engages in fighting, presumably during the mating season. Smaller circular scars round the mouth are almost certainly made by squid. Similar marks are found on great sperm whales.

Figure 82 (L. Gurr, Massey University, Palmerston North, New Zealand.)
A 7ft male pigmy sperm whale stranded near Otaki, New Zealand.

Stranded *Kogia* often have stomachs infested with roundworms, and the encysted larvae of the tapeworm *Phyllobothrium delphini* are embedded in the blubber (Johnston and Mawson 1939). The stomach contents have been examined by a number of workers, and the fact that a specimen taken in Holland had been feeding on small crabs suggests that the concept of the pigmy sperm whale as a pelagic species (Norman and Fraser 1937, Ratcliffe 1942) will have to be revised. A specimen stranded at Lyall Bay, Wellington, in November 1950 had been feeding on the locally common arrow squid *Nototodarus sloani* (Dell 1960).

The life-cycle of *Kogia* is not really known at all. We have odd scraps of information: the gestation period is about 9 months and the lactation period about a year (Slijper 1962); in the New Zealand region pregnant females have been recorded between January and April, and females with calves between April and August. We can assume that the majority of the calves in the South Pacific will be born between the beginning of February and the end of June. The inshore movement of the species might be associated with calving, but it is hard to reconcile the scanty biological evidence with the equally scanty evidence for migratory movements.

12. The Beaked Whales Family Ziphiidae

The Ziphiidae are among the least known Cetacea. They are toothed whales of moderate size, characterised among other things, by a snout which is frequently drawn out into a rostrum or beak and from which the group obtains its common name. Beneath the throat there is a pair of grooves arranged in a V-shape with an open apex; these grooves are diagnostic for the family. Ziphiidae also differ from Physeteroidea and Delphinoidea in the shape of the flukes; in the Ziphiidae there is no central notch between them.

The arrangement of the teeth is very variable. Males of *Berardius, Hyperoodon* and *Ziphius* have visible teeth only at the tip of the lower jaw, while males of *Mesoplodon* generally have only a single pair of visible teeth set half way along the lower jaw. The teeth of most female Ziphiidae never erupt, and remain buried in the gum. Only *Tasmacetus* has functional teeth in both upper and lower jaws.

The following genera are recognised; *Berardius,* Duvernoy, 1851, *Hyperoodon,* Lacépède, 1804, *Mesoplodon,* Gervais, 1850, *Tasmacetus,* Oliver, 1937, and *Ziphius,* Cuvier, 1823. There is, and will be for some time to come, considerable dispute about the relationships of these genera, and the exact number of species that exist. Very few specimens of some species of beaked whales have been examined (Moore 1966), and some specimens are represented only by skulls and a few other bones. Though most of the recognised species differ in a number of apparently sound diagnostic skull features, the samples are generally too small to be compared statistically with any great reliability. So far little commercial whaling interest has centred on these species, and until this happens, or it is possible to finance a scientific expedition to take large numbers of specimens, we must rely largely on stranded animals to increase our knowledge of the Ziphiidae. Experts differ on the identity of a number of *Mesoplodon* species; for example while Moore (1966) regarded *M.hectori* as a good species, McCann (1962b) forcibly argued that the known specimens were juveniles of *B.arnouxi.* It is worth mentioning the remarks by Fraser (1966b) on the plasticity of bone in cetaceans and the reliability of taxonomy based on purely skeletal characters: '. . . the earlier systematic cetologists had to depend very largely on skulls and other bones for their diagnosis of species, and, in general, their decisions have stood the test of time. J. E. Gray (1827, 1828-30, 1846, 1871a, b, 1874a, b, c) has been criticised for the number of species he erected, but much of the material available to him was inadequate for anything more conclusive than the recognition of differences, which he acknowledged specifically in the fashion of his times'. We are in a similar position now with Ziphiidae and Delphinoidea as Gray, in his time, was with Mystacoceti.

All the Ziphiidae are medium-sized cetaceans (McCann 1962a), and *Mesoplodon* species are generally less than 20 ft in length. Specimens of *Ziphius* may reach 30 ft (Norman and Fraser 1937) while *Berardius* and *Hyperoodon* frequently exceed 30 ft (Rice 1963b). Early descriptions and discussions of the beaked whale group were undertaken by Knox (1871), who referred to some

New Zealand specimens, and Flower (1872, 1879). A major account was given by Dr F. W. True in 1910, following a shorter paper by W. M. Turner (1872) on *Ziphius* and *Mesoplodon* in Scottish seas. The anatomy of various beaked whales was described by Forbes (1893), who discussed the development of the rostrum in *Mesoplodon* and made a major contribution towards separation of the species then known. Fraser (1936), examined the dentition in *Ziphius* and illustrated the mandibles of all the genera and a number of the most important species (Norman and Fraser 1937). Boschma (1950, 1951) discussed dentition in a number of Ziphioid genera.

Genus *Tasmacetus*
(Oliver 1937)

Only five specimens of this unique monotypic genus are known, all from New Zealand. The type specimen was described by Oliver (1937) from Wanganui beach.

SHEPHERD'S BEAKED WHALE
Tasmacetus shepherdi (Oliver 1937)

DESCRIPTION: This animal reaches lengths of up to 30 ft (Sorensen 1940). In appearance it strongly resembles a species of *Mesoplodon*, but differs from all other beaked whales in possessing functional small teeth in both jaws. The type specimen had 19 pairs of teeth in the upper jaw and 27 in the lower jaw, the pair at the tip of the jaw being particularly well developed. The colour of specimens so far recorded has been black on the dorsal surface, grey or greyish-yellow on the flanks and white beneath. The forehead in a photograph taken by Mrs J. Harrison of a specimen stranded on Stewart Island is quite bulbous (Gaskin 1968a, fig. 49). Although not clearly shown in the photograph, a distinct dorsal fin was present.

DISTRIBUTION: There are only five stranding records and these were summarised by the author (Gaskin 1968a). Two specimens were stranded at Mason's Bay, Stewart Island, in February 1933, one (the type specimen) at Wanganui in October 1933, and two more from the Avon-Heathcote estuary and Birdling's Flat in March 1951 and in 1962 respectively. The Dominion Museum also has in its collection another mandible of unknown origin.

The distribution of the strandings suggests the species is associated with the cool waters of the south side of the Subtropical Convergence, but nothing is known of the total range of its distribution, or of any movements. During the Marine Department whale survey, I sighted a small whale a few miles from the Kaikoura Peninsula which appeared to be this species. It was a beaked whale, somewhere between 15 and 20 ft in length, and a fleeting glimpse of the head left the impression of small teeth along the edge of the lower jaw. Although the animal swam very close to the boat for a few seconds there was no equipment on board with which it could be taken.

BIOLOGY: Nothing is known of the life-cycle or behaviour of this species.

Figure 83.
Tasmacetus shepherdi, Shepherd's beaked whale: 1. small but functional teeth present in upper and lower jaws.

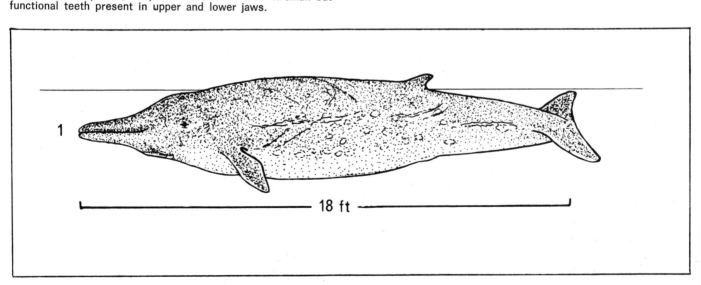

1

18 ft

106

Genus *Mesoplodon*
(Gervais 1850)

Some authorities, for example, Scheffer and Rice (1963) prefer to regard *Mesoplodon* as being two subgenera, *Mesoplodon* containing the small-toothed species such as *M.grayi*, and *Dioplodon* containing the large-toothed species such as *bowdoini, densirostris, layardi* and *stejnegeri*. Iredale and Troughton (1934) went further, and gave *Dioplodon* full generic status. I have doubts about the validity of either of these steps, when *Mesoplodon* species can be arranged in logical series showing either elongation of the medial teeth, or increase in size of these teeth. In this work I intend to follow Moore (1966) and consider *Mesoplodon* as a single genus.

In general appearance the species of *Mesoplodon* are very similar; they are of rather stout build, and the flippers are relatively short in proportion to the total body length, as in other Ziphiidae. Size and body colouration are not very useful for separating *Mesoplodon* species from one another (Norman and Fraser 1937). The known species do not appear to exceed lengths of about 20 ft at physical maturity, and they are all dark grey or blackish-brown dorsally, grey or yellowish-grey on the flanks, and white or yellowish-white on the ventral surface. Males, as in many odontocetes, are larger than females at maturity, and are readily separable from one another by tooth characteristics. The teeth rarely erupt in females, even at maturity, and are sometimes not visible in immature males. The size, shape and position of the teeth are very important criteria for recognition of species. A complete review of all the species known has recently been published (Moore 1968); a previous paper by Moore (1966) covers all the North American species. Earlier papers of a review nature include those of Flower (1879), Forbes (1893), True (1910) and Harmer (1924).

Eleven species are currently recognised in this genus, of which four occur in New Zealand waters; *M.layardi, M.hectori, M.grayi* and *M.bowdoini. M.bowdoini* has also been recorded from Western Australia (Glauert 1947), from Japan (Nishiwaki 1962a) and the Pacific coast of the United States (Hubbs 1946). However, the last two records are referred by Dr J. C. Moore to *M.stejnegeri*. *M.hectori* has also been recorded from the Falkland Islands (Fraser 1950), but as mentioned earlier, McCann (1962b) disputed the validity of this species. *M.layardi* has also been recorded from South Australia (Hale 1932a), and Victoria (Warneke 1963).

The seven species not recorded from the New Zealand region are: *Mesoplodon mirus* from the North Atlantic

(Raven 1937, Thorpe 1938, Allen 1939b, McKenzie 1940, Ulmer 1941, Moore and Wood 1957, Moore 1960 Moore 1966) and from South Africa (Talbot 1960 McCann and Talbot 1963); *Mesoplodon ginkgodens* from Japan (Nishiwaki and Kamiya 1958); *Mesoplodon stejnegeri* from the North Pacific and Bering Sea between latitudes 50° and 60°N (True 1885, 1910, Scheffer and Slipp 1948, Jellison 1953, Orr 1953, Nishiwaki 1962a, 1962b as *bowdoini,* Moore 1963a, 1966); *Mesoplodon bidens* from the North Sea and North Atlantic (Agassiz 1868, Allen 1869, True 1910, Harmer 1927, Stephen 1932, Fraser 1934, 1946, 1953, Berlin 1941, 1951, Jonsgaard and Hoidal 1957, Sergeant and Fisher 1957, Moore 1966); *Mesoplodon densirostris* from the North Atlantic (Allen 1906 as *bidens,* True 1910, Andrews 1914, Ulmer 1941, Hutton 1950, Moore 1958, 1966), the North Pacific (Galbreath 1963), the Seychelles (Norman and Fraser 1937, Moore 1966), Lord Howe Island in the Tasman Sea (Kreft 1870, Moore 1966), and a whale stranded in Tasmania may also belong to this species (Dr E. R. Guiler, in letter); *Mesoplodon carlhubbsi* from the Japanese (Nishiwaki and Kamiya 1959 as *stejnegeri*) and the United States' coasts of the North Pacific (Scheffer and Slipp 1948, Orr 1950, 1953 as *stejnegeri,* Moore 1963a, 1966, Roest 1964); and *Mesoplodon europaeus* from the North Atlantic, Caribbean and Gulf of Mexico (van Beneden 1888, Turner 1889, Allen 1906 as *bidens,* True 1910, Raven 1937, Ulmer 1941, Brimley 1943 as *mirus,* Rankin 1953, 1956, Aguayo 1954, Fraser 1955a, Twist and Twist 1956, Moore 1953a, 1960, 1966 and Moore and Wood 1957). A species *Mesoplodon gervaisi* discussed by Moore and Wood from the Gulf of Mexico appears to be synonymised by Moore in his 1966 paper with *M.europaeus,* although he does not mention this at all in his lengthy 1966 paper, making things rather confusing for the reader. Although McCann (1962b) synonymised *M.hectori* with *B.arnouxi,* I have left the records under that name as in an earlier publication (Gaskin 1968a), to avoid possible confusion. Similarly McCann's use of *stejnegeri* (1962a, 1964b) for New Zealand *bowdoini* is not being followed until an authority has completely reviewed the southern hemisphere *Mesoplodon.* However, the species *Mesoplodon pacificus* which Longman described from the coast of Queensland (Longman 1926) has been convincingly synonymised with *Hyperoodon planifrons* by McCann (1962c).

For general discussions on this genus the reader is referred to Norman and Fraser (1937), Moore (1953, 1966, 1968), and to a comparative key to the species in the New Zealand region by McCann (1962a). The genus in the New Zealand region has also been discussed by Gaskin (1968a), and complete lists of the known strandings of *Mesoplodon grayi, M.layardi, M.bowdoini* and *M.hectori* for New Zealand and the Chatham Islands are also given in the same paper, together with maps. The total number of *Mesoplodon* specimens recorded from these two archipelagoes is between 40 and 45. Not all of the known specimens were definitely identified before the remains were lost, and a few may have been mis-identified. In view of this, I do not want to be too dogmatic in my approach to the New Zealand and Chatham Islands material.

HECTOR'S BEAKED WHALE
Mesoplodon hectori (Gray 1871)
DESCRIPTION: There is no description of the external characteristics of *hectori*, since it is known from only three skulls. It is distinguished from all species of *Mesoplodon* except *mirus*, by having the single pair of relatively large teeth in the mandible set almost at the tip, as in *Ziphius* and *Berardius*. These teeth are triangular and almost flat. McCann (1962b) discussed its systematic position. Known distribution includes only New Zealand (two specimens) and the Falkland Islands (one specimen). Nothing is known of the life-cycle.

THE STRAP-TOOTHED WHALE
Mesoplodon layardi (Gray 1865)
DESCRIPTION: Males of this species may attain lengths of 18 ft, although 14 to 15 ft is the usual size. Females reach a maximum length of about 15 ft, and the beak is usually a more prominent feature in this sex than in the male. The usual colour pattern is dark purplish-brown above and white beneath, with grey intermediate

zones on the flanks. Specimens are frequently found with large irregular white patches on the back and flanks. The males of this species can be at once separated from any other Ziphiidae by the form of the single pair of visible teeth in the lower jaw. Their extreme elongation has earned this species its common name. In mature males the teeth curve upwards and backwards from their position midway along the mandible, and project upwards beside the upper jaws rather like the tusks of a boar. In the female the teeth rarely break the gum, and the same may also be true of immature males. In such cases the species can only be identified by skull characters, and determinations must be left to an expert.

DISTRIBUTION: The type specimen for this species was described from the Cape of Good Hope by Gray (1865) as *Ziphius layardi,* and at least one other South African specimen has been recorded ('M.C.-L.' 1960). The species has already been mentioned as occurring in South Australian waters (Hale 1932a), New South Wales (Marlow 1963) and from Victoria (Warneke 1963); it has also been recorded from Tasmania (Davies 1963b). However, most of the recorded strandings have occurred in New Zealand. No less than 19 specimens have been reported, and material from most of these is held in New Zealand museums (Gaskin 1968a). One of the New Zealand specimens came from the Chatham Islands. The strandings of this species have been very noticeably concentrated on the east coast of the South Island and in Cook Strait, the only exceptions to this being the solitary Chatham Island specimen and some remains found on Great Barrier Island (Oliver 1922c). The occurrence of this species in New Zealand has been discussed by Hector (1873a), von Haast (1877b), Waite (1912), Oliver (1922c, 1924a), McCann (1964b) and Gaskin (1968a).

Figure 84.
Mesoplodon layardi, Strap-toothed whale: 1. large strap-shaped pair of teeth in mature males.

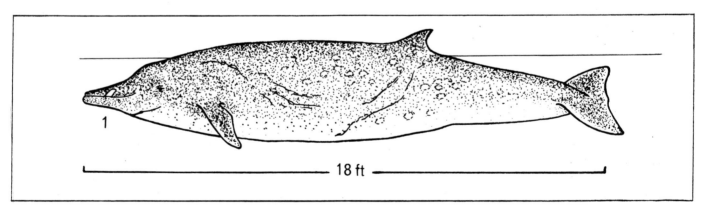

1

18 ft

However, we have little information on the overall distribution of the species even within the New Zealand region, except that the northern limit of range appears to be about 30°S. One male was recognised by the author from *Chiyoda Maru No. 5* in 1967 to the east of the Chatham Islands; the tooth on the left side of the head was clearly seen. This animal was one of a group of three, of which all were presumed to be of this species. Although this was the only one identified of 39 Ziphiidae seen by this vessel, it was noticed that all but one of the animals were observed in waters associated with the Subtropical Convergence, with a surface temperature range of about 10° to 16°C (Gaskin, in press[2], 1971).

BIOLOGY: McCann (1964a) described the reproductive system of a freshly stranded female of this species, but we know nothing of the life-cycle. Strandings on the coast of New Zealand predominated in summer months (Gaskin 1968a); twice, females with foetuses were recorded, both in winter, suggesting that calving occurs in spring or early summer.

ANDREW'S BEAKED WHALE
Mesoplodon bowdoini (Andrews 1908)

DESCRIPTION: This appears to be the smallest of the beaked whales occurring in the New Zealand region; the largest specimens recorded, both females, were 14 ft in length. In external appearance this is very similar to other species of *Mesoplodon*, but the males have a massive, broad pair of teeth situated behind the mandibular symphysis, quite characteristic in shape, and the species is also distinct from others on skull characters. The teeth do not always erupt in females.

DISTRIBUTION: The difference in opinion between authorities on the relationships between *M.bowdoini* and *M.stejnegeri* on one hand and *M.stejnegeri* and *M.carlhubbsi* on the other has been discussed in the generic

section. Moore (1963b, 1966) has presented evidence that all three are distinct species. He cites the range of *M.stejnegeri* as the North Pacific and Bering Sea, and the range of *M.bowdoini* as the South Pacific. The type specimen was described from New Zealand (Andrews 1908b), along with four others (Oliver 1922c, McCann 1962a, 1964b). Two teeth of this species were found at Campbell Island (Gaskin 1968a) in the Subantarctic, and Glauert (1947) recorded a specimen from Western Australia.

BIOLOGY: A female carrying a 5 ft foetus was stranded at the Manawatu River in late September 1937 (Gaskin 1968a). The size of the foetus strongly suggests that calving occurs in the late spring or early summer.

THE SCAMPERDOWN WHALE
or GRAY'S BEAKED WHALE
Mesoplodon grayi (von Haast 1876)

DESCRIPTION: This whale, which appears on the basis of known strandings (Gaskin 1968a) to be the commonest member of the genus in New Zealand waters, reaches a maximum length of about 18 ft for mature males and 16 ft for mature females. The dorsal surface is a dark brownish-grey in life, the flanks plain or mottled grey, and the ventral surface white. The teeth are triangular and relatively inconspicuous, and the visible pair in males is placed towards the posterior end of the mandibular symphysis. There are no external distinguishing features which can be used to reliably separate females and immature males from their equivalents in the other beaked whale species found in these waters. Occasionally a row of small maxillary teeth are present, so there is also the possibility of confusion with *Tasmacetus*.

DISTRIBUTION: This whale appears to have a circumpolar distribution in the southern hemisphere south of latitude 30°S. Gray's beaked whale has been recorded from Australia and the southern part of South America (Norman and Fraser 1937) and from South Africa (Bar-

Figure 85.

Mesoplodon bowdoini, Andrew's beaked whale: 1. broad and massive pair of teeth in lower jaw of males.

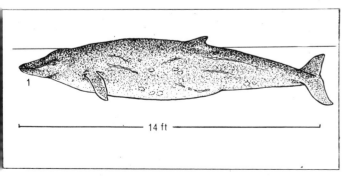

Figure 86.

Mesoplodon grayi, Gray's beaked whale: 1. inconspicuous pair of triangular teeth in the middle of lower jaw of mature males.

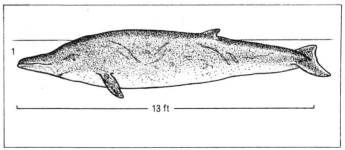

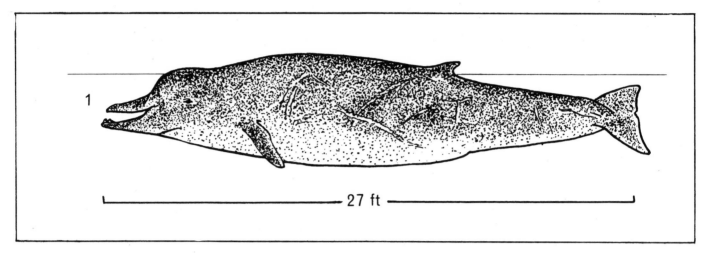

Figure 87.
Berardius arnouxi, large beaked whale: 1. two pairs of teeth at tip of lower jaw.

nard 1954). The type specimen was from New Zealand (von Haast 1876, 1877c). Other occurrences of this species in New Zealand waters have been discussed by Hector (1873b, 1875), Flower (1879 as *M.haasti*), Forbes (1893), Lillie (1915), Oliver (1922c), McCann (1964b), and summarised by Gaskin (1968a). Chatham Island records have been considered by Forbes (1893), Oliver (1922c), McCann (1964b) and summarised by Gaskin (1968a).

BIOLOGY: Nothing is known of the species' life-cycle.

Genus *Berardius*
(Duvernoy 1851)

A genus of relatively large beaked whales, with two species, *B.bairdi* from the North Pacific and *B.arnouxi* from the Southern Ocean, *B.arnouxi* was described first (Duvernoy 1851).

THE LARGE BEAKED WHALE
Berardius arnouxi (Duvernoy 1851)

DESCRIPTION: This species reaches lengths of over 30 ft at maturity, and is the largest beaked whale in New Zealand waters. Specimens of *B.bairdi* are even larger, attaining lengths of 40 ft (Rice 1963b). The colour is variable; usually specimens are blackish-brown dorsally, grey on the flanks and white ventrally. The dorsal surface turns dark grey after death. Patches of pale colour, like those found on killer whales and *Ziphius cavirostris,* are not unknown. Stranded specimens are frequently found

to bear rows of parallel scars, which are believed to be caused during fighting and mating. The forehead of mature males has a distinct bulge, directed upwards and forwards rather than forwards, as in *Hyperoodon* and the pilot whale *Globicephala*. Immatures of both sexes, and even mature females, are very difficult to separate from their equivalents in other beaked whale species. Adult males can usually be distinguished from other species by two pairs of triangular teeth protruding from the gum at the tip of the lower jaw. In females these teeth often remain buried in the gum.

DISTRIBUTION: It is the North Pacific species, *Berardius bairdi* (Stejneger 1883), which is the better known of the two species. It occurs on both sides of the Pacific and in the Bering Sea (Omura, Fujino and Kimura 1955, Slipp and Wilke 1953, Scheffer and Slipp 1948, Scheffer 1949, Pike 1951, Kotov 1958, Betesheva 1962, Rice 1963b, Brownell 1964). The southern species *B.arnouxi* has been recorded from New Zealand (Flower 1872, Knox 1871, von Haast 1870, Hector 1878, Oliver 1922c, McCann 1964b) where the records have been summarised by Gaskin (1968a); from South Australia (Hale 1962b); and from the Falkland Islands and the Falkland Islands Dependencies (Norman and Fraser 1937). The species has been observed by whaling ships in the Antarctic, and Taylor (1957) recorded the species trapped in pools of sea ice on the edge of the Antarctic mainland.

We know nothing of the movements of this species. All the New Zealand strandings occurred between January and mid-March, and neither sex predominated. Three beaked whales more than 30 ft in length, and presumed to be of this species, were sighted moving north in

Figure 88. (S. C. B. Blake, British Antarctic Survey, with acknowledgements also to Proc. zool. Soc. London.) Two large beaked whales *(B.arnouxi)* trapped in a pool of Antarctic sea ice off the north east coast of the Antarctic Peninsula. The characteristic tube-like snout is clearly seen.

November 1964 off the Marlborough coast by the spotter plane of the Tory Channel Whaling Company.

BIOLOGY: Nothing is known of the life-cycle of the southern species. However, Ohsumi (1964b) studied the rate of accumulation of *corpora albicantia* in the ovaries of 13 females captured on the coast of Japan, and Omura (1958b) described an embryo of the species. *B.bairdi* is taken commercially on the coasts of Japan (Omura, Fujino and Kimura 1955) and California (Rice 1963b).

The stomachs of specimens examined on the coast of New Zealand contained the remains of squid, according to Mr C. McCann, formerly of the Dominion Museum, Wellington.

Genus *Ziphius*
(Cuvier 1823)

The single species in this genus, *Ziphius cavirostris,* was first discovered as a fossil by Cuvier in 1823, and in the years that followed, was shown to be distributed throughout most of the oceans of the world. In the nineteenth century many names were given to regional forms of *cavirostris,* including *novaezealandiae* by Hector (1877)

to a specimen stranded on the New Zealand coast, and *chathamiensis* to one from the Chatham Islands (Hector 1873a). Only one world-wide species is now recognised (Hershkovitz 1966, Mitchell 1968b) under Cuvier's original specific name.

CUVIER'S BEAKED WHALE
or THE GOOSE-BEAK WHALE
Ziphius cavirostris (Cuvier 1823)

DESCRIPTION: This attains lengths of 25 to 30 ft at maturity. Specimens are usually purplish-black on the dorsal surface, grey on the flanks and white on the ventral surface. However, pigmentation is very variable in *Ziphius,* and is no real guide to identification. Oval scars resembling those found on rorquals (Pike 1951) are quite common (Mitchell and Houck 1967). Longitudinal scars on the flanks (von Haast 1880) are probably the result of intraspecific fighting. Smaller scars around the mouth may be caused by the beaks and hooks of squid.

Mature males of this species often lack a well-defined beak, and have a pronounced forehead bulge. This bulge is frequently very slight in immature males and in mature females, in which the beak is more obvious. The distance between the tip of the upper jaw and the blowhole is

111

generally about one ninth of the total length of the animal. Male specimens of *Ziphius* can usually be separated from male specimens of *Mesoplodon* and *Berardius* by a single pair of visible teeth at the tip of the lower jaw. However, even this character is not completely reliable. In the female this pair of teeth does not usually erupt through the gum, even with advancing age. Determinations are best made on skull characters. Other vestigial teeth are present in the gums, as in other ziphoids (Boschma 1950, 1951), but very careful dissection is generally needed to discover them.

DISTRIBUTION: The general distribution of *Ziphius cavirostris* throughout the world was mapped by Moore (1963a). It has not yet been recorded from Arctic or Antarctic waters. Specimens have been recorded from the coasts of the U.K. (Turner 1872, Fraser 1946, 1966b), Ireland (Cabot 1965), France (Paulus 1962), Italy (Tamino 1957, Tortonese 1957), Spain (Uriarte 1943, Sala de Castellarnau 1945), the east coast of the USA (Backus 1961), South Africa (Allen 1939a, Barnard 1954), Puerto Rico (Erdman 1962a, b), Australia (Iredale and Troughton 1934), New Zealand (von Haast 1877a, 1880, Hector 1873a as *chathamiensis,* 1875, Scott and Parker 1889, Waite 1912, Oliver 1922c, Gaskin 1968a), Tasmania (Scott and Lord 1921), Japan (Omura, Fujino and Kimura 1955), Hawaii (Richards 1952), the Midway Islands (Galbreath 1963), the Aleutian Islands (Scheffer 1949, Kenyon 1961), Alaska (Stejneger 1884 as *Z.grebnizkii,* True 1910), Vancouver Island (Cowan 1945), British Columbia (Cowan and Hatter 1940, Cowan and Guiguet 1952), and the west coast of the United States (Scheffer 1942, Houck 1958, Hall and Kelson 1959, Roest, Storm and Dumas 1953, Hubbs 1946,

1951a, Norris and Prescott 1961, Daugherty 1965, Orr 1966, Mitchell and Houck 1967, and Mitchell 1968b). In the last paper Mitchell very usefully summarised data on this species for the whole of the eastern North Pacific region.

There is no evidence of migratory movement by this species in the New Zealand region. Strandings (Gaskin 1968a) have occurred in January, April, May, June, July, August, September, October and November, indicating that the species occurs in New Zealand waters all the year round. Validated records of New Zealand strandings have occurred between Poverty Bay in the north and Otago in the south, suggesting that the species may be distributed across Subtropical Convergence waters in this region. However, Mitchell (1968b) found known western North American strandings to extend over a wide range of latitude, from about 25°N to about 60°N, but decided that there was insufficient information to generalise on seasonality in the records, except that the species probably occurred off the west coast of North America all the year round. Similarly, Omura, Fujino and Kimura (1955) found that the species was present in Japanese waters all the year round; far more specimens were caught in summer (between late May and September) than in winter (October to April), as the whaling effort for these beaked whales is much less in winter, the catching being suspended between December and March.

BIOLOGY: This whale is taken commercially on the coast of Japan (Omura and others, 1955); consequently most information on the biology of the species has come from there. Ohsumi (1964b), studied the rate of *corpora albicantia* accumulation, but had insufficient material on which to base a detailed life-cycle study or age determina-

Figure 89.

Ziphius cavirostris, Cuvier's beaked whale: 1. single pair of teeth protruding at tip of lower jaw in mature males.

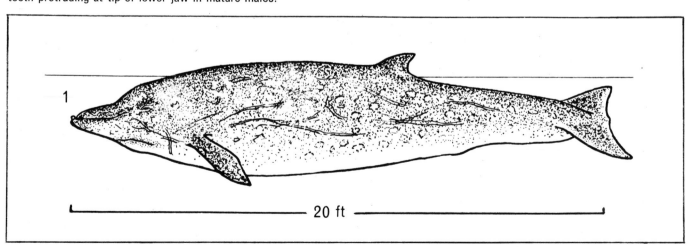

1

20 ft

Figure 90.   (Professor Warren J. Houck, Division of Biological
              Sciences, Humboldt State College, California.)
Cuvier's beaked whale *(Z.cavirostris)* stranded on the northern
coast of California in 1957.

tion analysis. Omura and his co-workers found that males of *Ziphius cavirostris* became sexually mature at about 17½ ft, and females at about 18ft in length. Calves are about 7½-8 ft in length at birth, based on Omura's observations on a 7 ft foetus which appeared to be nearly full-term. The only calf observed in New Zealand waters was stranded with its mother in July (von Haast 1877a), suggesting summer births in the southern hemisphere.

This species probably numbered among 39 Ziphiidae seen by the author and Dr Keiji Nasu from *Chiyoda Maru No. 5* in the 1966-67 Antarctic whaling season. Although these animals included at least some *Mesoplodon*, and probably *Berardius* and *Hyperoodon* as well, the largest school size seen was of only five animals. Mitchell (1968b) rightly questioned statements by Kellogg (1940) and Slijper (1962), neither with supporting data, that this species was found in schools of from 10 to 100 animals, usually in 'close-knit' herds of 30 to 40. However, this must be qualified by mentioning a sighting made by myself and co-workers of two groups of beaked whales on August 21, 1963, which we identified at the time as *Hyperoodon planifrons,* though I have become less sure of this with passing time. These animals were first seen

at a distance, creating a commotion in the water, and were thought to be pilot whales, since one observer reported he could see white patches behind the dorsal fins of some. However, they came very close to one boat, in two groups swimming quite near together, and it could then be seen that they certainly were not pilot whales. Some of the largest individuals were at least 25 ft in length, and these had very short beaks and bulbous foreheads. The total number in the school, which was first sighted about 20 miles south of Banks Peninsula, was estimated at 30-35 animals (Gaskin 1964a). A number of animals brought their heads clear of the water; no large teeth were visible, in fact no one recalled noticing any teeth at all. One other large school of beaked whales is recorded from the New Zealand region; 25 *Mesoplodon grayi* stranded near Waitangi in the Chatham Islands in 1875 (von Haast 1876), including among them the specimen chosen by Haast as the type of the species, which he at that time placed in a genus *Oulodon*.

Norris and Prescott (1961) described how an immature female was maintained in a Seaquarium for a few days on the west coast of the USA. The specimen was apparently dying when caught.

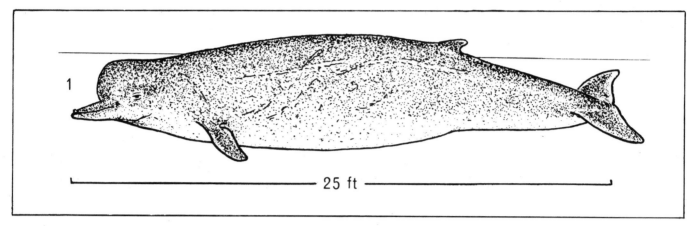

Figure 91.
Hyperoodon planifrons, Southern beaked whale: 1. single pair of teeth at tip of lower jaw in mature males.

Genus *Hyperoodon*
(Lacépède 1804)

Contains two species: *Hyperoodon ampullatus* in the northern hemisphere, and *Hyperoodon planifrons* in the southern hemisphere. While *H.ampullatus* has been taken extensively in commercial catches off the coast of Norway (Jonsgaard and Long 1959), *H.planifrons* is one of the least known Ziphiidae.

SOUTHERN BOTTLENOSED WHALE
Hyperoodon planifrons (Flower 1882)
DESCRIPTION: The following description is, of necessity, based on the external characters of the northern species. Whales of this genus possess a bulbous forehead which develops in size with age. The dorsal fin is small, and the body is stoutly cylindrical, resembling *Berardius* quite closely. Northern hemisphere *Hyperoodon* males attain lengths of up to 30 ft, and females to 25 ft. There is a short but well-defined beak, and a single pair of prominent teeth project through the gum of males at the tip of the lower jaw. The blowhole is placed further back than in *Ziphius*, but specimens are still very easy to confuse with that species. Colour is variable, generally dark brownish-black dorsally and greyish-white ventrally. White markings may be present on the back and flanks.
DISTRIBUTION: Very little is known of the movements of either *H.ampullatus* or *H.planifrons;* although Sanderson (1956) gave an account of the migration of bottlenose whales, it was without references, and a determination of the accuracy of the observations is impossible. In the northern hemisphere *H.ampullatus* ranges widely through arctic and subarctic regions (Gray 1882, Ohlin 1893). It

supported a substantial industry off the coast of Norway for many years (Southwell 1882, Kuekenthal 1887, Ruud 1937, Jonsgaard and Long 1959, Jonsgaard and Øynes 1952), and ranges as far south as the United Kingdom (Greenshields 1937, Ellison 1954a) and France (Renard 1954). The southern species *H.planifrons* has been recorded from northwest Australia (Flower 1882 type specimen), the Falkland Islands Dependencies (Fraser 1945), South Australia (Hale 1932a, 1939) and New Zealand (McCann 1961, Gaskin 1968a). The only positive records of this species in New Zealand waters have been from East Cape (correctly identified by Waite 1913, referred to as *Ziphius cavirostris* by Oliver 1922c), and South Canterbury, near Timaru.
BIOLOGY: Very little is known of the biology of *Hyperoodon.* Some anatomical aspects have been studied by Bouvier (1892) and Goudappel and Slijper (1958), the latter describing only the structure of the lungs of the northern species.

Akimushkin (1954) reported squid and fish as primary items of diet, as did Hale (1932a) with specific reference to *H.planifrons,* based on a study of a specimen stranded in South Australia.

Norman and Fraser (1937) reported school sizes as being up to a dozen animals, and recorded that *H.ampullatus* stranded frequently around the shores of the British Isles, and that from the records it was possible to estimate that the gestation period was about one year, and that calves are about 10 ft in length at birth. The breeding cycle seemed to be related to a migratory cycle, with the species wintering around England and Scotland and spending the summer in arctic waters. However, much more data is required before these facts can be fully confirmed.

13. The Great Dolphins
Delphinoidea: Globicephalidae

The old family Delphinidae of nineteenth century cetologists (Flower 1883, True 1889) is today considered to be a heterogeneous group containing a number of distinct families within three superfamilies. The first, the Platanistoidea or river dolphins, includes the Platanistidae. The second, the Monodontoidea, contains the narwhal *Monodon* and the beluga *Delphinapterus,* following revision by Fraser and Purves (1960). The third, the Delphinoidea, is by far the largest, and contains one family of porpoises, the Phocoenidae, and three families of dolphins, the Globicephalidae, the Stenidae and the Delphinidae. For more detailed discussion of the taxonomy of these animals the reader is referred to Fraser and Purves (1960), Nishiwaki (1963, 1964) and Fraser (1966a). The systematic position of a number of groups is not beyond dispute at this time.

The Globicephalidae are characterised as follows (Nishiwaki 1963, 1964): the rostrum is characteristically short and blunt, there are less than 15 pairs of teeth in the upper jaw; the atlas, axis and third cervical vertebrae are fused, and the body length at sexual maturity is generally more than about 8 ft. There are six recognised genera in the family (van Bree and Cadenat 1968): *Globicephala* Lesson, 1828, *Orcinus* Fitzinger, 1860, *Pseudorca* Reinhardt, 1862, *Orcaella* Gray, 1866, *Feresa* Gray, 1870, and *Peponocephala* Nishiwaki and Norris, 1966. One species of each of the first three genera listed has been recorded in New Zealand waters. *Orcaella, Feresa* and *Peponocephala* are subtropical and tropical genera, although *Feresa* has been recorded from Japan

(Nishiwaki 1966b), and *Peponocephala* on the coast of Australia (van Bree and Cadenat 1968).

Many sightings and stranding records of pilot and killer whales around the New Zealand coast are known (Gaskin 1968a); more so than for the Ziphiidae. Even so, these are nothing like sufficient to give a clear picture of the life-cycle, distribution and movements of these two species. Relatively few records of the false killer whale are known, and the stranding records are widely scattered.

THE COMMON PILOT WHALE
Globicephala melaena (Traill 1809)
Southern sub-species *G.m. edwardi* (Smith 1834)
DESCRIPTION: The forehead of this species is characteristically rounded and bulges out to form a 'melon', a term used by North Atlantic and Pacific whalers. There are 10 pairs of peg-like teeth on the anterior part of each jaw. The flippers are long and slender in proportion to the total body length. The dorsal fin is large and very broadly triangular, and is one of the features by which the pilot whale can be easily recognised at sea. Adults of this species attain body lengths of 10 to 20 ft.
DISTRIBUTION: There is still considerable dispute concerning the taxonomic status of the described forms of the pilot whale other than the originally designated species *Globicephala melaena*, which is distributed through the temperate regions of the North Atlantic (Joensen 1962) to Iceland and Southern Greenland, and is also found in the Mediterranean (Tamino 1953a, 1954), the North Sea and the Baltic (Davies 1960). A southern

115

form, allied to *G.melaena,* and given the sub-specific name *edwardi* by Davies, occurs in southern temperate and subantarctic waters with extensions of range into the north-flowing cold waters of the Humboldt and Benguela currents, according to the same author.

These two species are separated by subtropical and tropical waters in which occurs the Tropical or Short-flippered pilot whale *Globicephala macrorhyncha* which is recorded from the Caribbean (Caldwell & Erdman Indian Ocean and the Indonesian Archipelago (Weber 1923, Davies 1960). The recorded range of this species leads one to assume that it is distributed within the Tropical Convergences, and therefore probably occurs on the Queensland coast of Australia as well as the Indian Ocean coast of that continent, and may even have been recorded in northern New Zealand waters as *G.melaena.* The name *G.indica* used by Norman and Fraser (1937) for the form present in the Indian Ocean and the Bay of Bengal is a synonym of *G.macrorhyncha.* Similarly the name *G.brachyptera* used by the same authors for another form of the pilot whale is now regarded as a synonym of *G.macrorhyncha* (Sergeant and Fisher 1957). *G.scammoni,* which is the North Pacific form (Orr 1951, Norris and Prescott 1961, Brownell 1964, Ficus and Niggol 1965), appears to be a valid species.

In a 1962 paper Sergeant compared series of specimens from the recognised species and sub-species of the pilot whale, and could find no morphometric differences between the northern species *G.melaena* and the southern sub-species *m.edwardi.* However, Davies (1960) pointed out that the ranges of the two forms were quite distinct and probably have not overlapped since Pleistocene times. Sergeant found *G.macrorhyncha* to be distinct from the northern and southern species in that the flippers grew at the same rate as the body throughout life, while in the two sub-species of *melaena* the flippers grew allometrically, or faster than the rest of the body. However, the North Pacific *G.scammoni* was found to be intermediate in this respect. Davies examined the status of the southern pilot whale, which had been given full specific rank by Rayner (1939) as *G.leucosagmaphora* on the basis of colour pattern differences from *G.melaena.* Southern pilot whales frequently have a small white patch above and behind the eye and a white saddle-shaped marking behind the dorsal fin. However, a careful examination of the skulls and colour patterns of southern animals recorded by other authors (Oliver 1924b, Scott 1942, Cabrera and Yepes 1960, and also Barnard 1954) led Davies to conclude the *melaena* and *leucosagmaphora* could not be justifiably regarded as separate species, and hence proposed the southern sub-specific name *m.edwardi,* on the grounds that although no reliable morphological differences could be found, the two populations were still demonstrably isolated from one another.

Pilot whales, or 'blackfish' as they are known locally, are regarded as common animals by most New Zealand coastal authorities, and much potentially useful information has been lost when stranded specimens have been buried or towed out to sea without reference to a biologist. A list of strandings and sightings recorded to 20 January 1965 in the New Zealand area has been previously published (Gaskin 1968a). Probably in the same period at least as many strandings on beaches remote from human habitation went unreported. Since publication of the above list two more mass strandings have occurred, one at Campbell's Beach north of the Kaikoura Peninsula, and the other at McLeod's Bay, Whangarei, in November 1968.

Most of the specimens examined by reliable observers, or of which photographs were taken, appear to possess the white saddle-shaped mark behind the dorsal fin, except those around Whangarei, where many of the animals stranded have been all black except for a white mark under the chin extending posteriorly in the ventral midline. There is some suggestion in the data published earlier (Gaskin 1968a) that the species makes migratory movements in New Zealand waters, since pilots occur in Whangarei harbour and other nearby inlets much more often in spring and early summer than at other times of year. Behaviour which appeared to include mating among pilot whales sighted near Raoul Island in the Kermadecs has been reported to the Marine Department. However, until it is known whether one or two species of *Globice-*

Figure 92.

Globicephala melaena, pilot whale: 1. bulbous head with peg-like teeth in anterior parts of both jaws, white patch under chin, 2. long pointed flippers, and white streak on belly in many specimens, 3. large broad and deep dorsal fin, white streak from fin to tail stock in many specimens.

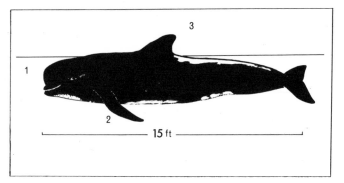

Figure 93. (*Northern Advocate*, Whangarei, New Zealand.)
School of pilot whales *(Globicephala)* stranded in shallow water in McLeods Bay, Northland, New Zealand, November 1968. Note the white pattern on the ventral surface. Note the large dorsal fin and the sharply elbowed flippers.

Figure 94. (*Northern Advocate.*)
Head and back of pilot whale stranded in McLeods Bay, Northland, showing the characteristic swollen forehead or 'melon' and the large dorsal fin.

phala occur in the New Zealand region, one northern and one southern, it would be idle to speculate on migratory movements. However, it is of interest to note that the officers of the whale chaser *Orca* sighted large numbers of *Globicephala* and *Pseudorca* (in separate schools), moving south in offshore waters near the Kaikoura Peninsula in January and February 1964.

There are certain regions in the western South Pacific in which pilot whale strandings appear to occur more frequently than anywhere else. Tasmania is a case in point (Scott 1942), although in comparison to Australia a relatively large number of people have been interested in the cetaceans of Tasmanian waters, for example Scott and Lord (1920a, 1920b), Pearson (1936), Scott (1942), Davies (1963), Davies and Guiler (1957), Guiler (1961). Perhaps a better example is the east coast of Northland, especially near Whangarei (Gaskin 1968a). At least 10 schools have been reported as stranding there, and more probably went unreported. The immediate thought is that the whales visit the area for breeding, but the fact that the strandings have been scattered over a period of 7 months makes this seem rather improbable. The region is, on the other hand, rich in fish and probably squid as well, so the whales may spend much time feeding in the area. The complex system of bars and banks around Whangarei harbour and the associated inlets probably increases the likelihood of whales being trapped and stranded.

BIOLOGY: The most detailed study of the pilot whale to date was made on the population found near Newfoundland (Sergeant and Fisher 1957, Sergeant 1962a, 1962b). Sergeant's work provides valuable information on morphometrics, colour patterns, growth rates, age determination methods, the reproductive cycle, feeding, parasites, and behaviour. The behaviour of the North Pacific *G.scammoni* has been studied in the wild and in captivity by Brown and Norris (1956), Brown (1960, 1962), Caldwell, Brown and Caldwell (1963), and by Caldwell and Caldwell (1966), and that of North Atlantic *G.melaena* by Starrett and Starrett (1955).

The breeding season of the pilot whale near New Zealand is not known, but in the coastal waters of Newfoundland (Sergeant 1962b) births were scattered over a six-month period with a maximum in August, just after the northern mid-summer, and the period of gestation was between 15 and 16 months. Calves at birth were about 5′ 9″ long, with males slightly longer (5′ 10″) than females (5′ 8½″). Maximum pairing occurred in spring, between April and May. There was evidence of a definite male sexual cycle lasting from April to September, with activity declining in August and September. Lactation was protracted, lasting 21 to 22 months, so that the whole reproductive cycle extended over a period of 3 years 4 months. The age of pilot whales in Newfoundland waters was determined from examination of tooth laminations, on a basis of one lamination representing one year's

117

growth. Most females appeared to bear up to nine calves, becoming sexually mature at a body length of about 12 ft. Males became mature at 15 to 16 ft. Maximum age, if the lamination reading scale was correct, was 40 years for males and 50 for females. Harrison (1949) studied the female reproductive organs of this species.

The bulk of the stomach contents of Newfoundland pilot whales consisted of squid of the genus *Illex,* and cod when these were not available. Neither species occur in New Zealand waters, although similar species occupy similar ecological niches in the local marine environment. In the wild state the species is very gregarious, travelling in tightly grouped schools of from 15 to 200 animals. The size and concentration of schools on the feeding grounds appeared to vary with the concentration of food available. The species is thought to be polygamous, with one or a few males controlling a harem (Sergeant 1962b) at least during the breeding season, although small schools containing either only mature males or senile females are not unknown. In the samples taken by Sergeant, females outnumbered males by about three to one, and the scarred nature of the integuments of mature males suggested that fighting was commonplace in the breeding season.

Norris and Prescott noted that the ' melon ' was used in intra-specific fighting (1961) of the North Pacific pilot. In fact, in the last decade, much has been learned of the behaviour of this species, both in the wild and in Marineland of the Pacific, where it has been found that the learning ability of the pilot whale is on a par with that of the bottlenosed dolphin *Tursiops truncatus.*

THE KILLER WHALE
Orcinus orca (Linnaeus 1758)

DESCRIPTION: The male has the proportionately largest dorsal fin of any whale, reaching fully 5 ft high in a mature specimen. The fin of the female is also well developed, but not as large in relation to body size as that of the male, and the posterior margin of the female fin is more curved than that of the male. Dorsally the killer is dark purplish-brown, and the ventral surface is white. Pale patches are present on the dorso-lateral regions of the flanks both behind the eye and behind the dorsal fin. The colour of these patches varies from pink to yellow as the animal ages. The head has a very short but distinct beak. The prominent teeth are oval in cross-section, and are fully developed in both jaws.

DISTRIBUTION: Present day authorities recognise only a single world-wide species, although in the past a number of names have been given to both the genus and the species. The very confused synonymy was discussed by

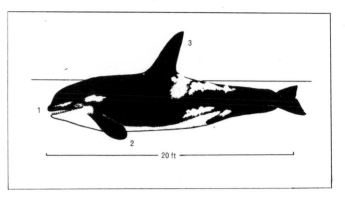

Figure 95.
Orcinus orca, killer whale: 1. rounded head with short beak, powerful teeth in both jaws, oval in cross section, 2. paddle-shaped flippers, with rounded apices, 3. large high dorsal fin, body patterned in black and whitish.

Iredale and Troughton (1933), who drew certain conclusions, but there is now general agreement on *Orcinus* as the generic name and Linnaeus' specific name *orca* (Hershkovitz 1966).

The killer whale has a nearly universal distribution in the seas of the world. It occurs in the North Atlantic and North Sea (Norman and Fraser 1937, Backus 1961), in the Mediterranean (Albert 1898, Daniel 1925), the South Atlantic (Barnard 1954), the Indian Ocean (Norman and Fraser 1937, Moses 1948, Barnard 1954), in the Indonesian Archipelago (Weber 1923), the North Pacific (Carl 1946, Nishiwaki and Handa 1958, Scheffer and Slipp 1948, Norris and Prescott 1961, Caldwell and Caldwell 1966, Brown and Norris 1956), off the coasts of South America (Clarke 1962, Cabrera and Yepes 1960), eastern and western Australia (Iredale and Troughton 1934), Tasmania (Pearson 1936, Davies 1963) and New Zealand (Oliver 1922b, c, Gaskin 1968a). It is frequently sighted in the Southern Ocean to the very edge of the pack ice (Sapin-Jaloustre 1953c), but since it is not taken commercially very little is known of its movements and population size.

Schools of 3 to 30 killer whales have been sighted frequently by vessels crossing the central Tasman Sea (Gaskin 1968a), and they also occur off the southeastern coast of Australia in some numbers (Dakin 1934). School sizes seem to be smaller in temperate than in high latitudes.

The species is especially common in the Southern Ocean between the Greenwich meridian and Drake Passage south of Cape Horn. The author saw two packs of killers two to three hundred strong between the South Shetland and South Orkney Islands in March and April 1962; these aggregations were sub-divided into smaller,

118

perhaps family groups, spaced several hundred yards to half a mile apart, all moving in the same direction. A similar phenomenon was observed by a New Zealand whale survey team north of Motiti Island in the Bay of Plenty in October 1963. On this occasion the total number of animals sighted was only 21 (Gaskin 1968a table 23), but the pack was split into four small groups spaced about a mile apart. All were travelling steadily northeast towards White Island, but it was impossible to say whether they were migrating or merely spread out in a search pattern for food. Packs of killer whales moving off the Californian coast appear to follow migrating schools of minke whales (Norris and Prescott 1961).

Killer whales have been sighted all the year round in New Zealand coastal waters, but most sightings have been made in the summer months, especially between October and January. Killers occur in the Bay of Plenty for most of the year, especially in the vicinity of the islands; they are frequently seen off the Kaikoura Peninsula and in the Marlborough Sounds, and around the mouth of Milford Sound in Fiordland. Why they tend to aggregate in these areas is unknown, except that all four possess abundant marine life, especially squid and fish. There is a slight suggestion in the sighting data (Gaskin 1968a) that killers are most common around northern New Zealand in winter and southern New Zealand in summer, but there is no direct evidence of migrations.

BIOLOGY: The killer whale's life-cycle in the western South Pacific is not known, but studies on North Pacific animals (Nishiwaki and Handa 1958) and the North Atlantic population (Slijper 1958, 1962, Jonsgaard and Lyshoel 1970) are not in accord and suggest that there may not be a fixed breeding season for the northern hemisphere. North Atlantic animals have a peak of pairing between November and January, while the peak in the North Pacific appears to occur between May and July. The gestation period in the North Pacific was found to be between 11 and 12 months. North Atlantic studies suggested a lactation period of about 12 months. Calves are about 7 ft in length at birth (Jonsgaard and Lyshoel 1970). Males are considerably larger than females at physical maturity; females are believed to be sexually mature at 15 ft, males at 20 ft or more. Males have been known to reach lengths of more than 30 ft at physical maturity. The estimated life span is about 20 years in both sexes, based on tooth laminations, but Dakin (1934) presented evidence, which if accepted, indicated that 'Old Tom' of Twofold Bay lived to be more than 50 years of age. This well-documented oddity of Australian whaling lore is discussed at the end of this section.

Figure 96.　(N. A. G. Heppard, British Antarctic Survey and with acknowledgements to Proc. zool. Soc. London.)
Three killer whales trapped in a pool of Antarctic sea ice off the northeastern coast of the Antarctic Peninsula.

Figure 97.　(N. A. G. Heppard.)
Killer whale trapped in a pool of Antarctic sea ice off the northeastern coast of the Antarctic Peninsula.

119

Figure 98. (D. E. Gaskin.)
Two killer whales (male with larger fin on right) off stern of FF *Southern Venturer* near the South Shetland Islands in March 1962, attacking tongues of baleen whales.

Much discussion has centred around the feeding habits of killer whales, and the possible danger to man of attack by these animals (Albahary and Budker 1958, Halstead 1959). The story of the 'attack' on Captain Scott's men is well documented, but the interpretation of the action of the whales can be challenged (Norman and Fraser 1937). Killers frequently attack the corpses of rorquals tied at the sterns of factory ships in the Southern Ocean, and in some areas near the South Shetland Islands the density of killers on the whaling grounds is so high that rorquals left 'in flag' for 24 hours have the blubber almost stripped away by the time they are finally towed to the factory ships. Usually however, killers appear to take only the tongue, and there are authenticated sightings of killers attacking mature animals. I have a reliable record of killer whales harrying a wounded sperm whale within sighting distance of the factory ship *Southern Venturer* in 1961, even though killers are reputed to attack sperm only very rarely. I observed them pulling at the mouth of the sperm when it had finally been killed and tied alongside one of the expedition chasers. I also witnessed, along with other crew members, two or three killer whales circling round a single large sei whale which may have had a calf with it. This was at a distance of about one mile from the whaling research vessel *Chiyoda Maru No. 5* in 1967. The final outcome was not observed. Killers have been recorded attacking schools of dolphins and seals (Norris and Prescott 1961, Brown and Norris 1956) on the coast of California well within temperate latitudes. A killer whale bleeding from a bullet wound in New Zealand waters was turned on by its companions and savagely attacked. Yet, as a 1967 Canadian Broadcasting Service television film shows, captive killer whales in the Seaquarium at Vancouver are just as amenable to training in captivity as dolphins such as *Tursiops*, and pilot whales. Newman and McGreer (1966) described the capture of the first killer whale for the Vancouver Oceanarium and its subsequent behaviour in captivity.

Under normal circumstances squid, fish, penguins and seals form the major part of the killer whale diet (Slijper 1958, 1962, Betesheva 1962, Caldwell and Brown 1964). When large rorquals are attacked, killers appear to concentrate on the extremities such as flukes, flippers and tongue. The officers of the *Southern Lotus* reported seeing a sei whale near the South Shetland Islands in 1962 with the left flipper completely missing, probably as the result of an attack by killers. Handcock (1965) reported killer whales killing and eating a minke whale. Little or nothing is known of the feeding habits of killer whales in New Zealand waters.

No discussion on the killer whale in New Zealand waters and the Tasman Sea would be complete without reference to 'Old Tom' of Twofold Bay, Eden, New South Wales. The story has been related in full by Dakin (1934), but the essential facts are as follows. Whaling was carried on for humpback whales on a small scale by the Davidson family at Twofold Bay from 1866 to 1930. Throughout this long period killer whales moved into the area before the start of each whaling season, prior to the arrival of right and humpback whales on migration north from the Southern Ocean. Many of the killers had recognisable individual characteristics, and the most famous, 'Old Tom', as he was locally known, was a very large male with a damaged dorsal fin. When a baleen whale came into the bay the killer pack would harry the animal to exhaustion, allowing it to be harpooned with ease by the whalers. It became a habit of the whalers to allow the killers to take the lips and tongue without interference, after which the carcase was towed back to shore to be rendered down.

Dakin (1934) pointed out that the details of this story have been so well documented by eye-witnesses and some of the Davidson family still living before the Second World War that there was no reason to doubt any part of it. The acceptance by a zoologist of apparently intelligent co-operation between whales and men would have been most unlikely even 30 years ago, but the work presently being carried out at the Marinelands of the Pacific and Florida on dolphin and pilot whale behaviour, and the use by the United States Navy of a porpoise to defend divers against shark attacks in the SeaLab Project, make the Twofold Bay story still interesting, but no longer remarkable.

Dakin was surprised to discover, by checking diaries of the settlers in the 1840s, that killer whales were operating in this bay even before the Davidsons began whaling there. I would therefore suggest that killer whales, with one generation learning from the previous one, had been using this bay for a long time as a convenient place in which to ambush baleen whales, and that the coming of the whalers meant only that they then had someone to do their killing for them. Providing the whalers make no move to attack the killers there is no reason why the two species, killer whale and man, should not form a temporary commensal relationship to exploit a common prey.

The effect of shooting or harpooning one killer out of a school has been recorded in two credible reports (Daniel 1925, Caldwell and Caldwell 1966). In both cases, the earlier one in the Mediterranean and the other in Puget Sound, other members of the school came to the aid of the injured or trapped animal and physically assaulted the boat with flukes or by attempting to crush it between them. In the Puget Sound incident a killer had been lassooed for transport to Marineland of the Pacific, and had not even been injured. However, neither of these could by any stretch of the imagination be interpreted as unprovoked attacks on man. A number of other examples of nurturant behaviour by this species are also outlined by Caldwell and Caldwell (1966), and a general account was given by Stephens (1963).

THE FALSE KILLER WHALE
Pseudorca crassidens (Owen 1846)

DESCRIPTION: To use the adjective 'false' is really unsuitable for this species, as it is nearly impossible to confuse it with the piebald killer. 'False pilot whale' would be a better name, as the two species are frequently confused in stranding records when trained biologists or experienced whalers are not able to go to the scene. In the water the false killer is easily distinguished from the killer whale by its dorsal fin and colouring. In general the adult false killer is a much smaller animal than the true killer. Specimens sometimes reach 18ft (Norman and Fraser 1937), but in the North Pacific 7 to 10 ft is more usual (Norris and Prescott 1961). The head of the false killer is much more tapered than that of the pilot whale, and also lacks the characteristic short beak of the killer. The dorsal fin is not only shorter than that of the killer whale, but it is much less broad than that of the pilot.

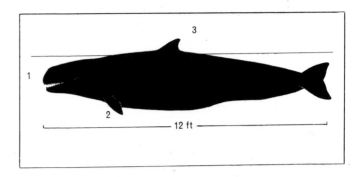

Figure 99.
Pseudorca crassidens, false killer whale: 1. blunt bulbous head, with powerful teeth, circular in cross-section, 2. pointed flippers, 3. relatively small dorsal fin compared to pilot and killer; body all black.

The flippers are pointed, as in the pilot whale, but not nearly so long in proportion to the total body length; the flippers of the killer whale are apically rounded and broadly paddle-shaped.

False killers are almost always black all over, with any white patches in the form of small scars. Only in badly decomposed animals might this species be difficult to recognise at sight. In such cases identification can be made from the teeth. The pilot whale has only an anterior semicircle of small peg-like teeth in each jaw, and the upper teeth are not prominently developed. In the killer and false killer whales the teeth are numerous, and fully developed in the upper jaw. However, the teeth of the killer are oval in cross-section, while those of the false killer are circular in cross-section.

DISTRIBUTION: Reliable sightings and strandings of the false killer whale have been very sporadic, both in New Zealand and elsewhere. It is very gregarious (Harmer 1931, Norman and Fraser 1937, Slijper 1939), and in the North Pacific a school containing 300 animals has been reliably observed (Norris and Prescott 1961). Almost all the information we have on this little known species has been obtained from very irregular and infrequent mass strandings. Owen described the type specimen in 1846, remarkably enough, from fossil or sub-fossil remains! A school of 150 animals stranded in the Dornoch Firth, Scotland in 1927 (Sanderson 1956, Slijper 1962); 100 stranded on the Chatham Islands in 1906 (Oliver 1922c, Gaskin 1968a); 167 at Velenai, Ceylon (Slijper 1958); 54 on the island of Zanzibar in 1933 (Slijper 1962); 29 at Napier, New Zealand in 1936 (Gaskin 1968a); 300 in South Africa in 1935 (Barnard 1954); 75 on the British coast in 1936 (Fraser 1946); 58 on the island of St Helena in 1936 (Slijper 1958); and 19 at Warrington, Otago, New Zealand in 1962.

Figure 100. (Dr Hideo Omura.)
False killer whale jumping for fish at Sea Life Park, Hawaii,
February 1966.

Smaller numbers have been recorded from the coast of Argentina (Marelli 1953); South Australia (Hale 1939); India (Silas and Pillay 1962); Formosa (Nishiwaki and Yang 1961); the Egyptian Mediterranean coast (Wassif 1956) and from the coasts of the United States (Stager and Reeder 1951, Bullis and Moore 1956, Mitchell 1965). A number of validated strandings of single animals have occurred both on the New Zealand archipelago and on the Chatham Islands (Gaskin 1968a). The total number of animals recorded from these two regions is only 152 in 90 years. The species has also been recorded several times by the whale chaser *Orca* (Gaskin 1968a) and by the Tory Channel whaling company spotter plane at the eastern end of Cook Strait and off the Marlborough coast. While we have no detailed knowledge of the movements of this species near New Zealand, all known strandings occurred between March and July with only one exception (Gaskin 1968a). The exception was a decomposed carcase from which Hector (1873b) obtained a skull near Wellington, in January 1870. All the schools seen by Tory Channel whalers were moving south, be-

tween January and March 1964 and again in December 1964, and twice were fairly close to schools of pilot whales.

BIOLOGY: A detailed examination of the anatomy of the false killer was made by Slijper (1939), and the female reproductive system has been examined by Comrie and Adam (1938). Data on animals stranded on the coasts of the USA have been summarised by Miller (1920) and Bullis and Moore (1956). Nothing is known of the reproductive cycle in the southern hemisphere.

Dr N. A. Wakefield (1968) was able to study the stomach contents of a specimen stranded on the coast of Victoria, and found the remains of both squid and fish, so there is no reason to expect the diet to differ from the typical toothed-whale pattern.

This species has been recorded making echo-location clicks, and appears to have a considerable repertoire of squeals, whistles and short cries (Poulter 1968). False killer whales have been studied in captivity at Hawaii and Marineland of the Pacific (Brown, Caldwell and Caldwell 1966).

14. The True Dolphins
Delphinoidea: Delphinidae

There are three families of relatively small cetaceans in the Delphinoidea; the Phocoenidae or true porpoises, the Stenidae, which have no common name, and the Delphinidae or true dolphins. Neither Phocoenidae nor Stenidae have been reliably reported from the New Zealand region. The majority of Phocoenidae are restricted to the northern hemisphere, although the genus *Neomeris* (finless black porpoises) occurs around the coasts of Southeast Asia and may penetrate through Indonesia to northern Australia. A single Stenid, *Steno bredanensis* is mentioned in older literature (Hector 1873a, Oliver 1922c) but there is in fact no evidence that a specimen has ever been taken in New Zealand (Gaskin 1968a). The classification of the Stenidae is at present in some confusion (Fraser 1966), but three tropical and subtropical genera appear to be more or less distinct; *Steno*, typified by *Steno bredanensis* the rough-toothed dolphin, which has been recorded in the South Pacific no nearer New Zealand than the Admiralty Islands (Mohr 1923) and the Galapagos Islands (Orr 1965), *Sotalia* which seems to be a genus restricted to South and Central American waters (Fraser 1966a), and the old-world tropical genus *Sousa*, a species of which appears to occur on the Queensland coast and which has been held in captivity for some years at the Tweed Head Marineland.

All the dolphins found in New Zealand waters belong to the Delphinidae. The genus with the largest number of described species within the Delphinidae, *Stenella*, has not been reliably reported from New Zealand waters (Gaskin 1968a, 1968c), although there is some suspicion that one species may occasionally pene-trate into northeastern coastal regions. Six genera have been recorded from New Zealand or the surrounding area; *Lagenorhynchus*, with two species, *L.obscurus* and *L.cruciger*, and *Cephalorhynchus, Delphinius, Tursiops, Grampus* and *Lissodelphis* with one each (Gaskin 1968a, 1968c).

Genus *Lagenorhynchus*
(Gray 1846)

Lagenorhynchus virtually lacks a visible beak or rostrum typical of other delphinid genera; the head cuts down sheer toward the tip of the snout. All species are bluish-black dorsally and have varying amounts of white on the flanks and ventral surface. The genus is in need of much more study (Fraser 1966a), and there is still some doubt if *L.obscurus* should not be placed in a separate genus from *L.cruciger* and *L.australis*. A number of species have been recognised in this genus apart from those already mentioned; there are *L.fitzroyi, L.acutus, L.albirostris, L.obliquidens, L.wilsoni, L.superciliosus, L.electra, L.latifrons* and *L.thicolea* (Norman and Fraser 1937). The status of a number of these is in doubt. Iredale and Troughton (1934) synonymised *obscurus, latifrons, fitzroyi* and *cruciger*. Scheffer and Rice (1963) also synonymised *L.obscurus* and *cruciger*, but Fraser (1966a) has demonstrated the distinctness of these two species very clearly. In a more recent paper (Gaskin 1968c) I presented evidence which shows that the range

Figure 101. (Professor Warren J. Houck.)
Head view of white dolphin, *Sousa* sp. in captivity in Queensland. This may be a species or subspecies new to science, for which perhaps the name *Sousa queenslandensis* might be considered. However a thorough systematic comparison between this population and Asian *Sousa* has so far been precluded by lack of specimens.

of the two species in the New Zealand region has virtually no overlap. However Fraser concurs with the synonymy of *fitzroyi* with *obscurus*.

L.obliquidens is the North Pacific species (Scheffer and Slipp 1948, Brownell 1964, Ficus and Niggol 1965), while *L.acutus* and *L.albirostris* are well-founded species occurring in the North Atlantic (Sergeant and Fisher 1957). *L.australis,* otherwise known as Peale's porpoise, is found around Patagonia (Fraser 1966a). *L.electra,* which has been taken on the coast of Australia (Dr W. H. Dawbin, in letter), should be referred to the genus *Peponocephala* in the Globicephalidae (Nishiwaki and Norris 1966, van Bree and Cadenat 1968). The status of the remaining species, *L.thicolea, L.superciliosus, L.latifrons* and *L.wilsoni* cannot be determined at this time.

THE DUSKY DOLPHIN

Lagenorhynchus obscurus (Gray 1828)

DESCRIPTION: Adults are from 4 to 7 ft in length, and have about 30 pairs of teeth in each jaw. The eye is characteristically surrounded with a rather diffuse ring of dark pigmentation. The species is dark bluish-black dorsally and white ventrally, and both black and white on the flanks, the pattern of black and white being quite

distinct from that of *L.cruciger*. The flukes and flippers of *L.obscurus* are black. A greyish band extends from the side of the head above the eye and merges posteriorly with the greyish-white zone on the anterior flank. Below the eye a dark streak runs to the anterior flipper base. The jaws are black, but the black pigment on the head, apart from that already described, is more or less limited to a triangular dorsal patch, though there are greyish areas within the white. The white area of the anterior flank merges without a break into the white on the ventral surface. Posteriorly the dorso-lateral dark zone divides into three streaks, one going to the vent, the second running with a break towards the flukes, and the third along the lateral crest of the back to the flukes. The dark pigmentation of the ventral surface of the flukes extends forward nearly to the vent. Boundaries of most of these markings merge into grey before white.

DISTRIBUTION: Sightings of this species in the New Zealand region are closely associated with the Subtropical Convergence (Gaskin 1968c), and rapidly become rare north or south of it. Observations during the Marine Department whale survey in 1962-64 indicated that *L.obscurus* occurs all along the North Canterbury coast in both summer and winter (Gaskin 1968a), and the observations of Mr Frank Robson of the Marineland of New Zealand showed the species to occur in the Hawkes Bay area generally only in winter and spring. This distribution pattern is probably related to the presence of a cold tongue of water from the Canterbury Current which pushes further northwards along the east coast of the North Island in winter than in summer, sometimes as far as Gisborne (Brodie 1960). The continuous presence of this species in the inshore waters of the North Canterbury coast throughout the year is almost certainly connected with the stable, cold, north-flowing Canterbury Current.

Figure 102.
Lagenorhynchus obscurus, the dusky dolphin: 1. ploughshare-shaped head, 2. large dorsal fin, and black streaks invading the white flanks of the caudal region.

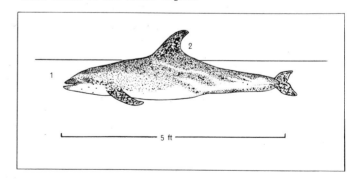

The sporadic occurrence of *L.obscurus* near Wellington in Cook Strait probably reflects the irregular local movements of water masses and variations in temperature in this semi-enclosed region (Garner 1961).

The southern limit of distribution of this animal is not known. It has been recorded as far south as Stewart Island and among the islands of Foveaux Strait, and the southern-most stranding record is from the Spit, Otago (Gaskin 1968a). Although Lillie (1915) reported the species at 58°S south of New Zealand there could easily have been confusion between *obscurus* and *cruciger*. A sighting of *cruciger* at this latitude would be in accord with data collected by the present author and Dr Keiji Nasu from *Chiyoda Maru No. 5* (Gaskin 1968c); a sighting of *obscurus* would not. Since Lillie did not apparently take specimens, I disregard this record.

The worldwide distribution pattern of *Lagenorhynchus* has been discussed by Slijper (1936), Bierman and Slijper (1947, 1948), Brownell (1965) and Fraser (1966). *L.obscurus* has been reliably reported from South Africa (Barnard 1954, Fraser 1966), and from Kerguelen Island at about 48°S (Paulian 1953). As *L.fitzroyi* it has been recorded from Tasmania (Davies 1963) and southern South America (Kellogg 1940). Brownell (1965) gave the distribution of *L.obscurus* as 'circumpolar to 30°S', but I cannot accept on present evidence that the species is circumpolar through the temperate South Pacific, or that in the New Zealand region the species ranges anywhere near latitude 30°S. Mr F. Robson's sightings, based on many years experience on the east coast of New Zealand, give Whitianga (about 37°S) as close to the absolute maximum northward summer or winter range of the species. Mr W. Skrzynski of the Fisheries Research Division, Wellington, recorded a multiple stranding (six specimens) on the Chatham Islands in October 1966, and a photograph of the animals was taken by Mr ·B. Marshall. One school was sighted by *Chiyoda Maru No. 5* to the southeast of the Chatham Islands (Gaskin 1968c), but there is no evidence that *obscurus* extends further east into the Pacific than the coastal shelf around the Chatham Islands. None were sighted during extensive cruising by the above vessel to longitude 140°W under ideal spotting conditions with surface temperatures and latitudes presumably favourable to the species. The Subtropical Convergence appears to become less well defined towards the Chatham Islands in surface layers (Garner 1959), and this would be consistent with the apparent limit of distribution of *L.obscurus* into the western South Pacific.

In New Zealand coastal waters, the stronghold of the

Figure 103. (Marineland of New Zealand.)
Two dusky dolphins *(L.obscurus)*, performing at Marineland of New Zealand, Napier.

species is the east coast of the archipelago from East Cape down to Timaru or Otago, although in summer Cape Palliser seems to be about its northward limit. This is consistent with a southward movement of cool water masses in summer (Garner 1961). Records from the Tasman Sea are non-existent. Five stranding records from the west coast of the North Island are all clustered within the Taranaki Bight (Gaskin 1968a). Fishermen have reported the species between Tasman Bay and Cape Egmont in the winter and very occasionally during summer. It also occurs sporadically as far south as Westport and has been known even to reach Greymouth. However, there is not one reliable record of *obscurus* south of Greymouth to West Cape in Fiordland. On this coast it appears to be largely replaced in the north by *Cephalorhynchus hectori* and in the south by *Tursiops truncatus*. Extensive cruising by *Chiyoda Maru No. 5* in the south-central Tasman Sea did not produce a single sighting that Dr Nasu could confirm as this species. There thus appears to be some evidence that the distribution of *L.obscurus* in the southern hemisphere may be discontinuous. Further study along the Subtropical Convergence waters in the Tasman Sea might show a link with the Tasmanian population, but this has to be proved.

BIOLOGY: Although this is one of the most common cetaceans in waters within easy reach of Christchurch and Wellington, we know nothing of its biology. The remains of small squid (*Nototodarus sloanei*) have been found in the stomachs of specimens stranded near Wellington, according to Mr C. McCann, formerly of the Dominion Museum, Wellington. I have also watched *obscurus* hunting surface shoals of small fish near the Kaikoura Peninsula. On another occasion I observed a school of 10 *obscurus* in the same locality playing round the edges of a small shoal of fish. As far as could be seen the dolphins were not feeding, merely playing. One by one, or in groups of two or three, dolphins breached from the water, turning complete somersaults before falling back.

School sizes vary through the year. Schools numbering hundreds of animals are seen during the summer months in the offshore waters of eastern Cook Strait, and less frequently near Cloudy Bay and Port Underwood on the southern side of the Strait within a few miles of shore. In winter, schools containing 6 to 20 dolphins are the rule in Cook Strait. I have records of over 20 schools recorded in the winter months close to Wellington, the largest of which contained 19 dolphins.

There is no definite information on the breeding cycle of *obscurus*. However, I suspect that mating off the east coast of New Zealand occurs in late summer, before the relatively large, temporary aggregations seen in offshore Cook Strait and 20 miles off the Marlborough coast in summer break up and scatter for the winter. I have observed apparent mating behaviour in *obscurus* between February and May in schools sighted off the Marlborough coast in 1963 and 1964, twice in South Bay, Kaikoura and once about 15 miles from shore in the same region.

This dolphin accompanies boats quite freely, riding the bow waves. For this reason it is easy to catch, and was the second species to be held successfully in captivity at the Marineland of New Zealand at Napier. Despite some problems the species has proved very amenable to training, and shares the same pools as *Delphinus delphis* without any overt signs of animosity. Like *D.delphis,* the dusky dolphin in captivity has been noted by its trainers to be most playful towards evening, and in warm weather when rain is threatening.

What little is known of the biology of other species of *Lagenorhynchus* in the northern hemisphere has been discussed and summarised by Bierman and Slijper (1947, 1948), Scheffer (1950), Wilke, Taniwaki and Kuroda (1953), Sergeant and Fisher (1957) and Norris and Prescott (1961). Sokolov (1961) studied the female reproductive cycle of *acutus*.

SOUTHERN WHITE-SIDED DOLPHIN
Lagenorhynchus cruciger (d'Orbigny 1847)

DESCRIPTION: *L.cruciger* reaches lengths of 6 to 7 ft. The species has a black circular mark around the eye. The jaws and top of the head are black, and a thin stripe of dark pigmentation runs from the angle of the jaw to the eye. The dorsal surface and dorsal fin are black, and the ventral surface white, except for the flippers, flukes and caudal stalk, which are also black. On the flank a broad, angular white zone extends posteriorly from the side of the head above the eye, ending abruptly just above the base of the flukes. This white band is interrupted by

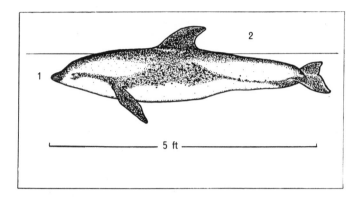

Figure 104.

Lagenorhynchus cruciger, Southern white-sided dolphin: 1. short beak, 2. flanks with an irregular band of black from base of flipper to tail stock.

a large black area intruding just anterior to the dorsal fin position. The latter is continuous with a large black patch which extends from the base of the flipper posteriorly, tapering below the base of the dorsal fin but widening again towards the flukes.

The colour pattern is quite distinct from that of *L.obscurus.* The simplest point for an observer at sea to look for is the prominent dark zone behind the flipper, which is present in *cruciger* and absent in *obscurus.* The second major distinction is the pure white *dorso*-lateral flank band of *cruciger.* The flank band in *obscurus,* both anterior to and posterior to the dorsal fin position, is ventro-lateral. Thirdly, the head of *cruciger* tapers less sharply than that of *obscurus,* and the dorsal fin of *cruciger* is somewhat broader than that of *obscurus.*

DISTRIBUTION: This dolphin's movements are restricted to the Southern Ocean, where it may be circumpolar in relatively high latitudes. Fraser (1966) was able to examine a specimen taken at 56°20'S 40°9'E south of the Prince Edward Islands, and 68 specimens seen from *Chiyoda Maru No. 5* between January and March 1967 were observed between 53°S and 61°S and between longitudes 165°W and 138°W to the southeast of the Chatham and Antipodes Islands (Gaskin 1968c). I have previously published (Gaskin 1968a) a possible sighting of six *L.cruciger* off the Marlborough coast during the winter of 1963. This was the first time I had seen this species; having since seen more from *Chiyoda Maru No. 5,* I can confirm this sighting as *L.cruciger,* at 42°20'S 174°5'E, moving south. Surface temperatures in this region can be as low as 9°C in winter (Garner 1961). Specimens recorded by *Chiyoda Maru No. 5* were in waters with surface temperatures ranging from 2°C to 9.5°C. The species appears to be widely distributed through the higher latitudes of the Southern Ocean, with apparent concen-

126

trations across the Antarctic Convergence region; in winter it approaches the southern side of the Subtropical Convergence when this is associated with surface temperatures of about 10°C (Garner 1959). On the basis of the Kaikoura sighting we could postulate a more northern distribution in winter than in summer, associated with the movement of both convergence zones, but the latitudinal movement of these is not great around New Zealand, and it would be wrong to consider the species as having a definite migratory behaviour without far more study.

BIOLOGY: Nothing is known of the breeding cycle of *L.cruciger;* no young were observed among the 68 specimens seen in 1967. Similarly the feeding habits and diet are unknown. One interesting and previously unrecorded facet of behaviour of *L.cruciger* was noted by Dr Keiji Nasu and myself from *Chiyoda Maru No. 5*. While riding the bow waves of this vessel, *cruciger* individuals consistently showed spinning behaviour, similar to that observed in Burmeister's porpoise *Phocoena spinipinnis* (Norris and Prescott 1961). Despite the fact that the species only came to ride the bow waves of the vessel for any length of time in very rough weather, I was able to record several feet of cinefilm of their behaviour. Individuals also performed short vertical jumps through the crests of particularly large waves as they came parallel to the bows.

The specimen described by Fraser (1966a) was dissected by Tomlinson and Harrison (1961). They discovered that this species, unlike most cetaceans, possessed only a single posterior vena cava in the blood system, and lacked an hepatic sinus. The histology of the kidney and liver was described by Cave and Aumonier (1962).

HECTOR'S DOLPHIN
Cephalorhynchus hectori (van Beneden 1881)

DESCRIPTION: This species reaches a length of about 6 ft. There is only a very short and poorly defined beak, which in most specimens merges into the head with little perceptible dorsal indentation. There are about 30 pairs of teeth in each jaw. The dorsal surface of living animals is very dark grey with a brownish tinge, sometimes with transient purplish hues. The ventral surface is white, and on much of the flanks the two zones of pigmentation have a narrow, greyish boundary between them. The forehead is either grey or white. A narrow belt of black connects the flipper bases across the belly. The flippers, dorsal fin and the flukes are the same colour as the back, but on some specimens there is a small whitish patch just behind the flippers. Several variable white streaks are present on the flanks posterior to the dorsal fin. The

dorsal fin is bluntly rounded in adults, but juveniles often appear to have a broadly triangular fin; its shape may thus change with age. A line of calluses has been observed on the anterior edge of the dorsal fin of some specimens. The flippers have more rounded tips than those of *Lagenorhynchus*.

DISTRIBUTION: While quite a number of nominal species have been described, the four generally accepted *Cephalorhynchus* species are *C.hectori* from New Zealand waters, *C.heavisidei* from South Africa (Barnard 1954), *C.eutropia* from the coast of Chile between latitudes 33° and 40°S (Rice and Scheffer 1968), and *C.commersoni* from waters around southern South America (Marelli 1953) and perhaps from Kerguelen Island (Davies 1963), where a species of this genus certainly occurs. Commerson's dolphin is a strikingly marked black and white species (Norman and Fraser 1937, Kellogg 1940), and it is possible that a piebald specimen recorded by Oliver (1946) in the Marlborough Sounds was of this species. On the other hand it may have been a semi-albino variety, as suggested by Oliver. *C.hectori* has not been confirmed as occurring outside New Zealand coastal waters. A report of the species from Sarawak (Rice and Scheffer 1968) is likely to be an error of identification not attributable to these authors.

Around New Zealand *hectori* tends to be a rather local species, often semi-estuarine and coastal in habit (Graham 1953). It is rarely seen in schools of more than four or five animals, although Hector recorded a specimen from Cape Campbell as 'one shot from a large school'. All stranding records are of single animals (Gaskin 1968a).

Although recorded sporadically from most parts of New Zealand except the extreme north and south it is only common in the vicinity of Cook Strait and Tasman Bay,

Figure 105.
Cephalorhynchus hectori, Hector's dolphin: 1. bluntly pointed snout, 2. rounded dorsal fin, body patterned with shades of grey.

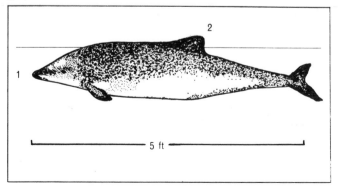

and on the adjacent east and west coasts for about one hundred miles north of the Strait and two hundred miles south. This suggests it is associated with the mixed waters of the Subtropical Convergence and the coastal currents, which are complex in the region of the Strait.

It is generally an inshore species, often penetrating some distance into the Clarence, Wairau and Grey River estuaries, into the Marlborough Sounds as far as Picton and Pelorus, and is also seen close to Nelson. Dr R. A. Falla, formerly director of the Dominion Museum, Wellington, told me that the species can nearly always be found close to the cement works at Cape Foulwind, Westport, and off the mouth of the Wanganui River.

I have recorded *hectori* outside the Grey River bar, in Wellington harbour and Palliser Bay, near Port Underwood and in Cloudy Bay, off the mouth of the Clarence River and Banks Peninsula, in the Marlborough Sounds at several localities and at Kaikoura. The frequency with which small schools are seen at the latter place suggests the animals are semi-resident. Whale survey observations in 1963-64 indicated that the species is fairly common on the west coast of the South Island to about Jackson's Bay, but as the shoreline becomes progressively more sheer *hectori* becomes rarer. The species is less common around the North Island. A record by Oliver (1922c) for the Bay of Islands seems to be extreme for the normal range of the species. The dolphin collectors of the Marineland of New Zealand regard it as a very rare visitor to Hawkes Bay. I have no information on its distribution on the northwest coast of the North Island north of about Cape Egmont.

There is no evidence of extensive migratory behaviour. Like other species of the genus it is essentially a cold water species, and seems to stray only rarely north of 40°S.

Figure 106.

Tursiops truncatus, bottlenosed dolphin: 1. short beak, 2. prominent dorsal fin; body colour grey above, shading to whitish on belly.

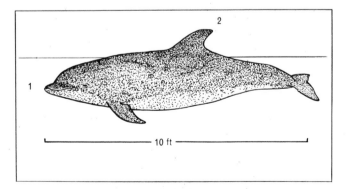

One interesting point is that while most species of cetacean avoid turbid water, *hectori's* characteristic fin can sometimes been seen breaking the surface of dirty brown water in the Marlborough Sounds when the sea has been mixed with mud by land run-off after heavy rain.

BIOLOGY: Very little information is available on the life history, feeding habits and behaviour of this species. There is no information on the breeding cycle. The stomachs of two specimens recorded by Dominion Museum staff (Gaskin 1968a, table 30) contained small quantities of shellfish, crustaceans, small fish and squid remains.

THE BOTTLENOSED DOLPHIN OR COWFISH
Tursiops truncatus (Montagu 1821)

DESCRIPTION: This is a large dolphin, reaching lengths of up to 15 ft at physical maturity. The beak is only a few inches long, but usually marked off quite distinctly from the rest of the head. There are 20 to 22 pairs of teeth in each jaw. The dorsal surface colour is variable, but is always grey or grey-green. The ventral surface is white, as is the lower jaw, and dorsal and ventral pigments merge indistinctly in the ventro-lateral region of the flanks. The flippers are slender, in contrast to the generally robust build of the animal. The dorsal fin is large in proportion to its body size.

DISTRIBUTION: The genus *Tursiops* is in great need of a detailed revision, so that the nomenclature of the various species or forms found in various parts of the world can be stabilised. The Delphinidae were last considered in detail by Flower (1883) and True (1889), and this is only one of many genera which require critical re-examination. The situation of *Tursiops* was further confused by the discovery by Dr F. C. Fraser (1940) of three dolphins taken from the Irish coast which could be arranged to give an orderly series connecting *Tursiops* to *Grampus griseus*. Fraser considered these to represent the results of localised inter-generic breeding. This may not be as surprising in the case of these two species as was once thought. While earlier authors have placed *griseus* in the Globicephalidae on skull characters and the short snout, more recent and detailed study of the internal anatomy, especially that of the ear (Fraser and Purves 1960, Fraser 1966a) has indicated that this species is a delphinid, closely related to *Tursiops,* and that the short snout seems to be a convergent character, rather than homologous with the condition found in globicephalids.

Five names are in common use for the genus *Tursiops:*

Figure 107. (Dr D. K. Caldwell.)
Adult female bottlenosed dolphin *Tursiops truncatus* swimming. Photographed at the Marineland of Florida.

T.truncatus, T.aduncus, T.catalania, T.gilli and *T.nuuanu*. The last name has been used to describe a form found in the North Pacific (van Gelder 1960). Although all five have been variously given specific status by authors such as Iredale and Troughton (1934), Norman and Fraser (1937), Scheffer and Rice (1963), *T.catalania*, a name used for a widespread tropical form, is almost certainly synonymous (Barnard 1954) with *T.aduncus*, unless later work shows justification for naming distinctly the forms from each major region of the world.

At this stage of our taxonomic knowledge it is probably best to recognise three species. *Tursiops gilli* is found on the American North Pacific coast (Mayer 1950, Kenyon 1952, Hubbs 1953, Norris and Prescott 1961, Scheffer and Rice 1963, Caldwell and Caldwell 1966), and the coast of Mexico (van Gelder 1960), and in the Hawaiian Islands (Rice 1960).

Tursiops aduncus (=*catalania*) has been recorded from Natal (Barnard 1954), Sumatra, Java and Solor (Weber 1923), and the coast of Queensland (Longman 1926). It is taken commercially in Ceylon (Medcof 1963).

In the northern hemisphere *Tursiops truncatus* occurs in the North Atlantic (Fraser 1953, 1946, Scott 1960, Cadenat 1959, Sergeant and Fisher 1957), in the North Sea (Flower 1880, Fraser 1946), in the Mediterranean (Postel and Mayrat 1956, Heldt 1954, Norman and Fraser 1937), the Black Sea (Slijper 1962), the Caribbean and the Gulf of Mexico (Moore 1953). In the southern hemisphere the same species has been recorded from

Argentina (Marelli 1953), southern New Zealand (Oliver 1922b, c, Gaskin 1968a) and Tasmania (Scott and Lord 1920a, Pearson 1936). The latter author found no significant differences between Tasmanian specimens and descriptions and measurements of those from European waters.

These distributions must be accepted with considerable reservation until the genus has been studied in more detail. However, although the specific identity of the Queensland *Tursiops* as *catalania* by Longman, could be wrong, the photograph in this book by Mr A. Dobbins clearly shows that a species of *Tursiops* is held in captivity in the Tweed Head Marineland.

Tursiops aduncus could prove to be a warm-water form or sub-species of *T.truncatus*, though *aduncus* is stated to have up to 28 pairs of teeth, about 6 more pairs than *truncatus* (Norman and Fraser 1937).

The Subtropical Convergence roughly bisects New Zealand in the latitude of Milford Sound and Oamaru (Garner and Ridgway 1965), and all year round, cold subantarctic water masses wash the southern coasts of the archipelago, and waters of subtropical origin the northern coasts. Because of this division of the New Zealand region marine environment, the *Tursiops* found on the northwest coast of the North Island, may belong to the *aduncus* form, a population distinct from the larger animals found in the southern Fiords and Foveaux Strait.

During the late spring and early summer this species is found both on the east and west coasts of the South Island, especially in the two localities mentioned above. Local residents of Milford Sound have observed that some *Tursiops* remain there all the year round, although the species is most common there in summer. *Tursiops* is almost always seen very close inshore in New Zealand waters, but in 1963 and 1964 the whale chaser *Orca* reported four small schools between 10 and 20 miles from the Marlborough coast.

According to Dr R. A. Falla, large *Tursiops* come into Wellington harbour during March and early April, but apparently not at any other time of year. Picton boatmen have reported the species in late summer in the Marlborough Sounds, and very occasionally at other times of year. There is evidence from these observations to suggest some migratory movement of *Tursiops* on the west coast of New Zealand. However, on the broad scale these movements are perhaps only local. It would be interesting to carry out tagging experiments to see if any relationship existed along the Subtropical Convergence across the Tasman Sea between the *Tursiops* populations of southern New Zealand and those of Tasmania.

Very little is known of the distribution of *Tursiops* on the west coast of the North Island north of Cape Egmont, except that the famous 'Opo' of Opononi was a *Tursiops,* and that during the 1962-64 Marine Department whale survey fishermen reported the species in or near the estuaries of the Mokau and Waikato Rivers, and in Kawhia, Aotea, Raglan, Kaipara and Hokianga harbours.

BIOLOGY: *Tursiops truncatus* has been studied more intensively than other cetaceans, especially in captivity on the Atlantic coast of the United States. Many papers have been published on the biology of the animal, on its breeding cycle, age determination, feeding and behaviour. Since very little research has been done on *Tursiops* in New Zealand most of the information in this section deals with observations made elsewhere in the world.

The gross anatomy of *Tursiops truncatus* was examined by Braun (1905), and more recently the fine structure of the brain of *T.gilli* has been studied by Kruger (1966). Earlier dissections of the brain and elucidation of the different functions of the cerebral cortex in *truncatus* were made by Langworthy (1931, 1932), and a special study of the thalamus was made by Kruger (1959). The structure and function of the inner, outer and middle ear of *truncatus* have been studied in great detail by Fraser and Purves (1960), Purves and van Utrecht (1963), and Purves (1966).

The swimming ability of *Tursiops* has been examined by Essapian (1955) and Lang (1966), and its ability to thermo-regulate by Tomilin (1951) and Kanwisher and Sundnes (1966).

The anatomy of the reproductive system has been considered by Wislocki and Enders (1941), and Slijper (1962 and 1966). Pregnancy and the gestation period is dealt with by Slijper (1962) and McBride and Kritzler (1951). Parturition is also discussed in both these papers, and in a very useful review by Slijper (1966).

A very large number of papers have been published on the behaviour of *Tursiops* in captivity. Among the most important of these are discussions by Kellogg and Rice (1966) on the species' problem-solving ability and visual discrimination; studies of echo-location and communication by Wood (1953), Evans and Prescott (1962), Lilly (1961, 1966), Dreher (1966) and Busnel (1966); and examination of social behaviour in the wild and in captivity by McBride and Hebb (1948), McBride and Kritzler (1951), Hubbs (1953), Moore (1953), Tavolga and Essapian (1957), Siebenaler and Caldwell (1956), Norris and Prescott (1961), Caldwell and Caldwell (1964, 1966) and Tavolga (1966).

The North Atlantic *T.truncatus* breeds in late winter and early spring, mating and calving between February and April. The gestation period is one year, and the lactation period 16 months. Sexual maturity is attained at an age of about six years, and the life span is about 20 years. Two years elapse between each successive birth (Slijper 1962).

Local residents of Milford Sound have made observations over a number of years on the behaviour of this species. In some summers small schools come up to the very end of Milford Sound near the Milford Hotel, into what is virtually fresh water. Very young calves accompany the adults. During the day most of the adults appear to go out to sea, presumably to feed; if not going out to sea they certainly move well beyond the first elbow of the sound beyond the Mitre. One adult always remains with the collection of calves in the blind end of the Sound, apparently acting as nursemaid. There were reports of calves attempting to suckle at the teats of this adult. None of the observers was able to ascertain if the adult which remained behind was always the same individual, or whether it had a calf of its own. This behaviour continued for several months of each summer. Towards autumn the calves, by then about two-thirds the size of the adults, started to join the adults on their daily expeditions, finally leaving the end of the Sound completely. During the winter only the occasional adult would be seen in the lower parts of the Sound.

These reports may be contrasted with observations made on a captive colony of *Tursiops truncatus* by Tavolga (1966). For the first few weeks following birth, mothers kept their infants close to them, and weaning did not seem to be completed for 18 months, the infants being dependent on the mothers throughout this period. Definite social order was discernible; the adult male was dominant over all others, and the descending order included adult females above juvenile males above infants. The male leader tended to spend much time alone, and among the adult females the individuals could be placed in a definite order of aggressiveness. The juvenile males stayed together as a sub-school, and often attempted to mate with the adult females. While the older infants also formed a sub-school this tended to be temporary, and they would all return to their mothers at intervals.

However, co-operation between adults at birth in *Tursiops* has been recorded by McBride and Hebb (1948) and Caldwell and Caldwell (1966). In these instances pairs of adult females supported the new-born infant at the surface until it was breathing by itself. McBride and Kritzler (1951) noted that in one case the mother and another adult female supported the infant at the surface by swimming close together and lodging it between their dorsal fins. The umbilicus in *Tursiops* does not have to be bitten through; its structure is such that it breaks when a relatively small strain is placed on it (Slijper 1966).

Other authors (Siebenaler and Caldwell 1956) reported wild *Tursiops* and captive *Tursiops* coming to the aid of injured individuals.

Aspects of *Tursiops* behaviour which have virtually become folklore in New Zealand concern the two individuals called 'Pelorus Jack' and 'Opo' of Opononi. The famous 'Pelorus Jack' of French Pass, who used to accompany certain ships through the narrows in this part of the Marlborough Sounds (Downs 1914, Parker 1933), was identified as a Risso's dolphin by the then director of the Dominion Museum, Dr Waite. However, Dr R. A. Falla informed me that the disappearance of the famous dolphin in 1913 coincided with the discovery of a large *Tursiops truncatus* dead on the rocks near the mouth of the pass after a gale. The body was reported by the local lighthouse keeper, who was very familiar with 'Pelorus Jack', and was convinced that the corpse was his, since there was a pattern of easily recognisable scars on the dolphin's back. Even if the question of identity cannot completely be settled at this distance in time, the behaviour of the dolphin in accompanying ships in the same place close inshore would be quite in keeping with the behaviour that could be expected of *Tursiops* in New Zealand waters. Secondly, *Tursiops* is a fairly common animal on this coast, while the only positive record of *Grampus griseus* in New Zealand waters is based on a single jawbone in the Dominion Museum (Gaskin 1968a). There are no reliable sight records of *Grampus* in this region, either. Reports by a fisherman in Tasman Bay of a school of *Grampus griseus* were shown to be *Tursiops* as soon as a full description was obtained.

'Opo' was a partly grown female *Tursiops* (Alpers 1963, Higham 1960), believed to be the offspring of a female shot by a local youth. The dolphin was first sighted in the winter of 1955, and it soon began to follow boats, and seemed particularly attracted to the sound of outboards. Despite an incident in which the animal was gashed by a propeller, it soon became very tame, moving right into the surf and playing with children in the waves, even allowing one particular child to ride on its back. The animal in fact became the greatest tourist attraction Opononi had ever known, but unfortunately was found dead on the rocks on 8 March 1956, apparently killed by a gelignite blast used in some illegal tidepool fishing.

Figure 108. (Northland Photography, Whangarei.)
'Opo' frolicking with holiday-makers.

The diet of this dolphin is varied, consisting of fish and squid (Slijper 1962). It is an active predator; Brown and Norris (1956) recorded in captivity the deliberate harrying and persecution of fish before they were finally eaten. No detailed analysis of stomach contents have been made of New Zealand *Tursiops*, but in June 1964 I filmed several schools of this species in the Acheron Passage and Bligh Sound in Fiordland. The dolphins were swimming in groups of from 4 to 12 very close to the shore, sometimes under overhangs of the bush-covered cliffs. Usually they ignored the boat unless it came towards them. They would then come in beneath the hull, rising to ride in the bow wave for a few minutes before turning away and resuming their purposeful sweeping along the shoreline. They appeared to be grubbing for bottom-living fish or shellfish. On one occasion, very close to the shore of Resolution Island we could see one animal rolling over and over in a spiral dive, following the contours of the rocky shelf about 15 to 25 feet under the surface.

Sergeant (1959) studied the teeth of *Tursiops* for age-determination purposes, and found regular growth layers

131

in the postnatal dentine, each consisting of a narrow, clear zone and a broad, opaque zone. Examination of an animal reared in captivity showed that one layer was laid down each year.

RISSO'S DOLPHIN
Grampus griseus (Cuvier 1912)

DESCRIPTION: Risso's dolphin is the least known of the species positively recorded in New Zealand waters. Superficially it resembles the pilot whale, and has in fact until recently been placed in the Globicephalidae (Nishiwaki 1963, 1964). The forehead rises directly from the tip of the upper jaw as in the pilot, but *griseus* lacks the short beak of the latter, and the bulge of the forehead is nowhere near as pronounced as in that species. The flippers are relatively long, but not as long in proportion to the total body length as in the pilot whale. The dorsal fin is prominent, but not as deep or proportionately large as that of the pilot. The dorsal surface, flippers and flukes are dark grey (Norman and Fraser 1937), while the undersurfaces of flippers and flukes, and the ventral surface of the body, are pale grey or white.

Stranded Risso's dolphins can be positively distinguished from pilot whales by an examination of the teeth. In the pilot, and other related species such as the false killer, functional teeth are present in both jaws, but Risso's dolphin almost invariably lacks teeth in the upper jaw, and those in the lower are confined to the anterior margin only. Body length of adult Risso's ranges from 8 to 14 ft.

DISTRIBUTION: Risso's dolphin has been reported from the North Atlantic, including the coast of the United Kingdom

Figure 109.

Grampus griseus, Risso's dolphin: 1. bulbous head with very short beak, forehead with slight 'melon', 2. prominent dorsal fin, 3. flippers of moderate length. In this species the teeth are confined almost entirely to the lower jaw, as in the sperm whales.

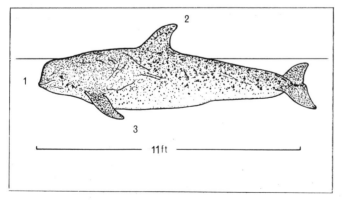

Figure 110. (Professor Warren J. Houck.)
Risso's dolphin, in captivity in Japan.

and France (Norman and Fraser 1937, Fraser 1946, Lami 1955, 1961), the Mediterranean (Tamino 1953b, Norman and Fraser 1937), from the North Pacific (Hall and Kelson 1959, Ficus and Niggol 1965), from South Africa and Mauritius (Barnard 1954, Dr P. B. Best, in letter) and the tropical Atlantic (Cadenat 1959). Mr J. LeClare-Eggleston (in letter), reported that the Marineland of Australia staff has sighted several specimens in New Guinea waters, supplying descriptions and a drawing which left no doubt of the animals' identity. There is only one record of *griseus* in New Zealand waters. A mandible was taken from a specimen stranded on the Manawatu beach north of Wellington in September 1867 and deposited in the Dominion Museum collection (Hector 1873a).

The famous New Zealand dolphin 'Pelorus Jack', called a Risso's dolphin in many publications, was almost certainly a specimen of *T.truncatus*.

BIOLOGY: No detailed study of this animal has yet been possible. Norman and Fraser (1937) recorded that in British waters the food appears to consist largely of small cuttlefish.

Among 45 animals stranded in United Kingdom waters (Fraser 1946) was a female, discovered in December, with a full-term foetus, suggesting that calves are born in winter. Nothing is known of the biology or behaviour of the species in the southern hemisphere. The anatomy of the adult was examined by Murie (1871), and that of the foetus by Le Danois (1912), both working on northern hemisphere specimens.

THE SOUTHERN RIGHT WHALE DOLPHIN
Lissodelphis peroni (Lacépède 1804)

DESCRIPTION: As the name suggests, the right whale dolphins, of which two species are recognised, have no dorsal fin. *L.peroni* attains a length of about 6 to 8 ft. The dorsal surface is purplish-brown and the ventral surface, flippers and undersurface of the flukes are glossy white, faintly tinged with pink. The teeth are small and numerous; 43 pairs in each jaw. The absence of a dorsal fin and the distinctive colour pattern enable this species to be distinguished even at a distance from any other southern hemisphere dolphin. In the illustrations of it shown by some writers, for example Sanderson (1956), the dorsal surface of the flukes is shown as white. While this may be a variable character, all the specimens I have seen had dark upper fluke surfaces.

DISTRIBUTION: The northern species of *Lissodelphis*, *L.borealis*, occurs in the North Pacific. The type specimen was recorded at latitude 2°N near New Guinea, and specimens have been taken off the coast of the United States (Scheffer and Slipp 1948, Brownell 1964, Ficus and Niggol 1965). The white zones in *borealis* are much less prominent than those of *L. peroni*, being largely restricted to a white streak along the ventral midline.

The type specimen of *L. peroni* was recorded by Lacépède at latitude 44°S, longitude 141°E, to the south-east of Tasmania. Other specimens have been sighted in the same area (Davies 1963). The distribution of the species was discussed by Fraser (1955b).

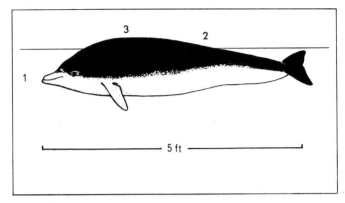

Figure 111.

Lissodelphis peroni, Southern right whale dolphin: 1. blunt head, 2 & 3. no dorsal fin, body blackish dorsally, white ventrally.

At present it appears that the range of *L. peroni* is restricted to temperate and subantarctic waters of the South Pacific and Tasman Sea, and waters directly South of Australia. There are no stranding records of this species available; it seems to lead an entirely pelagic existence, rarely venturing close to shore. Sightings made from *Chiyoda Maru No. 5* in the summer of 1967 suggest that it favours a distribution largely south of the Subtropical Convergence but north of the Antarctic Convergence (Gaskin 1968c); *L. peroni* has been recorded in the western South Pacific, including offshore coastal waters of New Zealand, at surface temperatures ranging from 9.8°C to 17.0°C.

Two specimens of *L.peroni* were recorded by Lillie (1915) during the *Terra Nova's* cruise in New Zealand waters, at latitude 47°04'S, longitude 171°33'E, to the southeast of the southeastern tip of the South Island. Until very recently the two sightings were the only records of the species in these waters.

However, on 3 October 1963 officers of the whale chaser *Orca* operating off the Marlborough coast, sighted five animals at latitude 42°36'S longitude 173°51'E, and on 6 October three more at latitude 42°27'S longitude 173°50'E. On the 11 January 1964 the same vessel sighted three more groups of *L.peroni*, all moving south, the largest being a huge aggregation of scores of small schools totalling over 1,000 animals. Between February and December 1964 seven more schools or aggregations of from 4 to 20 individuals were sighted, two more in January 1965 numbering over 200 animals each. The locations of the major sightings have been shown in a previous publication (Gaskin 1968a). All sightings were several miles from shore.

133

In February 1965 Mr F. Robson of the Marineland of New Zealand reported a single *L.peroni* in offshore waters of Hawke Bay south of the Mahia Peninsula at about latitude 39°50'S.

During the cruise of *Chiyoda Maru No. 5* a total of 86 *L.peroni* were sighted between latitudes 42°S and 54°S and longitudes 180° and 156°W between January and the end of March 1967. School sizes varied from 4 to 20 individuals (Gaskin 1968c).

BIOLOGY: Nothing is known of the biology of this species, except that in the summer months of 1967 the schools sighted from the *Chiyoda Maru No. 5* contained some half-grown animals. Their behaviour was indistinguishable from that of the adults with which they kept close company. None of the 12 schools of right whale dolphins seen by this vessel came to ride the bow wave at any time, though some schools came within fifty yards of the stern.

Scheffer and Slip (1948) and Scheffer (1953), recorded squid remains in the stomachs of specimens of the North Pacific right whale dolphin. Fitch and Brownell (1968) found that myctophids (lantern fish) were the most common item of diet of the northern species of *Lissodelphis*.

THE COMMON DOLPHIN
Delphinus delphis (Linnaeus 1758)

DESCRIPTION: Adults of this species reach lengths of 6 to 8 feet. The common dolphin has a relatively slender build, with a tapering head and distinct beak, the latter clearly delineated from the rest of the head externally. In most fresh specimens a dark ring is noticeable round the eye, but this fades after death. The dorsal surface, frequently described as black in books, is actually a rich purplish-brown in life, fading to greyish-black after death. The ventral surface is yellowish-white, and the two zones abut on angular areas of pale brown and yellowish-grey on the flanks. A greyish streak also runs from the base of the dark flipper towards the vent. The pale flank markings, which are indistinct in new-born specimens, are interrupted directly below the dorsal fin in the lateral midline by a dark triangular patch from the dorsal surface. The flukes and the ventral surface of the caudal stalk are dark brown.

DISTRIBUTION: *Delphinus delphis*, which has frequently been sub-divided into a number of geographic species with little real justification (Dr F. C. Fraser, in letter), has been recorded from all the temperate and warm waters of the globe. It is common on both sides

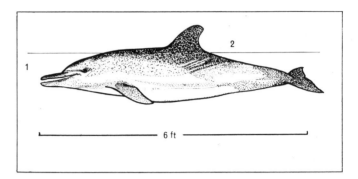

Figure 112.
Delphinus delphis, Common dolphin: 1. distinct beak, 2. purplish-brown back, shading to yellowish and whitish on flanks and belly.

of the North Atlantic (Flower 1880, Fraser 1946, Moore 1953), reaching the limit of its range in waters of about 11°C off the coast of Newfoundland (Sergeant 1958). It has also been recorded in the Black Sea (Sleptsov 1940), in the Red Sea and the Mediterranean (Norman and Fraser 1937), in the Gulf of Mexico (Caldwell 1955), off West Africa (Cadenat 1959), off South Africa (Barnard 1954), Ceylon (Medcof 1963), South Australia (Wood-Jones 1925a), Victoria (Wakefield 1968), Tasmania (Scott and Lord 1920b, Pearson 1936, Davies 1963), in Indonesia (Weber 1923), and around most coasts of the New Zealand archipelago (Gaskin 1968a). The *Delphinus* found in the North Pacific (Scheffer and Slipp 1948, Scheffer 1953, Guiguet 1954, Norris and Prescott 1961, Brownell 1964) is regarded as a separate species, *Delphinus bairdi*.

In the western South Pacific *D.delphis* appears to be distributed largely north of the Subtropical Convergence in surface temperatures of 14°C or above (Gaskin 1968c). The species has been reliably reported as far south in summer as southern Stewart Island, where water temperatures at that time of year are generally about 12° to 13°C (Garner and Ridway 1965). There are indications in records collected during the 1962-64 Marine Department whale survey that the southward distribution of this species is more restricted in winter than in summer. A map of over 4,000 sightings given in an earlier publication (Gaskin 1968a fig. 72) is shown here as two maps, one showing summer records only, the other winter records only, with spring and autumn sightings omitted. Observations made by the M.V. *Ocean Star* in the Hauraki Gulf (Gaskin 1968a fig. 73) indicate that the species is more common in this semi-enclosed region in the spring than at other times of year. This perhaps should be regarded as inshore-offshore movement rather than north-south movement. Further studies might reveal correlations with some of the movements of water masses recorded in the Gulf (Paul 1968).

The distribution of this species around New Zealand may be discontinuous. It is only rarely seen off the west coast of the South Island, and then only during the summer months. In the coastal waters of Marlborough, which are dominated by the cold north-flowing Canterbury Current (Brodie 1960) it is rarely seen, being replaced almost completely by *L.obscurus*. Similarly it is not seen in the inshore waters of Palliser Bay or Cloudy Bay, but is more common than *obscurus* in the offshore waters of the Marlborough coast and in eastern Cook Strait. Only near Hamauri Bluffs, south of Kaikoura, has it ever been seen in large numbers within three miles of shore. On the other hand I have seen this species playing just outside the surf line on the Wairarapa coast on a number of occasions.

During the summer months, between late December and February, large aggregations of schools congregate in relatively restricted areas about 15 miles off the Wairarapa coast. Sometimes such aggregations include over 1,000 animals. We observed such groups during the Marine Department aerial survey in 1963 off the east coast of Cape Palliser, off Castlepoint, and off Cape Turnagain and Cape Kidnappers further north. These large concentrations were apparently only temporary. None has been seen in the winter months. As winter approaches the large schools seem to break into smaller units of from 5 to 30 animals which take up residence in relatively restricted feeding areas. For example, one school appears to be semi-resident near the entrance to Gisborne harbour through the autumn and winter, being observed in the same locality day after day, not seeming to forage more than a few miles in each direction. A school of about 20 animals also appeared to be resident in the mouth of Wellington harbour during the winter of 1963, despite the fact that temperatures in Cook Strait in winter can fall as low as 9°C (Garner 1961).

The number of recorded strandings is very small for such a common species in New Zealand waters (Gaskin 1968a). This dolphin seems to strand singly rather than in schools, the only records of the latter in this country being on the north coast of Great Barrier Island. Single stranded specimens often have quantities of black foam in and around the blowhole, suggesting respiratory disease. This animal is so common that many strandings probably go unreported by coastal dwellers, but

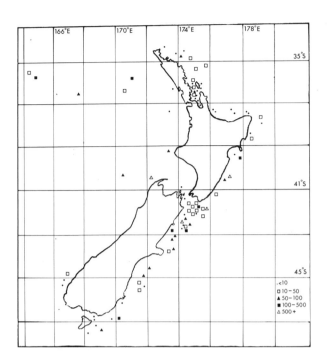

Figure 113. (Modified from Gaskin, 1968a.)
Sightings of the common dolphin *Delphinus delphis*, around New Zealand between 1962 and 1964 through the Marine Department whale survey. Summer sightings; November-April (inclusive).

Figure 114. (Modified from Gaskin, 1968a.)
Sightings of the common dolphin. Winter sightings; May-October (inclusive).

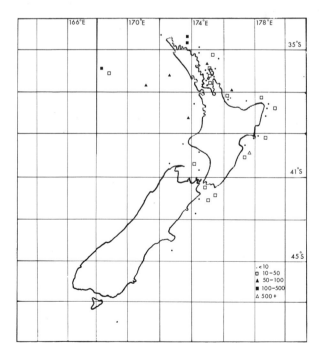

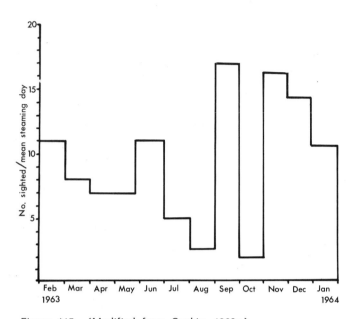

Figure 115. (Modified from Gaskin, 1968a.)
Seasonal variation in the concentration of common dolphins *Delphinus delphis* in the Hauraki Gulf, based on records of the MV *Ocean Star.*

Figure 116. (Marineland of New Zealand.)
Common dolphin *(Delphinus)* (right) with calf, born in captivity at Marineland of New Zealand, Napier. Note that the pigmentation pattern of the calf is not fully developed.

the small number of strandings may also reflect the species' familiarity with the coastal shallow water environment.

BIOLOGY: The only detailed study of the biology and life cycle of the common dolphin to date was carried out in the Black Sea (Sleptsov 1940, Klumov 1954, Kleinenburg 1956, Nikolov 1963), although Sergeant (1958) published details of observations on a single specimen taken off Newfoundland, and Essapian (1962) studied mating and courtship of this species. In the Black Sea the young are born in summer, after a gestation period of 11 months. There is apparently a resting year between every three pregnancies. The species attains sexual maturity at three years of age, and a body length of about five feet. The maximum life span was estimated to be about 25 years.

Sergeant found the remains, in the form of beaks and pens, of about 20 small squid in the Newfoundland specimen. In other regions cuttlefish and small fish form the bulk of the diet (Norman and Fraser, 1937), although echinoderm spines have also been recorded. Fitch and Brownell (1968) found the fish *Engraulis*

Figure 117. (D. E. Gaskin.)
Common dolphin *(Delphinus)* jumping for fish at Marineland of
New Zealand, Napier.

mordax and *Merluccius productus* to be the commonest
items of diet of *Delphinus* in the North Pacific. In
the Bay of Plenty and in the Hauraki Gulf I have
watched this dolphin chasing flying fish.

The behaviour of this species has been less extensively
studied than that of *T.tursiops,* although Sleptsov (1940)
discussed birth and the care of the young in the Black
Sea, and succorant behaviour has been reported by
Tomilin (1955) and Slijper (1958). *D.delphis* frequently
rides the bow waves of craft of all sizes, sometimes
accompanying them for considerable distances. In the
Bay of Plenty in October 1963 a small school of
D.delphis containing four individuals, rode in the bow
wave of HMNZS *Paea* from White Island to Hicks Bay
on East Cape, a distance of about 60 to 70 miles.

In very calm weather, generally just before rain, this
species may be observed making leaps of 10 or more
feet into the air, turning one or more somersaults before
falling back. This behaviour seems to represent sheer
exuberance, and is in marked contrast to the rapid porpoising noted when the animal is chasing a school of
small fish. *D.delphis* is quite audibly vocal, especially
at twilight and after dark. The noises which can be
easily detected by the human ear seem to be variations
of a single squeak, sometimes short, sometimes prolonged, sometimes rising, and sometimes falling. This
noise is made through the blowhole, not through the
mouth. At present it is not known if this sound is
employed in communication between individuals.

The common dolphin was the first species to be held
in captivity in New Zealand. Two animals were kept
in a temporary pool for a short time in late August

1964 at Takapuna, Auckland. This venture was not
successful, and the dolphins were released. A single
specimen of this species, 'Daphne', in a specially built
pool 50 feet across and 11 feet deep marked the modest
but thorough beginnings of the Marineland of New
Zealand at Napier early in 1965. This now has a complex of three tanks on the Napier Marine Parade, the
main public performance pool having sub-surface
viewing pools and a capacity of 250,000 gallons of sea
water. The species has proved very amenable to training, although overseas organisations such as the Florida
Marineland have tended to use *Tursiops truncatus,* as
has the Tweed Head Marineland in Australia. The
repertoire of tricks of the four star common dolphin
performers at Napier increased considerably during the
training programme. In addition to *delphinus, L.obscurus,*
and for a short time a pigmy sperm whale *K.breviceps,*
have also been held at Napier.

The Tauranga Marineland was begun in 1966 and
opened for public display in October of the same year.
In this organisation there is a complex of six pools of
varying sizes, the two largest divided by an underwater
viewing corridor and containing about 500,000 gallons
of sea water. To date only the common dolphin has
been held there as a cetacean exhibit, although a number
of Pinnipedia are also exhibited.

15. The 'Earless' Seals Phocoidea: Phocidae

The most obvious differences between the seals of this family and those of the Otariidae are the lack of external ears in the Phocidae and the methods of movement on land in the two groups. While the otariids can turn their hind flippers forward and use them to walk with a basically four-footed gait, the phocids cannot do this, and progression over land is accomplished by a hitching or caterpillar type of motion in which the fore flippers may or may not be used to provide some motive power.

The Phocidae are generally split into three subfamilies by present authorities (Scheffer 1958, King 1964). These are basically distinguished by skull characters and tooth arrangement, but the Phocinae differ from the Monachinae and the Cystophorinae in having the digits of the hind limbs almost equal in length and bearing long claws. In the other two subfamilies the first and fifth digits are longer than the other three, and except in the genus *Cystophora* the hind digit claws are reduced.

The subfamily Monachinae contains five genera: *Monachus* with three very rare species, *M.monachus* (Hermann, 1779) the Mediterranean Monk Seal, *M.tropicalis* (Gray, 1850) the Caribbean Monk Seal which may be extinct, and *M.schauinslandi* (Matschie, 1905) the Hawaiian Monk Seal; *Leptonychotes* with a single species *L.weddelli* (Lesson, 1826) the Weddell Seal; *Hydrurga* with a single species *H.leptonyx* (Blainville, 1820) the Leopard Seal; *Ommatophoca* with a single species *O.rossi* (Gray, 1844) the Ross Seal; and *Lobodon* with a single species *L.carcinophagus* (Hombron and Jacquinot, 1842) the Crabeater Seal. Monk

seals are confined to the northern hemisphere, but all the other species have been recorded from the New Zealand region or the Ross Sea pack ice.

The subfamily Cystophorinae contains only two genera: *Cystophora* with a single species *C.cristata* (Erxleben, 1777) the Hooded Seal; and *Mirounga* with two species *M.angustirostris* (Gill, 1866) the Northern Elephant Seal, and *M. leonina* (Linnaeus, 1758) the Southern Elephant Seal. Only the last named species is found in the New Zealand region.

The subfamily Phocinae contains six genera: *Erignathus* with a single species *E.barbatus* (Erxleben, 1777) the Bearded Seal; *Halichoerus* with a single species *H.grypus* (Eabricius, 1791) the Grey Seal; *Phoca* with a single species *P.vitulina* (Linnaeus, 1758) the Harbour Seal; *Pusa* with three species *P.hispida* (Schreber, 1775) the Ringed Seal; *P.siberica* (Gmelin, 1788) the Lake Baikal Seal, and *P.caspica* (Gmelin, 1788) the Caspian Seal; *Pagophilus* with a single species *P.groenlandicus* (Erxleben, 1777) the Harp Seal; and *Histriophoca* with a single species *H.fasciata* (Zimmerman, 1783) the Banded Seal. All Phocinae are northern hemisphere species.

Subfamily Monachinae—Tribe Lobodontini (The Antarctic Pelagic Phocids)

The Lobodontini differ from the other tribe of the Monachinae, the Monachini, in the following ways; the Monachini are northern hemisphere animals, have smooth vibrissae, four mammary teats and a black natal pelage, while the Lobodontini have beaded vibrissae, two mammary teats and a pale grey natal pelage (King 1964). Of the four species found in this region, the Ross Seal, the Crabeater, the Leopard Seal and the Weddell Seal, relatively little is known of their biology and movements, although the Weddell Seal has been studied in more detail than the others by research teams in the Ross Sea area. The Ross Seal is probably the least known of the Antarctic Pinnipedia.

THE ROSS SEAL
Ommatophoca rossi (Gray 1844)

DESCRIPTION: The Ross Seal is relatively stout with a short snout, small mouth and bulging eyes, the extraordinarily large size of which is not apparent until the eyelids have been removed. There is a little individual variation in the basic colour pattern in the form of pale longitudinal markings on the neck and back. The body is typically dark grey dorsally, with the dark pigmentation shading gradually into paler grey on the flanks and then to white on the ventral surface (King 1964). However, Polkey and Bonner (1966) described a specimen taken in the Weddell Sea as sandy-brown. The species possesses large, powerful flippers which give it great manoeuvrability in the water. Males are slightly larger than females at physical maturity. Females range from about 8 to 8½ ft in length, while males may attain lengths of up to 10 ft.

Figure 119. (Mr J. Terhune, University of Guelph.)
'White-coat'. The pup of a harp seal *Pagophilus groenlandicus*, on the ice field south of the Magdalen Islands in mid-March 1970. This species is the basis for an industry of great economic importance to the poorer fishing centres of eastern Canada.

Figure 120. (Mr J. Terhune.)
Adult and pup harp seals, *Pagophilus groenlandicus* on the ice field in Northumberland Strait in mid-March 1969, between New Brunswick and Prince Edward Island.

DISTRIBUTION: *Ommatophoca rossi* was first described by Gray (1844) from two specimens collected by the *Erebus* and *Terror* expeditions in the Ross Sea between 1839 and 1843. Subsequently the range of the Ross Seal has been shown to be nearly circumpolar, although records are widely scattered. The species appears to have a low population density. The only recorded sighting north of the winter pack ice limit is from just south of Heard Island (King 1964). There are no records for the subantarctic islands south of Australia or New Zealand. In this sector of the Antarctic most specimens have been sighted southeast of the Balleny Islands, although King (1964) showed a number of records from the coast of Wilkes Land and Marie Byrd Land. In other sectors the Ross Seal has been seen on the Queen Maud Land coast, in the Scotia Sea, on the coast of the Antarctic Peninsula (King 1964) and in the South Orkney Islands (Wilson 1907).

BIOLOGY: The Ross Seal is the most rarely observed of all the Antarctic seals, and little is known of its biology or habits. The reproductive cycle is unknown. There are no published records of the species concentrating either ashore or on pack ice for breeding or moulting, although an Argentinian expedition did take a few young animals in one area near the South Orkney Islands in 1903 (Wilson 1907). The pelage structure was described by Polkey and Bonner (1966).

The species feeds along the edges of the heavier pack ice. Squid are the major item of diet, but King (1964) noted that euphausids and fish have also been recorded. The Ross Seal is host to the usual parasites which attack Pinnipedia. Lice have been found in the pelage, and the digestive tract is usually infested with nematodes. Cestodes have also been reported (King 1964).

As it lives among the heaviest and least accessible pack ice this species probably has few enemies. Some animals have been seen with long scars on the flanks which may have resulted from escapes from killer whales or Leopard Seals. However, these may just as well be the results of intraspecific fighting during the mating season (King 1964).

The species is said to be unusually vocal for a seal; a variety of noises have been recorded, including cooing sounds, trilling, and sharp warning clucks (Ray 1966).

THE CRABEATER SEAL

Lobodon carcinophagus (Hombron & Jacquinot 1842)
DESCRIPTION: Adults are dark grey dorsally, shading to light grey on the flanks and white on the ventral surface.

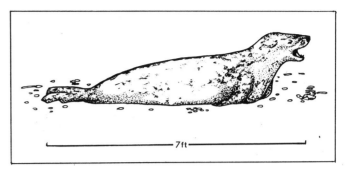

Figure 121.
Lobodon carcinophagus, Crabeater Seal.

The back and flanks are also marked with reddish-brown circles and irregular-shaped patches, mostly around the shoulders. There are interesting colour changes in the pelage through the year. As the summer progresses the coat of the Crabeater Seal becomes paler until almost white. At this stage it is moulted for a new, dark pelage. King (1964) noted that the general colour even of the dark hair becomes paler as the animals age. At maturity males and females are of approximately the same size, attaining lengths of over 8 ft and weights of 500 lbs (King 1964). Although not as lithe as the Leopard Seal, the Crabeater is swift and agile both on the ice and in the water.

DISTRIBUTION: The Crabeater Seal has a circumpolar distribution in the pack ice regions of the Southern Ocean (Bertram 1940). However, each year a few stragglers are reported as far north as South America (Berg 1898); South Africa (Courtenay-Latimer 1962); southern Australia (Hall 1903, Wood-Jones 1925b), LeSouef 1929, Brazenor 1950); Tasmania (anon, 1946, Davies 1963b) and New Zealand (Oliver 1921, McCann 1964c). Occasionally odd specimens have been encountered by survey parties on the Antarctic mainland as much as thirty miles from the coast and at altitudes of up to 2,700 ft (Wilson 1907, Péné, Rivard and Llano 1959, Stirling and Rudolph 1968), but this is exceptional, and the species is generally found on the shore ice, the deep-sea icefields or in the sea itself.

Crabeater Seals are semi-gregarious (Rudmose-Brown 1913, Bertram 1940, Haga 1961, Holdgate 1963), and the species is one of the most abundant of the Antarctic seals. Scheffer (1958) gave a total population estimate of two to five million animals for the whole Southern Ocean.

The species is migratory, and between April and June part of the population moves into subantarctic waters between latitudes 55°S and 65°S (Scheffer 1958).

Specimens which stray to the southern hemisphere temperate zone landmasses are usually immature individuals and are mainly found in the winter months.

BIOLOGY: The anatomy, morphology and pathology of the adult Crabeater Seal have been discussed by Lindsay (1938), Laws and Taylor (1957), King (1961), and Nishida and Amemiya (1962), and of the pup by King (1957). The growth rate of the species and methods of age determination were described by Laws (1958).

The reproductive cycle of this species is not completely known, since the colonies form and pupping is at its height early in the spring, while daylight hours are still very short and before the ice near the Antarctic has started to break up sufficiently to allow the easy passage of supply and survey ships (King 1964). Consequently information on the early part of the breeding cycle is lacking. Moulting occurs on the pack ice in January and February. The adults have never been observed mating, but the condition of the testes of males shot in October and November suggests very strongly that mating takes place at this time of year (Bertram 1940). Males and females appear to reach sexual maturity at about 2½ to 3 years of age. Pups of this species have only rarely been seen. Most are born between early September and late November (Bertram 1940). They are about 4 to 4½ ft in length at birth, and weigh 50 to 60 lbs (King 1957, 1964). Lactation may last as long as five weeks (Bertram 1940), or only a few days (Poulter 1968); more information is required. When weaning is complete the pups shed their natal pelage and take to the water for the first time.

The Crabeater Seal is ill-named, as euphausid crustaceans, the diet of the whalebone whales, are almost the only food eaten. This seal has a set of teeth with perhaps the most intricate arrangement of cusps found in any living mammal, to facilitate the capture of euphausids at sea. The animal appears to swim into a shoal of krill with its mouth open and the breathing passages closed. When the mouth shuts the water is squeezed out between the complicated cusps of the cheek teeth by raising the tongue, leaving a mouthful of euphausids to be swallowed (King 1961).

Adult crabeaters have few natural enemies except killer whales (King 1964), and at the present time the species is not commercially exploited. Greatest mortality occurs among young animals, with bad weather and starvation being probably the biggest killers. Tapeworm parasites are relatively rare, but nematodes are usually abundant in the digestive tract; they probably do little harm to their hosts (King 1964). Lice are common in the pelage (Hopkins 1949). Laws and Taylor (1957) reported the death of over a thousand Crabeater Seals during the winter of 1955, off the east coast of the Antarctic Peninsula. The cause of death was not positively discovered, but was believed to be a contagious virus.

THE WEDDELL SEAL
Leptonychotes weddelli (Lesson 1826)

DESCRIPTION: Adults of this species are dark grey or black dorsally, paler grey on the flanks, and light grey ventrally. In summer the basic colour changes from dark grey to dull brown. The body is covered with irregular pale streaks and patches. Adults attain lengths of 9 to 10 ft and weights of up to 900 lbs at maturity (King 1964). Fully grown females are usually a little larger than males of the same age. General descriptions of this species have been given by Moore (1963b) and King (1964).

DISTRIBUTION: The Weddell Seal has a circumpolar distribution in the Southern Ocean, and generally frequents the coast of the Antarctic mainland, staying close to the shore ice and hauling out on beaches where these occur (Arseniev 1959, Scheffer 1958). Weddell Seals have been known to make prodigious journeys over land, and carcases have been found more than 30 kilometres inland and at elevations of 2,000 feet (Wilson 1907, Korotkevich 1958). Quite recently radio-isotope dating methods have shown that some of these mummified remains are several hundreds of years old. Tracks found far inland in the Taylor Dry Valley are also apparently ancient (Claridge 1961).

The Weddell Seal ranges further south under normal circumstances than any other species, reaching to about latitude 80°S in the Ross Sea. It is a semi-gregarious species occurring in colonies in scattered localities. The groups are usually relatively small. Recent estimates of the total population in the southern hemisphere

Figure 122.
Leptonychotes weddelli, Weddell Seal.

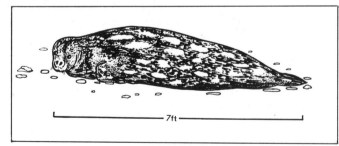

7ft

have ranged from about 250,000 to 500,000 (Scheffer 1958, King 1964). An earlier estimate of 800,000 by Laws (1953a) may be too high. The species was known early to breed around the Ross Sea in considerable numbers (Wilson 1907). In recent years it has been intensively studied in this region by scientists working at McMurdo Sound (Lindsay 1937, Heine 1960, Littlepage and Pearse 1962, Littlepage 1963, De Vries and Wohlschlag 1964, Ray and Lavallee 1964, Featherston 1965, Ray 1965, Smith 1965, 1966a, b, Schevill and Watkins 1965, Murray, Smith and Soucek 1965, Flyger, Smith, Damm and Peterson 1965, Stirling 1966a, b, c, 1967, Ray and Smith 1968, Poulter 1968).

Weddell Seals have also been recorded, sometimes abundantly, on the Adelie Coast (Sapin-Jaloustre 1953a, b); the Prince Harald Coast (Haga 1961); the Kemp Coast (Scheffer 1958); the South Shetlands and South Orkneys (Bertram 1940, Sivertsen 1954, Mansfield 1958); and in the Weddell Sea (Rudmose-Brown 1913) from the subantarctic and temperate latitudes. Gray (1837) recorded specimens from the coast of Patagonia, and other specimens from the same region were described by Allen (1905), Cabrera and Yepes (1960), Osgood (1943) and Vaz Ferreira (1956a). Hamilton (1945) reported the species from the Falkland Islands, and Matthews (1929) noted that South Georgia had a small breeding colony. Small numbers have also been recorded from Kerguelen Island (Paulian 1957b); Heard Island (Law and Burstall 1953); the southern coast of Australia (Iredale and Troughton 1934, Carter, Hill and Tate 1945, Brazenor 1950, and Troughton 1951); from Enderby Island in the Auckland group and from New Zealand (Turbott 1949, 1952, McCann 1964c). The New Zealand records, three in number, were all from the west coast of the North Island; one from Muriwai beach and the other two from just north of Wellington. All were immature animals, and all seen in the winter or early spring.

Recently, the movements of Weddell Seals which breed near the Antarctic bases in McMurdo Sound have been studied in detail (Heine 1960, Littlepage and Pearse 1962, Littlepage 1963, Smith 1965, Stirling 1966a, 1967). Definite seasonal changes in density were discovered. The overwintering population consists of less than 300 animals; the seals move northwards in autumn to avoid the heaviest ice of winter. In the spring and early summer the population rises steadily until nearly 2,700 seals are present in the Sound. There are two quite distinct phases to this influx; the first wave in October consists almost entirely

Figure 123. (S. E. Csordas.)
A rare visitor to Macquarie Island in 1957, a Weddell Seal (left), with a damaged jaw and a bad flesh wound in its back, lies among kelp close to a young Elephant Seal.

of pregnant females, while the second wave a month or so later is made up of males, and females which have already pupped further north. The seals begin to leave McMurdo Sound in late February, the newborn, weaned pups leaving first. Once it starts the emigration is very rapid. Immature animals of both sexes stay in the pack ice north of McMurdo until they are old enough to breed.

BIOLOGY: Bertram (1940) made a detailed study of the biology of this species on the east coast of the Antarctic Peninsula, and Mansfield (1958) undertook a similar study of the South Orkney Island population. In these localities the cows come ashore from the end of August onwards and form aggregations on the breeding beaches. The pups are born during September and October, about two weeks after the mother comes ashore. The mothers are very vocal after the birth of their pups (Poulter 1968). The pups are about 4 to 4½ ft in length at birth and weigh about 60 lbs. They are covered with a thick, grey natal pelage with a dark longitudinal stripe down the middle of the back. This coat is moulted at the age of about two weeks, the actual moulting process taking about a month to complete. The pups are very active at this time, and may take to the water even before the moult is complete.

Lactation lasts for about six to seven weeks, small crustaceans supplementing the mother's milk in the later stages. Cows with young pups are very aggressive towards intruders (King 1964), unlike fur seals which

143

will often abandon their offspring if danger threatens. During lactation the cows do not feed, and stay close to the pups at all times. Once the pups are weaned the temporary colonies rapidly break up as soon as the cows have been mated again by the bulls. Females mature at an age of 3 years, and the total length of the breeding life of a cow is thought to be about 9 years.

Unlike other species Weddell Seals do not lie up and fast while moulting. The process may begin at any time during the summer, and does not appear to affect the metabolism in the drastic way observed in Elephant Seals, where the eyes run, the nasal passages are clogged with mucus and the animals appear to run a temperature.

During the winter months the Weddell Seals spend much of their time in the sea beneath the ice, presumably because in the water the temperature fluctuates less widely than on land, and the lowest extreme is only at or slightly below freezing. Under these conditions thermo-regulatory problems are kept to a minimum (Ray and Lavallee 1964, Ray 1965, Ray and Smith 1968). The animals rest on ledges beneath the ice and cluster in places where the ice is folded and split and air pockets are continually renewed (Perkins 1945). Over-wintering scientists have heard Weddell Seals calling beneath the ice throughout the winter. The animals use their teeth to keep breathing holes open, and the canines are often found to be broken or badly worn (Lindsay 1937, King 1964).

Poulter (1968) described the results of recording the underwater sounds produced by this species. He noted a considerable variety of sounds, but the most usual one a call which started with a high frequency and high pulse rate and then descended to a very low frequency and pulse rate. Other signals were deciphered as being composed of isolated parts of the total range of this common call. Single clicks, which could be heard echoing and re-echoing under the ice were identified as echo-location finding signals. Other studies on the range of underwater calls made by the Weddell Seal were described by Schevill and Watkins (1965).

The seals feed along the edge of the shore ice. Fish is the major item of diet, but squid, octopus, crustaceans and holothurians (sea cucumbers) are also eaten (King 1964). Experiments carried out at McMurdo Sound with pressure-recording apparatus strapped to specimens have shown that Weddell Seals are capable of submerging for at least 10 minutes and diving to depths of more than 350 metres (De Vries and Wohlschlag 1964).

The Weddell Seal has few enemies. Killer whales probably take a number of adults at sea and among leads in the shore ice. Starvation and bad weather, especially moving ice, together with Leopard Seals, probably kill numbers of young.

The digestive tracts of Weddell Seals are usually infested with cestodes (Markowski 1952, Featherston 1965) and nematodes (King 1964). The pelage is commonly infested with the louse *Antarctophthirus ogmorhini* (Murray, Smith and Soucek 1965). These lice are only occasionally found on mature bulls, and the parasite is transmitted solely by the movement of adult lice from mothers to their pups. Infestations are heavy on immature animals, and become lighter as they age. The lice are generally concentrated on the caudal region, sometimes around the genital apertures. They feed while the seals are at sea, and reproduce on animals which are hauled out on the ice, or on beaches. Eggs of this louse are capable of developing and hatching at the remarkably low temperature range of 0 to 4°C.

Figure 124. (S. E. Csordas, Melbourne.)
A young Weddell Seal ashore on Macquarie Island in 1955.

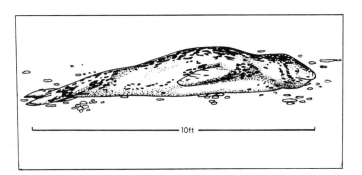

Figure 125.
Hydrurga leptonyx, Leopard Seal.

THE LEOPARD SEAL
Hydrurga leptonyx (Blainville 1828)

DESCRIPTION: Adult Leopard Seals are grey dorsally, merging into lighter grey on the flanks and yellowish-grey or white on the ventral surface. The pelt of most specimens is also generously spotted with dark grey and whitish markings, the latter being subject to considerable individual variation. Females are slightly larger than males at any given age, also heavier. When fully grown a female may weigh 1,000 lbs and attain a length of 12 to 12½ ft. Males weigh about 600 lbs and attain lengths of about 10 ft at maturity.

The Leopard Seal is very lithe and graceful in the water, and is capable of strong bursts of speed. It can make prodigious jumps from the water to the surface of ice floes, sometimes jumping to heights of more than 6 ft. A Leopard Seal was photographed at South Georgia just after it had jumped from the water on to the back of a dead whale tied near a shore station, a height of several feet. The species is also surprisingly nimble on land, where it moves by an undulating motion of the body, arching the back and pushing itself forward using the pelvic region. The fore flippers are not used (O'Gorman 1963).

DISTRIBUTION: This species ranges widely through the Southern Ocean from the outer limits of the Antarctic shore ice to the southern coasts of the inhabited regions of the southern hemisphere (Scheffer 1958, King 1964). It does not range as far south as the Weddell Seal, although there is one record from McMurdo Sound at latitude 78°S (Dearbon 1962).

The Leopard Seal has been recorded from Victoria Land, the Ross Sea and the Antarctic Peninsula (Allen 1905, Turbott 1952, Stonehouse 1965a); the Adelie Coast (Sapin-Jaloustre 1953b); South Georgia (Matthews 1929); the South Orkneys and South Shetlands (Scheffer 1958); Heard Island, where larger concentrations have been seen than anywhere else in the world (Law and Burstall 1953, Gwynn 1953a, Chittleborough and Ealey 1953, Howard 1954, Brown 1957); Macquarie Island (Gwynn 1953a, Carrick 1957, Csordas 1962b); Kerguelen Island (Angot 1954, Paulian 1953, 1955, 1960); Amsterdam Island (Paulian 1957b); Bouvet Island (Sivertsen 1954), and Campbell Island (Wilson 1907, Bailey and Sorensen 1962).

Both immature seals and adults migrate northwards in the autumn and winter, and sporadic observations of the species have been made on the coast of Patagonia (Wilson 1907, Cabrera and Yepes 1960); South Africa (Roberts 1951, Courtenay-Latimer 1962); Western Australia (Serventy 1948); New South Wales (Troughton 1951); Victoria (Brazenor 1950, Wakefield 1963); Tasmania (Davies 1963b); Lord Howe Island (Hamilton 1939a); New Zealand (Hutton and Drummond 1923, Thomson 1921, McCann 1964c); and the Cook Islands (Berry 1961). The latter record at the island of Rarotonga, latitude 21°S, is the present northern limit recorded for the species.

The Leopard Seal is a solitary animal for most of the year. It is widely distributed but nowhere really common. Only at Heard Island have large numbers (about 900) been seen in one place. Laws (1953a) estimated the population of the Falkland Islands' Dependencies region to be about 40,000, while Scheffer (1958) placed the total southern hemisphere population as somewhere between 100,000 and 300,000 animals.

BIOLOGY: Very little is known of the breeding habits of

Figure 126. (S. E. Csordas.)
Very young Leopard Seal resting on beach on Macquarie Island.

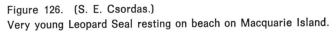

Figure 127. (S. E. Csordas.)
Head view of Leopard Seal. Taken on Macquarie Island.

Figure 128. (S. E. Csordas.)
Leopard Seal showing its teeth in threatening gesture. Taken on Macquarie Island.

this species. The pups are thought to be born between September and January. A female was killed at King Edward Point, South Georgia, in July 1925, and was found to be carrying a three foot foetus, and very young Leopard seals six to seven feet in length have also been seen on South Georgia beaches (Hamilton 1939a). In one experiment research workers at Heard Island prevented a pregnant female from leaving the island, with the result that the pup was born on the beach instead of on the pack ice some hundreds of miles away. At birth the animal was a little over 5 ft in length and weighed 65 lbs. The eyes were open and the tips of the canine teeth visible protruding through the gums. This pup was born in November (Brown 1952).

Females are believed to reach sexual maturity at about 2½ years and give birth to their first pup in their third or fourth year. Males do not become sexually mature until their fourth year (Paulian 1955). The duration of

lactation is not known with great accuracy, but is thought to last about two months (Paulian 1960). The adults moult in February (Hamilton 1939a).

Pregnant female Leopard Seals are known to drink quantities of sea water. The reason for this is unknown (Brown 1952), but it may be a method of balancing body fluid salts diminished during pregnancy

The species is migratory, and the younger animals tend to wander further afield than adults of breeding age. Most of the specimens recorded in New Zealand were young animals, some apparently in the last stages of exhaustion; specimens hauling out on the Southland coast, on the east coast of the South Island, around the shores of Cook Strait. A young specimen which came ashore near Wellington in 1964 was in very poor conditions, possibly through near-starvation, or parasitic infestation (Charles McCann, in letter). The species is subject to parasitism by cestodes (Markowski 1952).

Somatic tumours, especially bony growths in the nasal passages (King 1964), are also found, especially in older animals. An 11 ft female taken by Massey University staff on a beach near the mouth of the Manawatu River in 1966 had suffered a large gash in the side of the face, perhaps by a boat propeller. The wound was deep, and the jaw fractured in several places.

The Leopard Seal is an active carnivore, with no predatory enemies other than man and perhaps killer whales. It feeds mainly on penguins and fish, although pups of other seal species are almost certainly taken as well (King 1964). A specimen taken on the South African coast had the remains of two terns in its stomach. The species is often seen cruising in the vicinity of penguin colonies, and the sight of a Leopard Seal's head breaking the surface causes great agitation among the birds. However, once a seal has gorged itself and hauled out on the ice or on a beach penguins will approach it without any sign of fear.

Leopard seals have been observed to wait for Adelie penguins returning from feeding at sea (Poulter 1968). The seal dives and turns very swiftly in the water to catch swimming penguins. If the birds are running across thin ice the seals rise sharply to fracture the ice and knock their prey into the water. Once a penguin is taken it is brought to the surface, being held by the loose scruff of the belly skin. The bird is then bitten and shaken so vigorously that it is literally shaken out of its skin. The carcase is then eaten and the skin and feathers discarded (King 1964). King also noted that pieces of whale meat have been found in the stomachs of Leopard Seals, though these were probably taken from half-flensed whale carcases rather than from living animals.

This species has a reputation for ferocity (King 1964) and generally responds to provocation more rapidly than any other Antarctic seal. It has been accused of making unprovoked attacks on man, but entry of a human into the temporary territory of a Leopard Seal may be construed as a provocation, even if the human did not consider it as such. Attacks on humans can occur when a man walks between a Leopard Seal and the sea, thus appearing to cut off the animal's escape route and making it very nervous (Mr T. Bruce, pers. comm.).

The Leopard Seal is more vocal underwater than on the surface, and has been recorded making a range of noises, including glottal throbbing, prolonged hisses made through the nostrils with the mouth closed, and sharp grunts. Some of these noises have been interpreted as warning sounds. Underwater calls have been taped in the McMurdo Sound area, and include echo-locating

Figure 129. (Logan Print Ltd.)
'Brutus', adult Leopard Seal (*H.leptonyx*) taking a fish from trainer at Mt. Maunganui Marineland, New Zealand.

clicks, broken or interrupted high-pitched or low-pitched sounds of constant frequency, sweeping sounds similar to those produced by Weddell Seals, and low-frequency chirping noises. At least some of the animal's more complex calls seem to be used for communication (Poulter 1968).

The Mount Maunganui Marineland at Tauranga is one of the few seaquaria in the world to have held a Leopard Seal in captivity. 'Brutus' was a 12 ft male, captured at Ohiwa near Opotiki in the Bay of Plenty during 1967. This animal spent most of the day in its pool, surfacing to breathe and feed, but usually hauling out on to a rock at night to sleep. About 30 pounds of fish were eaten every day.

Subfamily Cystophorinae
Elephant and Hooded Seals

The species of this family share the common character of inflatable nasal passages in the male. Males are generally much larger than females at any given age, except shortly after birth. The anterior nares (nostrils) are nearly vertical, and in the jaws there is only one lower incisor on each side. There is a single species in the New Zealand region, the Southern Elephant Seal.

THE SOUTHERN ELEPHANT SEAL
Mirounga leonina (Linnaeus, 1758)

DESCRIPTION: The male is the largest of the southern Pinnipedia, and at maturity may attain lengths of 18 to 20 ft and weights of up to 8,000 lbs (King 1964). Males are dark grey or dark brown dorsally, lighter on the flanks and ventral surface. Bulls are pugnacious during the breeding season, and generally become extensively scarred while fighting for possession of territory and females. The scarring is concentrated about the neck and shoulders, and the skin becomes very cornified and thickened there. The most prominent feature of the male is the inflatable proboscis. This is not fully developed until the animal is about eight years of age, and fully mature sexually. The proboscis can be raised in aggressive display by a combination of inflation, blood pressure and muscular action, and at full display the snout is elevated into a large, ridged swelling divided by two transverse grooves which become more prominent with age (Harrison and King 1965).

Females weigh about 2,000 lbs at maturity and attain lengths of 10 to 13 ft; they do not possess the characteristic proboscis of the male.

On land the Elephant Seal uses its front flippers to gain purchase while hauling its body forward in a hitching movement. It is far more at home in the water, where the hind flippers are used to supply movement, in combination with sinuous thrusts of the body. The propulsive driving stroke is made with the digits of the hind flipper expanded, presenting a large webbed surface. On the recovery stroke the digits are closed to reduce the surface area. The front flippers may be used for short sculling strokes at slow speeds, but in general their function is that of stabilisers.

DISTRIBUTION: The main breeding range of the Elephant Seal lies in the circumpolar subantarctic region. The species has been known to breed well into the southern temperate region, but past over-exploitation has resulted in its extermination in regions where it was formerly quite common.

It has been exterminated at the type locality, the Juan Fernandez Islands (Osgood 1943). Cabrera and Yepes (1960) indicated that it was also killed out on the coast of Chile before the beginning of the twentieth century. On the Atlantic coast of South America it is slowly recovering in numbers (Santiago Carrara 1952, Vaz Ferreira 1956a).

Further south and southeast the Elephant Seals of the Atlantic Antarctic sector are concentrated on South Georgia (Matthews 1929, 1952, Hamilton 1940, Bonner

1955, 1958a, Laws 1952, a, b, c, 1953b, 1956 a, b, 1962). This population scatters southward in the summer months, reaching the South Shetland Islands and South Orkney Islands (Liversidge 1950), where small breeding colonies are present. There are also scattered records from the coast of Antarctica (Ingham 1957, 1960, Eklund 1960). Small numbers are found on the South Sandwich Islands (Matthews 1952) and on Gough Island (Swales 1956, Holdgate, LeMaitre, Swales and Wace 1956). Stragglers, probably from the Gough Island colony, have been observed at Tristan da Cunha (Elliott 1953), and historically have even been reported from St Helena (Scheffer 1958, King 1964). Dr R. M. Laws (1953a) estimated the total population of the Falkland Islands-Falkland Islands Dependencies region as about 300,000.

In regions south of the Indian Ocean a few Elephant Seals breed on Amsterdam and St Paul Islands (Paulian 1953, Laws 1956b), and about 40,000 breed on Heard Island (Chittleborough and Ealey 1953, Law and Burstall 1953, Howard 1954, Gibbney 1957). The species is abundant on Kerguelen Island and surrounding islets, and about 80,000 appear to breed in this region (Paulian

Figure 130.

Mirounga leonina, Elephant Seals. Male in aggressive posture with proboscis inflated.

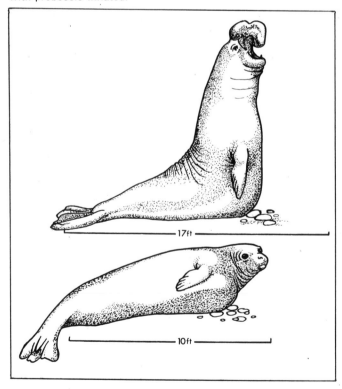

148

Figure 131. (M. W. Cawthorn.)
'Blossom'; young male Elephant Seal at Oriental Bay, Wellington, New Zealand in 1964.

1953, 1957b, Aretas 1951, 1958, Angot 1954). The species is also found on Marion Island (Rand 1955b, 1962) and has been reported from the Crozet Islands (Matthews 1952, Laws 1956a).

There is a breeding population on Macquarie Island (Carrick 1957, Carrick and Ingham 1960, 1962a, b, c, Carrick, Csordas, Ingham and Keith 1962, Carrick, Csordas and Ingham 1962, Ling and Nicholls 1963), which has increased from about 70,000 (Doutch 1952) to nearly 95,000 in recent years (Carrick and Ingham 1962). On Campbell Island the population is small (Sorensen 1950, 1951, Degerbol 1956), and in the late 1940s was only 417 (Sorensen 1950), but is now increasing (Bailey and Sorensen 1962). Small colonies have also been observed on the Auckland Islands (Sorensen 1951) and on the Antipodes (Turbott 1952).

The species used to occur in Bass Strait (Mr R. M. Warneke, in letter), but was exterminated by sealers in the last century. A few stragglers now appear to return from time to time. Immature animals stray north in the winter months, and have been recorded from Tasmania (Davies 1961) and South Africa (Kettlewell and Rand 1955), in both of which places pups have been born; from New Zealand (Thomson 1921, Hutton and Drummond 1923, McCann 1964c); and Mauritius (Vinson 1956).

BIOLOGY: The reproductive cycle has been studied in detail at South Georgia by Matthews (1929), Hamilton (1940), Bonner (1955) and Laws (1956a, b, 1960); at Kerguelen Island by Angot (1954); at Heard Island by Gibbney (1957) and at Macquarie Island by Carrick (1957), Carrick, Csordas, Ingham and Keith (1962) and Campbell Island by Sorensen (1950).

Breeding begins in late August and early September. The mature bulls come ashore at beaches which are habitual colony sites. These mature animals take up individual territories on the most favoured parts of the beaches, while younger bulls are forced to live temporarily on the fringes of the beaches, or among rocks and vegetation a short distance inland.

Sexually mature and pregnant cows come ashore in middle and late September and congregate into groups. Harems are formed by the end of this month; the number of females taken by each of the dominant bulls, now generally referred to as beachmasters, is initially about 10, but this number may rise to about 50. This is approximately the maximum number any bull can manage successfully; larger units tend to be split by other bulls taking females, or cows moving away while the beachmaster's attention is engaged elsewhere. The beachmasters mate with the sexually mature cows in their harems, and with the formerly pregnant females shortly after the pups are born. Younger, less mature bulls, are also sometimes successful in mating with cows on the periphery of a large harem.

If a beachmaster is challenged by one of these bachelor bulls he takes up an aggressive posture, roaring and inflating the proboscis. The warning sound generally consists of a deep cough and a throaty, bubbling belch. The bachelors, and occasionally a beachmaster past his prime, generally find it more discreet to retreat than fight when confronted with an obviously superior opponent. A serious fight for a harem generally only occurs when challenger and challenged are evenly matched for weight and experience. There is usually an 'eyeball to eyeball' phase where the two seals rear up chest to chest, probosces inflated, bellowing at very short range. If neither retreats they proceed to lunge and slash at each other, often inflicting deep gashes to the neck and chest. It is not unusual to see old bulls with one eye missing. Eventually one or other retreats from the harem area to the beach perimeter to join the other bachelor animals. Fights to the death are probably very rare.

Pupping occurs from five to seven days after the arrival of the cows on the beaches. Twin births are rare, but not unknown, although one nearly always dies. Lactation lasts about three weeks. The mother does not feed during this time and may lose as much as 500 pounds in weight. The pups are generally finished weaning about 21 days after birth, during which time they gain weight at a rate of about 20 pounds per day. The rate of growth in this species at Macquarie Island was studied by Bryden (1968).

149

The cows are ready to mate once more towards the end of lactation. After this they leave the young and the harem confines and return to the sea to feed. When the mating season is over the aggressive behaviour of the bulls also moderates, and by the end of November they have lost interest in the females and also return to the sea to begin feeding again.

Pups are about 4 to 4½ ft in length and weigh approximately 80 lbs at birth. A thick black natal pelage is present, which they begin to moult about 10 days after birth and complete the moult in about 25 days. As they suckle, the pups accumulate a thick blubber layer on which they subsist for a further five to six weeks. Towards the end of this time they begin to feed themselves successfully, returning to the sea more frequently after the moult is complete.

The earliest age at which bulls may achieve challenger and breeding status is not known; the bulls at the Macquarie Island colonies are all at least six years old.

However, at South Georgia, where the animals have been successfully culled (Laws 1952a), males are present in the breeding areas from the age of four years onwards and may control a harem at seven years of age.

Females become sexually mature in their third year and generally pup for the first time in their fourth year. The breeding life is not long, only about five years. A very high mortality rate appears to occur among breeding cows; from a third to a half of the cows breeding in a given year do not survive through the following two years. The maximum age of females is about 12 years, and the maximum number of pups born in a lifetime is about seven. Adult mortality factors are not well known. Disease and bad weather are probably responsible for many deaths.

During December, January and February adults of both sexes haul out on suitable beaches to moult. This process takes about 30 to 40 days, during which time the animals cease feeding and rarely go near the sea unless

Figure 132. (S. E. Csordas.)
Small Elephant Seal harem formation on a Macquarie Island Beach. In the middle the cows huddle around the beachmaster, and are in turn surrounded by bachelor bulls without harems. A few weaned calves lie in the foreground. To the left three bachelor bulls attempt to catch a cow in the surf.

Figure 133. (S. E. Csordas.)
Harem master or beachmaster Elephant Seal moving into position to drive off a challenger on a Macquarie Island beach.

Figure 134. (S. E. Csordas.)
Two young bull Elephant Seals fighting on Nuggets Beach, Macquarie Island. A harem can be seen in the middle background.

disturbed. Hair and skin moults off in large patches, and the irritation set up during the moult encourages the animals to wallow in any suitable muddy depression. They return to such wallows year after year, and swampy areas on the South Georgia coast are pocked with seal wallows several feet deep. Once the moult is complete the animals once again return to the sea to feed, and no more large concentrations of Elephant Seals are seen ashore again until the following breeding season.

Apart from man this animal has few enemies. Some pups may be taken by Leopard Seals and killer whales; starvation, being crushed by adults, and bad weather account for others. Only about 50 per cent of the pups born on any island survive their first winter. Pup deaths actually on the breeding beaches at Macquarie Island were estimated to vary between 2 and 16 per cent from all causes. Numbers of pups are killed during fights by beachmasters and bachelor bulls; the pups are simply crushed by the great bulk of the adults, who make little or no effort to avoid them. Cows sometimes accidentally roll on their own or other animals' pups. Skuas, gulls and giant petrels may badly peck young pups left unattended. Sometimes mothers are for some reason not able to relocate their own pups. Although lost pups will try to suckle at any teat within range, their advances are rarely accepted.

Elephant seals are parasitised by cestodes and acanthocephalans (thorny-headed worms) (Markowski 1952),

and by the louse *Lepidophthirus macrorhini*. The ecology of this animal has been studied in detail by Murray and Nicholls (1965) on Macquarie Island. These workers found that the louse requires a temperature of about 25°C for rapid multiplication, and that this temperature is about that of the hind flippers and general caudal region of Elephant Seals hauled out for breeding. Most louse mortality occurred if a seal stayed in the sea for an extended period, and during the seal moult. They found that the body temperature of the lice fell to nearly that of the sea, as did the skin temperature of the seal when it remained in the water for any lengthy period. The lice could only tolerate this condition for about one week without feeding. Lice were also unable to reproduce on older animals which moulted in deep muddy wallows. However, by burrowing into the *stratum corneum* layer many lice on younger animals were able to survive when the annual moult sloughed off the outer layer. Pups are infected by transmission of adult lice from their mothers or adjacent cows. Although lice occur all over the body they were thought to concentrate on the hind flipper region as there the temperature of the skin rose fastest after the seal emerged from the sea, offering most opportunities to feed.

Reproduction of the louse only takes place when the seals are hauled out on the beaches. The parasite does not oviposit in water, and eggs do not hatch if sub-

151

merged. About six to nine eggs are laid each day and the life cycle is completed in about three weeks only. Reproduction of the louse population occurs during the seal breeding season, but perhaps not when the seals haul out to moult.

The anatomy and physiology of the Southern Elephant Seal moult has been studied in detail in this region by Dr J. K. Ling (1965, 1966, 1968) and by Ling and Thomas (1967). These workers studied the natal moult, discovering that the hairs of the natal pelage were not connected to the underlying *stratum corneum* below the top layer of the skin, thus allowing them to be shed individually. Adult hair is rooted in the deep layer, and the moulting process of an adult is a much more drastic affair, leading Ling to discuss in some detail the thermoregulatory problems associated with hair loss on a massive scale in a subantarctic climate and the physiological mechanisms brought into play to compensate.

The feeding habits of the species have not been studied in detail in the New Zealand region, but fish, cephalopods and small crustaceans appear to be the major items of diet. The pups forage under rocks at the edge of the breeding beach where they were born as their natal moult progresses, and gradually begin to take small crustaceans. Sand and stones have been found in the stomachs of adults; the reason for this is not known. Some theories concerning this habit are discussed in the section on the New Zealand Fur Seal in the final chapter.

Only the Northern Elephant Seal's vocalizations have been studied in detail (Bartholomew and Collias 1962, Orr and Poulter 1965, Rice, Kenyon and Lluch 1965, and Sebeok 1965). In the period before the cows arrive on the breeding beaches the bulls make low-pitched, honking signals, generally repeated three times, with the end of each signal tapering off into resonating vibrations caused by the inflation of the thin-walled proboscis. Very often in this species the warning noises are sufficient to scare away bachelor bulls from a harem, even if the beachmaster responsible is in the sea at the edge of the beach at the time.

In general the cows and pups are much more vocal than the bulls. The cow is capable of a threatening screech as a warning sound; a more urgent warning call is a low-pitched, rapid bark. Under various circumstances the cows produce a very wide variety of sounds, some high-pitched, others low-pitched, some short and

Figure 135. (Logan Print Ltd.)
'Caesar', immature male Southern Elephant Seal at Mt. Maunganui Marineland, New Zealand.

abrupt, others long and drawn out into groans or squeals. There is also a long, blubbering whimper (which I have heard both on South Georgia and Campbell Island from cow Southern Elephant Seals) which seems particularly prevalent during the moult, and sounds exactly as if the animal was bewailing extreme discomfort. The mother seal also appears to have a specific, vibrating call which she uses for calling the pup; this is generally answered by a high-pitched, repeated yapping by the latter. In contrast to the great vocal activity recorded on the breeding beaches the Elephant Seal apparently has only a very limited range of underwater sounds (Poulter 1968).

The Mount Maunganui Marineland at Tauranga is the first seaquarium in the southern hemisphere, perhaps in the world, to hold Elephant Seals in captivity. Two young animals, each about seven weeks old, were brought to Tauranga by the Marineland Manager, Mr A. R. Piggales, from Campbell Island on the *Holmburn* in October 1967, during one of the regular charters used to take personnel and equipment to the weather station on the island. The two animals were thriving at the time of writing, and have become very tame, each eating from 20 to 25 pounds of fish per day.

16. The 'Eared' Seals
Otarioidea: Otariidae

In the latest revision of this group (Mitchell 1968a), two living sub-families are recognised within the Super-family Otarioidea. The Odobeninae (walruses) have a single living species *Odobenus rosmarus* (Linnaeus, 1758) characterised by: five distinct nails on the digits of the fore flippers, the external ears lacking pinnae, the tip of the tongue being rounded, the testes remaining internal and not descending into a scrotal sac, the tail being enclosed in a web of skin, and the canines being developed into long tusks, reaching their greatest size in the mature males. The Otariinae, the sea lions and fur seals, have 13 recognised living species, arranged in two sub-families. The Otariinae differ from the Odobeninae in having the nails of the fore flippers rudimentary, distinct pinnae on the external ears, the tip of the tongue medially notched, and the testes enclosed in a scrotal sac.

Apart from having pinnae on the external ears, the otarine seals can at once be distinguished on land from phocids by their method of locomotion. When a fur seal or sea lion walks on land the foreflippers are swung out at right angles to the body. The hind flippers are also drawn up and turned outwards, enabling the animal to walk using the flippers in pairs or alternately. In fast land movement the hind flippers are always used in unison because they are less mobile than the fore-limbs. An otarine can, when the situation demands, produce a surprising burst of speed, covering the ground with a galloping action. The head is swung back and forth on the long flexible neck at the same time, giving added impetus to the forward motion.

The otarine body is modified considerably for life in the aquatic environment. Most of the forelimb is actually within the body, the 'armpit' occurring between elbow and wrist. The hind flippers are free from the ankle only. In the water the fore flippers are used like oars, and the seal progresses with a slow, leisurely sculling motion; the fore flippers are also used for rapid acceleration. It is commonly held that unlike phocids, otarine seals use the fore flippers for high-speed swimming (Backhouse 1961, King 1964).

Without generalising about otariines, Mr R. M. Warneke (in letter) originally expressed some doubts about this to me some years ago. Since then I have studied some under-water film of *A. tasmanicus* made off the coast of Victoria, and watched *A. forsteri* on a number of occasions swimming at high speed near Kaikoura and it appears that southern fur seals do use the fore flippers for acceleration and turning. During swimming at constant high speed the main motive force appears to be supplied by the movement of the caudal part of the body, with the fore-flippers held back in a steep dihedral. New Zealand fur seals have frequently been observed 'porpoising' clear of the water at high speeds. The fore flippers were not seen to be used in this process.

The Sea Lions
Tribe Otariini

The two otariid tribes are the Otariini (sea lions) and the Arctocephalini (fur seals). The Otariini have a relatively blunt snout, a coarse outer coat of guard hair

Figure 136. (Dr Ian Stirling, Canadian Wildlife Service.)
Shark wound on adult male Australian Sea Lion, *Neophoca cinerea*.

with only a very little underfur, the first digit of the fore flipper longer than the second, and the inner digit of the hind flipper shorter than the outer ones. Five genera are presently recognised in this subfamily, each containing one species. The living sea lions are: *Eumetopias jubatus* (Schreber, 1776) Steller's Sea Lion, ranging from the latitude of Los Angeles right round the margin of the North Pacific to the eastern coast of Hokkaido (King 1964), also occurring in the Sea of Okhotsk and the Beaufort Sea; *Zalophus californianus* (Lesson, 1828) the Californian Sea Lion, found along the coast of California, Lower California and around the Galapagos Islands; *Otaria flavescens* (Shaw 1800) the Southern or South American Sea Lion, which ranges all round the coast of South America from the southern coast of Ecuador at about latitude 3°S in the Pacific to Santa Catarina on the southern coast of Brazil near latitude 27°S in the Atlantic, and is also found in large numbers on the Falkland Islands (King 1964); *Neophoca cinerea* (Péron, 1816) the Australian Sea Lion, found on the small groups of islands making up Houtman's Abrohos off the Indian Ocean coast of Western Australia at about latitude 33°S, and on the islands off the southern coasts of Western Australia and South Australia from the Recherche Archipelago a longitude

122°E to Kangaroo Island at longitude 137°E; and *Phocarctos hookeri* (Gray, 1844) Hooker's Sea Lion, which is more or less restricted to the Auckland Islands, although non-breeding animals are frequently reported from the Snares, Campbell Island and Macquarie Island, and each year a few stragglers are reported from the southern part of the South Island of New Zealand.

While the Otariini are not commercially as useful as the Arctocephalini with their dense layer of underfur, they are exploited. There have been some valuable scientific studies; Steller's Sea Lion populations have been examined by Kenyon and Scheffer (1955) on the American and Canadian Pacific coasts, and by Naumov (1933), Nikulin (1937), and Okada (1938) in the Bering Sea and in Siberian and Japanese waters. The 1955 estimate of the total size of the American population was 54,500. The size of the population in Asiatic waters is less well known, but may be more than 50,000.

The Californian Sea Lion, which is the species most commonly seen in captivity in both hemispheres, and which is currently held at the Marineland of New Zealand at Napier, is thought to contain at least two separate subspecies. The monotype *Z.c.californianus* has been studied by Bonnot (1951), Hubbs (1956) and Kenyon and Scheffer (1955), who have estimated the total population to be about 50,000 animals. The extreme northern limit appears to be about 49°N off the coast of British Columbia (Cowan and Guiget 1955). A small stock of an isolated subspecies given the name *Z.c.japonicus* (Peters, 1866), and numbering no more than a few hundred animals, occurs in Japanese waters (Scheffer 1958). Scheffer expressed doubts concerning the validity of this name. The other well known subspecies is *Z.c.wollebaeki* (Sivertsen, 1953), which numbers between 20,000 and 50,000 specimens (Scheffer 1958) and is confined to the Galapagos.

The Southern Sea Lion has been the subject of detailed studies by Hamilton (1934, 1939b), Matthews (1952) and Laws (1953a) in the Falkland Islands, and by Brandenburg (1938), Kellogg (1942), Osgood (1943), Vaz Ferreira (1950, 1956a) and Santiago Carrara (1952) on the South American mainland.

Relatively little is known about the Australian Sea Lion, compared to other species. The population was drastically reduced in the early sealing days, really before scientists had a chance to make detailed examinations. Limited information on the habits and range has been supplied by Iredale and Troughton (1934), Lewis (1942) and Sivertsen (1954) The New Zealand region representative, Hooker's Sea Lion, is now discussed in detail.

154

THE NEW ZEALAND,
OR HOOKER'S SEA LION
Phocarctos hookeri (Gray 1844)

DESCRIPTION: Adult males are dark brown with a well-developed blackish-brown mane reaching to the shoulders. They attain lengths of 8 to 9 ft, and weights of about 700 lbs. The females are lighter coloured than the males, varying from buff to creamy grey with darker pigmentation around the muzzle and flippers. Females reach a length of about 6 ft at maturity, and may weigh up to 300 lbs.

DISTRIBUTION: Almost the whole of the breeding population of this species is found on the coasts of the Auckland Islands, especially Enderby Island. A few specimens are seen each year at Campbell Island (Bailey and Sorensen 1962), and there is a recent record of the species at Macquarie Island (Csordas 1963a). Various New Zealand observers have reported sea lions on the small beaches at the Snares south of Stewart Island, and generally a few animals, presumably males, come ashore in winter on the coasts of Stewart Island and the southern part of the South Island, south of Dunedin (R. J. Street, in letter).

BIOLOGY: Sexually and physically mature males haul out on Auckland Island's beaches between late October and early November, establishing territories which they guard against the intrusion of other males (Bailey and Sorensen 1958). They do not feed during the period of active harem formation and maintenance, but live instead on the fat reserves accumulated over the previous months.

The first females begin to arrive about three weeks after the males, and are already in the final weeks of

Figure 137.
Phocarctos hookeri, Sea Lions.

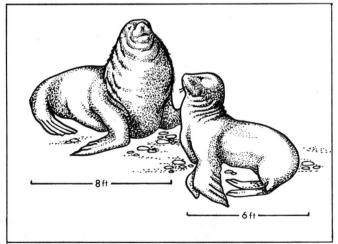

Figure 138. (S. E. Csordas.)
A Hooker's Sea Lion hauled out on Macquarie Island in 1955. Called 'Mr. Brown' by the survey party, this animal was recognisable by an odd-shaped scar. He was recorded on Macquarie during six consecutive seasons.

pregnancy. As they arrive they are gathered into harems by the beachmaster bulls, each bull controlling about a dozen females. Pups are born during November, December and January; they measure about 20 inches long at birth, and possess a thick natal coat of dark brown hair.

Mating takes place on land only a few days after the birth of the pups. The bulls then show a decrease in interest in the females, and allow them to make short trips to the sea. The females return at frequent intervals, usually every few hours, to feed the pups. Once mating is completed the harem bulls become noticeably less aggressive towards younger bachelor bulls, and the relatively tight structure of the harem during the early part of the short breeding season is harder to discern.

While the females are at sea the pups retreat beyond the high water mark into the tussock and scrub, making short trips down to play in pools in the littoral zone (Falla, pers.comm). The mother locates the pup on her return from the sea by uttering a call which appears to be individually distinctive to the pup, if not to human ears.

The period of lactation is apparently about 6 to 8 months, but the pups may stay with their mother for some months after weaning. The pup is introduced to the sea by the mother, and its initial days in the water are supervised. This behaviour is in marked contrast to the singular lack of attention shown by the mother fur seal.

155

By late February the harems have completely broken up and the cows and pups have taken to the sea. The bulls also return to the sea and begin to feed again after fasting for as long as four to five months. In April both mature and immature seals again return to shore, this time for the annual moult, which is completed in about three weeks to one month.

Hooker's Sea Lion feeds mainly on small fish, crustaceans, sea birds and penguins. It also swallows pebbles; the reason for this is not known, but the habit has likewise been observed in fur seals. Patches of basaltic pebbles on the Snares Islands (which are composed of granite) confused geologists until it was realised that they had been carried there and disgorged by sea lions (King 1964).

Hooker's Sea Lion is now a completely protected species. Funds have not been available for an accurate census, and population estimates ranging from 10,000 to 50,000 (Scheffer 1958) are largely guess work.

The Fur Seals
Tribe Arctocephalini

The Arctocephalini possess a relatively sharply pointed snout, a coarse outer coat of guard hairs overlying a dense inner fur layer, have the first digit of the fore flippers equal to or even shorter than the second, and the digits of the hind flippers about equal in length. Specialists recognise two living genera: *Callorhinus* with one species, and *Arctocephalus* with seven species. *Callorhinus ursinus* (Linnaeus, 1758) the Northern or Pribilof Fur Seal, ranges around the margin of the North Pacific Ocean from the eastern coast of Honshu Island in Japan to the latitude of San Diego (King 1964). However, the species is strongly migratory, and most breeding takes place on the Pribilof and Commander Islands in the southern Bering Sea. The genus *Arctocephalus* is largely southern hemisphere in distribution, although *A.philippii* (Peters, 1866) the Guadalupe Fur Seal is now known only from the island of Guadalupe off the coast of Lower California. A few stragglers have been recorded north of this point on the Santa Barbara Islands off Los Angeles (King 1964), and the subspecies *A.p.townsendi* (Merriam, 1897) now extinct, was once common on the Juan Fernandez group at about latitude 30°S. *A.australis* (Zimmerman, 1783) the South American Fur Seal, is found on the Galapagos Islands, and on the coast of the South American mainland it is found as two subspecies *A.a.galapagoensis* (Heller, 1904)

Figure 139. (S. E. Csordas.)
'Mr. Brown' in 1959. Note the thickening neck ruff as he reaches adulthood.

and *A.a.gracilis* (Nehring, 1887), the monotypic form *A.australis australis* (Zimmerman, 1783) being distributed around the Falklands. *A.pusillus* (Schreber, 1776) the South African Fur Seal, is found on the coasts of South West Africa and South Africa from Cape Cross at latitude 21°S on the Atlantic coast to Algoa bay near Port Elizabeth on the south coast of Cape Province (King 1964). *A.doriferus* (Wood-Jones, 1925) the Australian Fur Seal, occurs along the southern coast of Australia from the vicinity of Albany in Western Australia at about longitude 115°E to Kangaroo Island south of Adelaide near longitude 137°E. Recently King (1968) has synonymised *doriferus* with *forsteri,* but the evidence for doing this is not completely satisfactory. *A.tasmanicus* (Scott and Lord, 1926), the Tasmanian Fur Seal, ranges to the east of *doriferus,* being found from about the western border of the state of Victoria at approximately longitude 141°E, around Tasmania and throughout Bass Strait, and reaching a limital range a little north of Sydney on the coast of New South Wales near latitude 33°S. *A.forsteri* (Lesson, 1828) the New Zealand Fur Seal, has been recorded from all the subantarctic islands of the New Zealand region, and from

156

most rocky localities on the east and west coasts of Stewart Island, the South Island and the southern parts of the North Island. It is rare for the species to breed north of the Kaikoura Peninsula at about 42°S. The remaining species of this genus, *A.tropicalis* (Gray, 1872) usually called the Kerguelen Fur Seal, has a unique and complex distribution. Two subspecies are recognised; *A.t.tropicalis* is distributed to the north of the Antarctic Convergence, and is found on Tristan de Cunha, Gough Island, Marion Island, Prince Edward Island, the Crozet Islands, Amsterdam Island and St. Paul Island. The other subspecies *A.t.gazella* (Peters, 1875), is found south of the Antarctic Convergence on the South Shetland Islands, South Orkney Islands, South Georgia, South Sandwich Islands, Bouvet Island, Heard Island and Kerguelen Island.

Complete or partial protection of fur seals following the drastic depletion of the herds by uncontrolled sealing in the late eighteenth and early nineteenth centuries has resulted in the numerical recovery of a number of species. The pelts of all Arctocephalini are of great commercial value, and this has stimulated a number of scientific studies of both the northern and southern species.

Recently the populations of *Callorhinus ursinus* in the North Pacific and Bering Sea have been studied by Kenyon and Wilke (1953), Kenyon, Scheffer and Chapman (1954), Kenyon and Scheffer (1955), Taylor, Fujinaga and Wilke (1955) and Bartholomew (1959). Estimates by these authors indicate a total population size of a little under 2 million animals.

After being nearly exterminated by the depredations of sealers the Guadalupe Fur Seal population appears to be increasing again (King 1964), although there are still probably fewer than 500 animals in all (Scheffer 1958). This seal has been compared critically with related species by King (1954).

The South American Fur Seal population on the Falkland Islands numbers about 20,000; it has been examined by Laws (1953a). On the South American mainland the animal has been studied by Santiago Carrara (1952) and Vaz Ferreira (1956a, b); the population size is not known with accuracy but could be more than 100,000 (Scheffer 1958).

The South African Fur Seal, with a present population of about 400,000 individuals, is of considerable economic importance; at present it is the subject of an active research programme (Rand 1967). Notable contributions to the biology of this seal or useful reviews of the industry which it supports have been

Figure 140. (Dr Ian Stirling.)
Adult male fur seal *Arctocephalus forsteri* in normal sitting posture.

presented by Cross (1928), Black, Rapson, Schwartz and Van Rensburg (1945), Rand (1949, 1955a, 1956b, 1959) and Meschkat (1956).

At the time of writing little has been published on the biology of the Australian and Tasmanian Fur Seals, although the Victorian Fish and Wildlife Service has initiated an active programme on these species (Warneke 1966). Some brief facts on these animals have been summarised by Scott and Lord (1926) and Coleman (1951).

Population studies on the New Zealand Fur Seal have been greatly hampered by lack of funds. Some early information was given by Clarke (1873) and Chapman (1893), and the colonisation or recolonisation of Macquarie Island by this animal has been documented by Csordas (1958, 1963b) and Csordas and Ingham (1965). Street (1964) carried out a detailed study of feeding habits in relation to coastal commercial fisheries, while Falla (1953, 1962) discussed some of the problems of exploitation and conservation. Stirling (1968) described the movements of animals at the Kaikoura Peninsula colony. Sorensen (1969a, 1969b) has most usefully summarised the scattered, unpublished Marine Department and Dominion Museum reports on the distribution, population censuses and feeding habits of this species.

The total size of the very scattered populations of the

two subspecies of the Kerguelen Fur Seal probably does not exceed 12,000 animals. The two forms have been intensively studied with a view to future rational management when the stocks have recovered their numbers more fully. The population of St Paul and Amsterdam Islands has been considered by Clark (1875) and Paulian (1957a, 1964), that of Gough Island by Holdgate (1957), of Marion Island by Rand (1956a), of Heard Island by Gwynn (1953b), of Kerguelen Island by Peters (1875), of South Georgia by Bonner (1958b, 1963, 1964) and O'Gorman (1961, 1962). Csordas (1962a) provided a record from Macquarie Island, and King (1964) suggested that a few specimens had reached the South Island of New Zealand as stragglers. The best critical study to date of the northern and southern populations of this species has been made by King (1959a, b).

NEW ZEALAND FUR SEAL
Arctocephalus forsteri (Lesson 1828)

DESCRIPTION: New Zealand Fur Seal males attain lengths of 6 to 7 ft and weights of up to 450 lbs. Females are of slimmer build, and a weight of 208 lbs was recorded for an animal 6½ ft long. Weights of immature animals range between 100 and 300 lbs.

The colour merges from a dark grey-brown on the back to a lighter grey-brown beneath. Pups are light grey at birth and gradually become darker after their first

Figure 141.
Arctocephalus forsteri, New Zealand Fur Seal.

— 6ft —

moult. The underfur in this species is a rich chestnut colour, overlaid by coarser, darker guard hairs. These are tinged with white at the tips, so that a fur seal has a faint silvery sheen when dry. The body is well proportioned, and streamlined. The neck is long and mobile, and the ventral surface slightly flattened. The body tapers evenly back to the caudal region.

DISTRIBUTION: The New Zealand fur seal is found mainly around the coasts of the South Island and Stewart Island, although colonies also occur on the offshore islands and the Chatham Islands. In winter a few stragglers are found as far north as Hawke Bay (Mr F. Robson pers. comm.) and Cape Egmont (pers. obs.). It seems reasonable to assume that the geographical limits of distribution are largely defined by the waters of the Subtropical Convergence. There are semi-permanent colonies in the central part of Cook Strait, especially in the Marlborough Sounds and near Turakirae Head not far from the mouth of Wellington harbour; this marks about the most northerly point at which breeding is likely to occur. At the southern end of the geographical range temporary colonies have been reported on Campbell Island (Bailey and Sorensen 1958), which may number more than 2,000 seals (Dr R. A. Falla, in: Sorensen 1969b), and Macquarie Island (Csordas 1958, 1963b). At the latter locality 471 animals were counted in 1962. Interestingly enough, the species has only been recorded on Macquarie Island since the middle 1950s. It is possible that it was formerly present on this island, for a species of fur seal did occur there (McNab 1907), only to be exterminated by sealing parties in or about 1825. *A. forsteri* may be recolonising this island, or colonising it for the first time; there is no way of knowing. The species does not range much further south than this, and has not been recorded from Antarctic Convergence waters.

It has recolonised the Bounty Islands although the population is still probably less than 3,000 strong (Dr R. A. Falla in: Sorensen 1969b). It has also returned to the Antipodes Islands (Mr R. H. Taylor in: Sorensen 1969b), where it once occurred in large numbers (see chapter 7). Mr Taylor counted about 1,100 fur seals but saw no newly-born pups, and concluded that this was a visiting seasonal non-breeding population only. The Chatham Islands population is probably less than 2,000 (Dr R. A. Falla in: Sorensen 1969b), although Mr A. Wotherspoon noted that the tiny outlying Star Keys and the Forty-fours had breeding rookeries, the latter group being densely covered. Dr Falla also reported that a breeding population of about 3,000 occurred on the Snares, and about 1,000 more on Solander Island.

This species is generally confined to localised rocky areas on the coasts of the main islands. In addition to the small colonies at Turakirae Head and in the Marlborough Sounds there are small colonies at Makara on the west coast of the North Island not far from Wellington (B. R. Tunbridge in: Sorensen 1969a), at Island Bay just south of Wellington, on D'Urville Island (B. Tunbridge, in letter), at Kaikoura (Stonehouse 1965b, Stirling 1968), Akaroa, Shag Point, Cape Saunders, Nugget Point, Chaslands Mistake and Toetoes Bay (R. J. Street, in letter). There is a small colony at Cape Foulwind just west of Westport (pers. obs.), and larger aggregations occur on the Open Bay Islands in Jackson Bay, where with Mr M. W. Cawthorn, I counted 1,300 animals in the winter of 1964; off Cascade Creek (pers. obs.), Milford Sound, Caswell Sound (pers. obs.), and at several localities in Dusky Sound, Chalky Inlet and Preservation Inlet (Dr R. A. Falla in: Sorensen 1969b). Other local but relatively large colonies, each of several hundred animals, occur on Centre Island and similar places in Foveaux Strait and around Stewart Island, such as Ruapuke, Big South Cape Island, Bench Island, Poutama, Mogi and Codfish Island, and on the coast of Stewart Island itself, especially the west coast (Mr R. J. Street, in letter; and Sorensen 1969b).

Stirling (1968) studied the diurnal movements of animals at the Kaikoura colony. This was originally a temporary winter aggregation site, but some seals now remain there all year, although the colony size drops to about 60 animals in summer. In the winter months as many as 550 animals have been counted (Stonehouse 1965b). Stirling found that the daily pattern of movements was correlated closely with tidal rhythm. The number of fur seals was lowest in the morning and in mid-afternoon. The pattern of aggregation between 2 and 4 p.m., was most marked when this period coincided with low tide, but was still recognisable when high tide occurred then. Offshore rocks were preferred at low tide, and rocks right on the edge of the peninsula at high tide. Stirling suggested that a movement away from land at low tide was probably a precaution against land predators.

BIOLOGY: Very little is known of this animal's reproductive biology. The following short account is based on personal observations by Dr R. A. Falla, formerly Director of the Dominion Museum, Wellington, and is based on his observations before, during and after the 1946 open season (in: Sorensen 1969b), and published notes by Dr E. G. Turbott (1952). Dr R. A. Falla (in: Sorensen 1969b) has given the following average age-

Figure 142. (Dr Ian Stirling.)
Pup of fur seal *Arctocephalus forsteri;* one or two days old.

length relationships for the male of the species based on about 30 specimens collected in 1947 and 1948—1 week old: 16 inches; 1 year old: 36 inches; 2 years old: 42 inches; 3 years old: 50 inches; 4 years old: 56 inches. Age was determined by examination of skull changes.

Mature males establish territories in the Fiordland breeding colonies in late November-early December. Females arrive on the breeding grounds in an advanced state of pregnancy and are collected into harems of about 8 to 10 animals by the most aggressive bulls, who fight to defend their territories and cows. Pups are generally born in early December; they measure as much as 20 inches in length at birth and weigh about 10 lbs. The duration of lactation is not accurately known. Dr R. A. Falla (in: Sorensen 1969b) noted that females were still lactating in July, 6 to 7 months after the birth of the pups. He assumed that weaning occurred in or about September at an age of 8 to 9 months. The harem bulls mate with their females about one week after the birth of the pups, and shortly after this the harems break up, as the cows begin to make long feeding visits, sometimes of several days, back to the sea. Pups appear to enter the sea for the first time about one month after birth, but there is no precise information on this.

Between 1967 and 1969 Dr I. Stirling carried out detailed observations on the behaviour of this species at the non-breeding Kaikoura colony and the breeding colony on the Open Bay Islands, and also initiated a valuable tagging programme. He found that numbers on the Open Bay Islands did not vary greatly from one part of the year to another, but that there was considerable variation in population structure, with immature weaned animals being segregated from the breeding population. Ten-month old pups were abundant in September, but adult females were absent. Some signs of harem formation were detected at this time. Stirling pointed out that immature animals disappear at the non-breeding season at the same time as they disappear from the breeding colony; at this time it is not known where they go, though they may scatter throughout the length of available coastline to feed.

Breeding activity was at a peak on the Open Bay Islands between November and January. By the middle of February no adult males could be seen there, and relatively few adult females. At this time the number of sub-mature animals was noted to be increasing rapidly.

Stirling (1970) has also classified fully the behavioural postures seen in this species. 'Full neck display' is a high intensity threat posture in which the

Figure 144. (Dr Ian Stirling.)
Adult male fur seal *Arctocephalus forsteri* in full neck display.

animal puffs out the neck and chest to increase the apparent size of the body. This posture was recorded both in harem bulls and non-breeding animals. If a large seal postured at a smaller, the latter invariably backed down from any direct encounter. Only if the animals were very much of a size did the next posture, 'neck waving' take place. A distinctive 'submissive posture' was recorded by animals which did not wish to challenge or fight. This was a lowering of the head performed after full neck display or by an animal which was trying to pass through a group of seals. 'Open mouth display' with canine teeth showing, was used as a threatening gesture by a female with a pup to another female approaching too closely. 'Facing away', a motion of the head of a smaller animal away from a larger one, appeared to be a submissive gesture in males of breeding age.

In general, conflicts between fully mature adult males were short and sharp, but rarely with serious wounds resulting. At the end of these fights a clear dominance by one animal was observed. This was not the case when fighting occurred between sub-mature males; these

Figure 143. (Dr Ian Stirling.)
Adult male fur seal *Arctocephalus forsteri* giving full threat call from horizontal neck threat posture.

160

animals would fight for up to an hour at a time without any clear victory at the end. Stirling pointed out the advantages to the species of the development of these rituals; practice and establishment of a clear pattern of postures by maturity results in the elimination of unnecessary battles which would otherwise waste the energy of males controlling harems. There appears to be a clear correlation between large neck size and reproductive dominance among males of this species.

During the breeding season on Open Bay Islands the males were observed to keep alert, and only a few slept in the daytime. Once the peak of mating activity was over, on the other hand, the same males spent much of the day sleeping or grooming their pelage.

When a female returned from the sea a process of mother-pup recognition was observed by Stirling similar to that recorded for other species, especially the Northern Fur Seal and Californian Sea Lion by Bartholomew (1959), Peterson (1968), and Peterson and Bartholomew (1967). The mother began to give out a characteristic high-pitched call with the head and neck arched. The pup generally moved towards the mother, not vice versa. Final acceptance or rejection appeared to be based on the female smelling the pup. On a number of occasions hungry pups would approach females other than their mothers.

Seal pups suffer high mortality. Studies on the Northern

Figure 145. (Dr Ian Stirling.)
Two adult male fur seals *Arctocephalus forsteri* giving mutual full neck display and vocalizing.

Fur Seal *Callorhinus ursinus* on the Pribilof Islands showed that in any one year 85 per cent of the pups may die before attaining an age of three years. In this species hookworm and *Salmonella* bacillic dysentry are believed to be the major causes of death, although some pups are trampled by bulls in the congestion of the breeding colonies, and may even be squashed by their own mothers. Starvation in the post-weaning period is almost certainly another contributing factor to mortality. We have little or no information on causes of pup mortality in *A.forsteri*.

In New Zealand the fur seal has frequently been accused of depredations among commercially important fish. In fact a detailed investigation carried out by Mr R. J. Street for the New Zealand Marine Department showed that there was little or no justification for this view (Street 1964; Sorensen 1969a, 1969b).

Stomach contents of seals culled from colonies at the Kaikoura Peninsula, Banks Peninsula, Cape Saunders, Nugget Point and Bench Island in Foveaux Strait consisted mainly of octopus, squid and barracouta. These three items made up 90.8 per cent of the stomach contents by weight. Since the octopus is a major predator of crayfish, the presence of fur seals may even benefit the crayfishing industry. While curious sub-mature fur seals may take undersized, or moribund female crayfish thrown overboard by fishing boats and play with them, bringing them to the surface and throwing them in the air, there is no evidence at all that they take crayfish as a staple item of diet.

Fur seals feed mainly at night. There are few records of them feeding on the surface, although fishermen at Kaikoura have reported seeing seals breaking up ling *(Gerypterus blacodes)* and blue cod *(Parapercis colias)* at the surface before eating them. Street noted that 9 to 10 lbs was ingested at each meal, although one animal had four octopus in its stomach, with a total weight of 15 lbs. At the Cape Foulwind colony I noted not only octopus and squid mandibles, but also some bird feathers in the seal droppings on the rocks. While the fur seal may take sea birds from the surface, this does not, however, seem to be a common practice. Fur seals appear to be able to catch food at considerable depths; Street recorded one example of a seal snared in a trawl, apparently at a depth of 360 ft.

My former assistant, Mr M. W. Cawthorn, reported a young male seal at the Turakirae Head winter colony as taking and swallowing stones from the water's edge. The stones ingested were smooth pebbles about 1-1½ in. in diameter. The seal was also seen to take pebbles from

Figure 146. (Dr Ian Stirling.)
Adult female fur seal *Arctocephalus forsteri* (right) giving open mouth threat to adult male.

the bottom of a pool about three feet deep. It hauled out on a flat rock and dropped some pebbles there, playing with them for several minutes before finally swallowing them.

The reason for this stone-swallowing behaviour is not known. Possibly the pebbles may serve to adjust trim, or to aid in breaking up food in the stomach, or perhaps even to allay hunger pangs during fasting.

At the present time there is no reliable estimate of the numbers of the New Zealand Fur Seal. Scheffer (1958) gives an estimate of 5,000 to perhaps 20,000 animals, but no complete census work has been carried out. Combination of local censuses carried out by R. J. Street between 1962 and 1964 and Mr M. W. Cawthorn and myself in 1963 and 1964 suggests that the correct number may be about 15,000 to 20,000, and this is in broad agreement with the figures published recently by Dr R. A. Falla (in: Sorensen 1969b).

In summary, the best present evidence gathered by Falla, Sorensen and Taylor indicates that the populations of the New Zealand fur seal are distributed as follows: New Zealand mainland and Stewart Island, 8,000; Snares, 3,000; Solander Island, 1,000; Campbell Island, 2,000; Auckland Island, nil; Antipodes, 1,100; Bounty Islands, 3,000; Chatham Islands, 2,000; Macquarie Island 500; grand total about 20,600 animals. The influence of the crayfishing industry in disturbing the once large breeding colonies on the coast of Fiordland should be considered as a factor in any population estimate. While there have been some suggestions that the *A.forsteri* population is increasing (Stirling 1968), I would advocate caution in accepting this idea. We may well be witnessing only a redistribution of a relatively stable population both southwards (to Macquarie Island) and northwards (to the southern coasts of the North Island), possibly as a result of the influence of man on the coastal environment of the South Island and Stewart Island since the end of the Second World War.

162

Appendices

A. Tables

Table 1. Catches of Whales in the New Zealand Antarctic Sector, from latitude 40°S to the shelf ice.

Area V (130°E to 170°W)											
Season	Blue	Fin	Humpback	Sei	Sperm	Season	Blue	Fin	Humpback	Sei	Sperm
1931-40 (totals)	1,326	1,993	28	*	*	1955/56	15	1,494	194	*	*
						1956/57	0	23	0	*	*
1940-47 (totals)	693	478	0	*	*	1957/58	76	1,130	23	*	*
						1958/59	176	2,625	885	*	*
1947/48	1,467	1,591	0	*	*	1959/60	391	4,730	931	1,649	*
1948/49	1,013	2,489	0	*	*	1960/61	90	3,174	293	563	*
1949/50	1,494	1,664	903	*	*	1961/62	74	1,098	0	409	362
1950/51	2,211	2,198	162	*	*	1962/63	115	645	0	430	732
1951/52	1,462	2,768	146	*	*	1963/64	0	1,144	0	1,820	2,434
1952/53	512	1,709	505	*	*	1964/65	0	747	0	2,207	1,503
1953/54	83	654	1,097	*	*	1965/66	0	385	0	1,008	2,209
1954/55	561	3,021	171	*	*	1966/67	0	304	0	717	2,091

Table 2. Catches of Humpback Whales from the Antarctic Area Group IV Stock from opening of modern Australian shore whaling to the collapse of the industry by over-exploitation.

Catches from Group IV (70°E to 130°E)					
Date	Pt. Cloats (22.35S 113.40E)	Carnavon (24.53S 113.38E)	Albany (35.05S 117.56E)	Antarctic pelagic	Total
1949	190	—	—	0	190
1950	348	40	—	779	1,167
1951	574	650	—	1,112	2,336
1952	536	600	51	1,127	2,314
1953	603	600	100	193	1,496
1954	600	600	120	258	1,578
1955	500	500	126	28	1,154
1956	Trans. to Carnavon	1,000	119	832	1,951
1957	—	1,108	102	0	1,120
1958	—	885	82	0	967
1959	—	541	159	1,413	2,113
1960	—	440	105	66	611
1961	—	475	105	4	584
1962	—	503	40	56	599
Total Australia w. coast 12,312				5,868	18,180

Table 3. Catches of Whales at New Zealand Shore Stations

Year	Tory Channel Humpback	Tory Channel Others	Kaikoura Humpback	Kaikoura Others	Whangamumu (to 1931) Great Barrier (to 1962) Humpback	Others
1912	?	?	?	?	27	1 sei
1913	?	?	?	?	56	0
1914	?	?	?	?	57	0
1915	36		?		70	0
1916	57	6 right	?	1 right	25	0
1917	37		6		52	0
1918	40	0	9	0	41	0
1919	47	0	11	0	61	0
1920	43	0	20	0	44	0
1921	34		16		39	0
1922	17	1 right	6	1 right	35	2 right
1923	17	0	Operations ceased		62	0
1924	52	0			55	0
1925	48	0			48	0
1926	43	2 right			35	0
1927	53	0			74	0
1928	55	0			50	0
1929	49	0			53	0
1930	47	1 blue			31	0
1931	61	0			48	0
1932	18	0			Operations ceased	
1933	41	0				
1934	52	0				
1935	57	0				
1936	69	0				
1937	55	1 blue				
1938	75	1 blue 1 sperm				
1939	80	1 blue				
1940	107	2 sperm				
1941	86	0				
1942	71	0				
1943	90	0				
1944	88	0				
1945	107	0				
1946	110	0				
1947	101	1 blue 9 sperm				
1948	88	0				
1949	139	0				
1950	79	0				
1951	111	0				
1952	122	1 sei				
1953	109	0				
1954	180	0				
1955	112	0			Great Barrier Island	
1956	81	1 sei			62	13 Bryde's
1957	155	2 sperm			29	2 Bryde's
1958	183	0			—	—
1959	214	0			104	2 Bryde's
1960	226	0			135	0
1961	55	0			26	0
1962	27	3 sperm			8	0
1963	9	119 sperm			Operations ceased	
1964	0	129 sperm 1 fin 4 sei				
	Operations ceased					

Table 4. Catches of sperm whales in the Pacific Ocean north of latitude 40°S and south of 0° by pelagic factory ships.

Year	Catch	Year	Catch
1957	48	1963	1,219
1958	0	1964	1,779
1959	0	1965	400
1960	0	1966	160
1961	0	1967	36
1962	164		

Table 5. Catches from Group V Humpback Stock.

	Catches from Group V (130° to 170°W)					
Date	Tangalooma (27.11S 153.23E)	Byron Bay (28.37S 153.38E)	Norfolk Ind. (29.01S 167.58E)	N.Z. total	Antarctic pelagic	Total
1949	—	—	—	141	0	141
1950	—	—	—	79	903	982
1951	—	—	—	111	162	273
1952	600	—	—	122	146	868
1953	700	—	—	109	504	1.313
1954	598	120	—	180	0	898
1955	600	120	—	112	1.097	1,929
1956	600	120	150	143	194	1,207
1957	600	121	120	184	0	1,025
1958	600	120	120	183	0	1,023
1959	660	150	150	318	885	2,163
1960	660	150	170	361	931	2,272
1961	951	140	170	80	293	1,274
1962	68	105	4	32	0	209
Total Australian 8.307				2,155	5.115	15,577

B. Bibliography

Addison, W. H. F. 1915 On the rhinencephalon of the dolphin *Delphinus delphis* L. Anat Rec. Philadelphia, 9:45-6.

Agassiz, L. 1868 *Mesoplodon* from Nantucket, Massachusetts. Proc. Boston Soc. nat. Hist., 11:318.

Aguayo, C. G. 1954 Notas sobre cetaceos de aguas Cubanas. Circl. Mus. Bibl. zool. Habana, 13:1125-6.

Aguayo, L. A. 1963 Observaciones sobre la madurez sexual del cachalote macho (*Physeter catodon* L.) capturado en aguas Chilenas. Montemar, 11:99-125.

Akimushkin, I. I. 1954 (Basic food of *Hyperoodon rostratus*). Dokl. Akad. Nauk SSSR, 95:419-20.

Akimushkin, I. I. 1955 (The feeding of the cachalot). Dokl. Akad. Nauk. SSSR, 101:1139-40.

Albahary, C. and P. Budker 1958 Dangers des animaux aquatiques Encyclopédie médico-chirurgicale, 16078 C.I.O. Paris, pp. 1-9.

Albert, Prince of Monaco 1898 Some results of my researches on oceanography. Nature, 58:200-4.

Alderson, A. M. E. Diamantopoulos, & C. B. B. Downman 1960 Auditory cortex of the seal (*Phoca vitulina*). J. Anat. London. 94:506-11.

Allen, G. M. 1906 Sowerby's whale on the American coast. Amer. Natl. 40:357-70.

Allen, G. M. 1916 The whalebone whales of New England. Mem. Boston Soc. nat. Hist., 8:107-322.

Allen, G. M. 1939a A checklist of African mammals. Bull. Mus. Comp. zool. Harvard, 83:1-763.

Allen, G. M. 1939b True's beaked whale in Nova Scotia. J. Mammal., 20:259-60.

Allen, G. M. 1941 Pygmy sperm whale in the Atlantic. Publs Field Mus. nat. Hist., zool. ser., 27:17-36.

Allen, J. A. 1869 Catalog of the mammals of Massachusetts, with a critical revision of the species. Bull. Mus. Comp. zool. Harvard, 1:143-253.

Allen, J. A. 1905 The Mammalia of Southern Patagonia. Repts. Princeton Univ. Exped. Patagonia, 1896-99, vol. 3 (zool.), 1:1-210.

Allen, K. Radway 1966 Some methods for estimating exploited populations. J. Fish. Res. Bd. Canada, 23:1,553-74.

Alpers, A. 1963 A Book of Dolphins. J. Murray, London.

Anderson, H. T. (ed.) 1969 The Biology of Marine Mammals. Academic Press, New York, 511 pp.

Anderson, J. 1878 Zoological results of the two expeditions to western Yunnan in 1868 and 1875, London, I.417-551.

Andrews, R. C. 1908a Notes upon the external and internal anatomy of *Balaena glacialis* Bonn. Bull. Am. Mus nat. Hist., 24:171-82.

Andrews, R. C. 1908b Description of a new species of *Mesoplodon* from Canterbury Province, New Zealand. Bull. Am. Mus. nat. Hist., 24:203-15.

Andrews, R. C. 1909 Observations on the habits of the finback and humpback whales of the eastern North Pacific. Bull. Am. Mus. nat. Hist., 26:213-26.

Andrews, R. C. 1911 Shore Whaling: A world industry. Nat. Geogr. Mag. 22:5.

Andrews, R. C. 1914 Note of a rare ziphoid whale, *Mesoplodon densirostris*, on the New Jersey coast. Proc. Acad. nat. Sci. Philadelphia, 1914:437-40.

Andrews, R. C. 1916 Monographs of the Pacific Cetacea. II. The sei whale (*Balaenoptera borealis* Lesson.). Mem. Am. Mus. nat. Hist. (n. ser.), 1:289-388.

Angot, M. 1954 Observations sur les Mammifères marins de l'archipel de Kerguelen avec une ètude détaillée de l'elephant de mer, *Mirounga leonina* (L). Mammalia, 18:1-111.

Anon 1946 (*Lobodon carcinophagus* in Tasmania). Note in Pap. Proc. R. Soc. Tasmania, 1945:165.

Anthony, R. 1922 Recherches anatomiques sur l'appareil génito-urinaire mâle du *Mesoplodon*. Mem. Inst. Espanol. Oceanog., 3:35-115.

Anthony, R. and F. Coupin 1930 Recherches anatomiques sur le vestibule de l'appareil respiratoire (poche gutterale-hyoide larynx) du *Mesoplodon*. Mem. Inst. Espanol, Oceanog., 14:1-40.

Archey, G. 1926 The blue whale (*Balaenoptera musculus*). Two new records from New Zealand. N.Z. Jl Sci. Technol., 8:189.

Aretas, R. 1951 L'elephant de mer (*Mirounga leonina* L.) étude biologique de l'espèce dans les possessions françaises australes (archipel des Kerguelen). Mammalia, 15:105-17.

Aretas, R. 1958 Etude d'une société animale dans les terres australes, l'elephant de mer *Mirounga leonina* (L) C. R. 3rd Congr. Ass. Sci. Pays Océan Indien. (Sect. B. 1958: 39-44.

Arseniev, V. A. 1959 (Observations on the seals of the Antarctic). Bull. Soc. nat. Moscow, Biol., 62:39-44.

Arseniev, V. A. 1961 Lesser rorqual of the Antarctic, *Balaenoptera acutorostrata* Lac. Rep. Conf. Sea Mammals, 1959. Ichthyol. Commiss. USSR Acad. Sci., 12:125-32.

Backhouse, K. M. 1961 Locomotion of seals with particular reference to the forelimb. Symp. zool. Soc. Lond., 5:59-75.

Backus, R. H. 1961 Stranded killer whale in the Bahamas. J. Mammal., 42:418-9.

Backus, R. H. and W. E. Schevill 1966 *Physeter* Clicks. pp. 510-28, In: Whales, Dolphins and Porpoises, K. S. Norris (ed.) University of California Press, Berkeley, Los Angeles.

Bailey, A. M. and J. H. Sorensen 1962 Mammals of Campbell Island, pp. 48-83, In: Subantarctic Campbell Island. Proc. Denver Mus. nat. Hist., 10:1-305.

Baker, R. C., F. Wilke and C. H. Baltzo 1963 The Northern Fur Seal. Fish. Circ. U.S. Bur. Fish. 169:1-19.

Bannister, J. L. 1964 Australian whaling 1963: catch results and research. CSIRO Div. Fish. Oceanogr. Rep. 38:1-13.

Bannister, J. L. 1968 An aerial survey for sperm whales off the coast of Western Australia 1963-1965. Aust. Jl mar. Freshwat. Res. 19:31-51.

Bannister, J. L . 1969 The biology and status of the sperm whale off Western Australia — an extended summary of results of recent work. Intn. Whaling Comm. Rep., 19:70-76.

Bannister, J. L. & R. Gambell 1965 The succession and abundance of fin, sei and other whales off Durban. Norsk. Hvalfangsttid., 54:45-60.

Bannister, J. L. & J. R. Grindley 1966 Notes on *Balaenophilus unisetus* P.O.C. Aurivillius, 1879, and its occurrence in the southern hemisphere (Copepoda, Harpacticoidea). Crustaceana, 10:296-302.

Barnard, K. H. 1954 A Guide Book to South African Whales and Dolphins. South African Museum, Cape Town, 33 pp.

Barnett, C. H. R. J. Harrison & J. W. D. Tomlinson 1958 Variations in the venous system of mammals. Biol. Rev. 33:442-87.

Bartholomew, G. A. 1959 Mother-young relations and the maturation of pup behaviour in the Alaska Fur Seal. Anim. Behav., 7:163-71.

Bartholomew, G. A. & N. E. Collias 1962 The role of vocalization in the social behaviour of the Northern Elephant Seal. Anim. Behav., 10:7-14.

Bartholomew, G. A. & F. Wilke 1956 Body temperature in the Northern Fur Seal *Callorhinus ursinus*. J. Mammal., 37:327-37.

Bassett, E. G. 1961 Observations on the retractor clitoridis and retractor penis muscles of mammals. J. Anat., 95:61-77.

Bateson, G. 1966 Problems in cetacean and other mammalian communication. pp. 569-79. In: Whales, Porpoises and Dolphins, K. S. Norris (ed.) University of California Press, Berkeley, Los Angeles.

Baudrimont, A. 1956 Structure des veines pulmonaires et circulation fonctionelle du poumon du dauphin commun. Bull. Microsc. Appl. 2:57-78.

Beale, T. 1839 The Natural History of the Sperm Whale. John van Voorst, London. 393 pp. (2nd edit.).

Beauregard, H. 1894 Recherches sur l'appareil auditif chez les Mammifères. J. Anat. (Paris), 30: 367-413.

Beauregard, H. and R. Boulart 1882 Recherches sur le larynx et la trachée des Balénides. J. Anat. (Paris), 18:611.

Beddard, F. E. '1900 A Book of Whales. John Murray, London. 320 pp.

Beklemishev, C. W. 1960 Southern atmospheric cyclones and the whale feeding grounds in the Antarctic. Nature 187:530-1.

Belkovich, U. M. 1961 On the question of physical thermal regulation in Beluga (*Delphinapterus leucas*). Trudy Sovesh Ikhtiol. Komiss., Akad. Nauk SSSR, 12:50-9.

Benham, W. B. 1901 On the anatomy of *Cogia breviceps*. Proc. zool. Soc. Lond. 1901:107-34.

Benham, W. B. 1902a Notes on the osteology of the short-nosed sperm whale. Proc. zool. Soc. Lond., 1902:54-62.

Benham, W. B. 1902b An account of the external anatomy of a baby rorqual (*Balaenoptera rostrata*). Trans. Proc. N.Z. Inst. 34:151-5.

Benham, W. B. 1902c Notes on *Cogia breviceps*, the lesser sperm whale. Trans. Proc. N.Z. Inst. 34: 155-68.

Benham, W. B. 1935 The teeth of an extinct whale *Microcetus hectori* n.sp. Trans. Proc. Roy. Soc. N.Z. 65:239-43.

Benham, W. B. 1937a Fossil Cetacea of New Zealand ii. On *Lophocephalus* a new genus of zeuglodont whale. Proc. Roy. Soc. N.Z., 67:1-7.

Benham, W. B. 1937b Fossil Cetacea of New Zealand, ii. The skull and other parts of the skeleton of *Prosqualodon hamiltoni* n.sp. Proc. Roy. Soc. N.Z., 67:9-14.

Benham, W. B. | 1937c | Fossil Cetacea of New Zealand. iv. Notes on some of the bones of *Kekenodon onamata* Hector. Proc. Roy. Soc. N.Z., 67:15-20.

Benham, W. B. | 1939 | *Mauicetus:* a fossil whale. Nature, 143:765.

Bennett, A. G. | 1931 | Whaling in the Antarctic. Blackwood, London, i-x, 221 pp.

Benz, F. W. Schuler and A. Wettstein | 1951 | Adrenocortico-tropic hormone in the pituitary gland of the whale. Nature, 167:691-92.

Berg, C. | 1898 | *Lobodon carcinophagus* (Ho'b'rn and Jacq.) en el Rio de la Plata. Comun. Mus. Nac. Buenos Aires, Tomo 1:15.

Berlin, H. | 1941 | Nagra sydsvenska valfynd fran senare tid. Fauna och Flora, Uppsala 1941:241-51.

Berlin, H. | 1951 | Ett nytt fynd av Sowerby's nabbwal. *Mesoplodon bidens* (Sowerby) vid skanska kusten. Fauna och Flora, Uppsala 1951:1-10.

Berry, J. A. | 1961 | The occurrence of a leopard seal (*Hydrurga leptonyx*) in the tropics. Ann. Mag. nat. Hist. 13th ser, 1960 (1961):591.

Bertram, G. C. L. | 1940 | The biology of the Weddell and Crabeater seals, with a study of the comparative behaviour of the Pinnipedia. Brit. Mus. (Nat. Hist.) Sci. Repts. Brit. Graham Land Exped., 1934-1937, 1:1-139.

Best, P. B. | 1960 | Further information on Bryde's whale (*Balaenoptera edeni* Anderson) from Saldanha Bay, South Africa. Norsk Hvalfangsttid., 49:201-15.

Best, P. B. | 1967 | The sperm whale (*Physeter catodon*) off the west coast of South Africa. 1. Ovarian changes and their significance. Investl Rep. Div. Sea Fish. S. Afr., 61:1-27.

Best, P. B. | 1968 | The sperm whale (*Physeter catodon*) off the west coast of South Africa. 2. Reproduction in the female. Investl Rep. Div. Sea Fish. S. Afr., 66:1-32.

Best, P. B. | 1969 | The sperm whale (*Physeter catodon*) off the west coast of South Africa. 3. Reproduction in the male. Investl Rep. Div. Sea Fish. S. Afr., 72:1-20.

Best, P. B. and J. L. Bannister | 1963 | Functional polyovuly in the sei whale *Balaenoptera borealis* Lesson. Nature, 199:89.

Betesheva, E. I. & I. I. Akimushkin | 1955 | (Food of the sperm whale (*P.catodon*) in the Kurile Islands). Trudy Inst. Okeanol. 18:86-94.

Betesheva, E. I. | 1962 | (Food of whales of economic importance in the Kurile region). Trudy Inst. Morf. Zhiv. 34:7-32.

Bettum, F. | 1958 | The development of whaling and the present position. Norsk Hvalfangsttid., 47:485-502.

Beu, A. G. and P. A. Maxwell | 1968 | Molluscan evidence for Tertiary Sea temperatures in New Zealand: A reconsideration. Tuatara; 16:68-74.

Beverton, R. J. H. and S. J. Holt | 1957 | On the dynamics of exploited fish populations. Fishery Invest., London, ser. 2, 19:1-533.

Bierman, W. H. & E. J. Slijper | 1947 | Remarks upon the species of the genus *Lagenorhynchus*, I. Proc. K. ned. Akad. Wet., 50:1353-64.

Bierman, W. H. & E. J. Slijper | 1948 | Remarks upon the species of the genus *Lagenorhynchus*, II. Proc. K. ned. Akad. Wet., 51:127-33.

Black, M. M. W. S. Rapson, H. M. Schwartz & N. J. van Rensburg | 1945 | South African fish products Part 19. The South African Seal Fishery. J. Soc. chem. Ind., London, 64:326-31.

Boenninghaus, G. | 1903 | Das Ohr der Zahnwhales, zugleich ein Beitrag zur Theorie der Schalleitung. Zool. Jb. (Anat.), 19:180-360.

Boerema, L. K., D. G. Chapman, T. Doi, R. Gambell & J. A. Gulland | 1969 | Report of the IWC-FAO working group on sperm whale stock assessment. Intn. Whaling Comm. Rep., 19:39-69.

Bonin, W. & L. F. Bélanger | 1939 | Sur la structure ou poumon de *Delphinapterus leucas*. Trans. Roy. Soc. Canada, 33 (ser. 3):19-22.

Bonner, W. N. | 1955 | Reproductive organs of foetal and juvenile elephant seals. Nature, 176:982-3.

Bonner, W. N. | 1958a | Exploitation and conservation of seals in South Georgia. Oryx, 4:373-80.

Bonner, W. N. | 1958b | Notes on the Southern Fur Seal in South Georgia. Proc. zool. Soc. Lond., 130:241-52.

Bonner, W. N. | 1963 | Cropping the Fur Seals. New Scient. 20:557.

Bonner, W. N. | 1964 | Population increase in the fur seal *A. tropicalis gazella* at South Georgia. pp. 433-43 In: Antarctic Biology; Proceedings of the first SCAR Symposium in Antarctic Biology, Hermann, Paris.

Bonnot, P. | 1951 | The sea lions, seals and sea otter of the California coast. Calif. Fish. Game, 37:371-89.

Boschma, H. | 1950 | Maxillary teeth in specimens of *Hyperoodon rostratus* (Muller) and *M.grayi* von Haast stranded on the Dutch coasts. Proc. K. ned. Akad. Wet., 53:775-86.

Boschma, H. | 1951 | Rows of small teeth in Ziphoid whales. Zool. Meded., Leiden, 31:139-48.

Bouvier, E. L. | 1892 | Observations anatomiques sur l'*Hyperoodon rostratus*. Ann. Sci. nat. Zool., 13:259.

Boyden, A. & D. Gemeroy | 1950 | The relative position of the Cetacea among the orders of mammalia as indicated by precipitin tests. Zoologica, 35:145-51.

Braekkan, O. R. | 1948 | Vitamins in whale liver. Hvalrad. Skr. 32:1-25.

Brandenburg, F. G. | 1938 | Notes on the Patagonian sea lion. J.Mammal., 19:44-7.

Braun, M. | 1905 | Anatomisches und biologisches ueber den Tuemmler. Schr. phys. Oekon. Ges. Koenigsb., 46:136-41.

Brazenor, C. W. | 1950 | The Mammals of Victoria. Nat. Mus. Victoria, Handbook 1, Brown, Prior and Anderson, Melbourne, 125 pp.

Breathnach, A. S. | 1955 | The surface features of the brain of the humpback whale (*Megaptera novaeangliae*). J. Anat. London, 89:343-54.

Breathnach, A. S. | 1960 | The cetacean nervous system. Biol. Rev., 35:187-230.

Breathnach, A. S. & F. Goldby | 1954 | The amygdaloid nuclei, hippocampus and other parts of the rhinencephalon in the porpoise (*Phocaena phocoena*). J. Anat. 88:267-91.

Breschet, G. | 1836 | Histoire anatomique et physiologique d'un organe de nature vasculaire découvert dans les Cétacés. Bechet Jeune, Paris.

Brimley, H. H. | 1943 | A second specimen of True's beaked whale, *Mesoplodon mirus*, True, from North Carolina. J. Mammal., 24:199-203.

Brodie, J. W. | 1960 | Coastal surface currents around New Zealand, N.Z. Jl Geol. Geophys., 3:235-52.

Brodie, J. W. | 1964 | Bathymetry of the New Zealand Region. N.Z. Dept. sci. industr. Res. Bull. 161:1-55.

Brooks, S. A. | 1936 | Whaling in the Antarctic. The Seagoer 3:35-68.

Brown, D. H. | 1960 | Behavior of a captive Pacific pilot whale. J. Mammal., 41:342-49.

Brown, D. H. 1962 Further observations on the pilot whale in captivity. Zoologica, 47:59-64.

Brown, D. H. D. K. Caldwell, & M. C. Caldwell 1966 Observations on the behaviour of wild and captive false killer whales, with notes on associated behaviour of other genera of captive delphinids. Contr. sci. Los Angeles Mus. 95:1-32.

Brown, D. H. R. W. McIntyre, C. A. Delli-Quadri, & R. J. Schroeder 1960 Health problems of captive dolphins and seals. 4 pp separate. 1960.

Brown, D. H. & K. S. Norris 1956 Observations of captive and wild cetaceans. J. Mammal., 37:311-26.

Brown, K. G. 1952 Observations on the newly born Leopard Seal. Nature 170:982-3. ,

Brown, K. G. 1957 The Leopard Seal at Heard Island, 1951-54. Interim Rep. Aust. natl. Antarct. Res. Exped , No. 16:1-34.

Brown, S. G. 1963 A review of Antarctic whaling. Polar Rec. 11:555-66.

Brown, S. G. 1967 Report on the 4th Antarctic cruise of Umitaka Maru (Tokyo University of Fisheries) 1966/67. pp. 1-20. In: Report on Whale Observations in the Antarctic 1966/67, S. G. Brown & D. E. Gaskin, FAO Fish. Circl. 111:1-48.

Brownell, R. L. jr 1964 Observations of odontocetes in Central Californian waters. Norsk Hvalfangsttid., 53:60-66.

Brownell, R. L. jr 1965 A record of the dusky dolphin, Lagenorhynchus obscurus, from New Zealand. Norsk Hvalfangsttid., 54:169-71.

Bryant, J. 1927 Antarctic whales in peril of extermination. Illust. London News 1927:419.

Bryden, M. M. 1964 Insulating capacity of the subcutaneous fat of the Southern Elephant Seal. Nature, 203:1299-1300.

Bryden, M. M. 1968 Development and growth of the Southern Elephant Seal Mirounga leonina (Linn). Pap. Proc. R. Soc. Tasmania, 102:25-30.

Buchanan, T. 1828 Physiological illustrations of the organ of hearing. Consitt and Goodwill, London.

Bullen, F. T. 1928 (reprinted) The Cruise of the Cachalot Round the World after Sperm Whales. Dodd, Mead and Co. New York. 301 pp.

Bullis, H. R. and J. C. Moore 1956 Two occurrences of false killer whales, and a summary of American records. Amer. Mus. Novitates., 1756:1-5.

Burling, R. W. 1961 Hydrology of Circumpolar waters south of New Zealand. N.Z. Dept. sci. industr. Res. Bull. 143:1-66.

Busnel, R. -G. 1966 Information in the human whistled language and sea mammal whistling. pp. 544-68. In: Whales, Porpoises and Dolphins, K. S. Norris (ed), University of California Press, Berkeley, Los Angeles.

Busnel, R. -G. A. Dziedzic, and S. Andersen 1965 Seuils de perception du système sonar du Marsouin Phocaena phocoena L., en fonction du diamètre d'un obstacle filiforme. C. r. Acad. Sci. Paris, 260: 295-97.

Busnel, R. -G., and A. Dziedzic 1966 Acoustic signals of the pilot whale Globicephala melaena and of the porpoise Delphinus delphis and Phocaena phocoena. pp. 607-46. In: Whales, Porpoises and Dolphins, K. S. Norris (ed), University of California Press, Berkeley, Los Angeles.

Cabot, D. 1965 Cuvier's whale Ziphius cavirostris on Achill Island, Co. Mayo. Irish Natl. J. 15:72-3.

Cabrera, A. and J. Yepes 1960 Mamiferes Sud-Americanos. Compagnia Argentina de Editores, Buenos Aires, 160 pp (2nd edition).

Cadenat, J. 1959 Rapport sur les petits cétacés ouest-africains. Résultats des recherches entreprises sur les animaux jusqu'au mois de mars 1959. Bull. Inst. Fr. Afr. Noire, 21A:1367-1409.

Caldwell, D. K. 1955 Notes on the spotted dolphin, Stenella plagiodon, and the first record of the common dolphin, Delphinus delphis, in the Gulf of Mexico. J. Mammal., 36: 467-70.

Caldwell, D. K. 1960 Sperm and pigmy sperm whales stranded in the Gulf of Mexico. J. Mammal. 41: 136-8.

Caldwell, M. C., D. H. Brown and D. K. Caldwell 1963 Intergeneric behaviour by a captive Pacific pilot whale. Contrib. Sci. Los Angeles County Mus., 70:1-12.

Caldwell, D. K. and D. H. Brown 1964 Tooth wear as a correlate of described feeding behavior by the killer whale—with notes on a captive specimen. Bull. S. Calif. Acad. Sci. 63:128-40.

Caldwell, M. C. and D. K. Caldwell 1964 Experimental studies on factors involved in care-giving behaviour in three species of the cetacean family Delphinidae. Bull. Soc. Calif. Acad. Sci., 63:1-20.

Caldwell, M. C. and D. K. Caldwell 1966 Epimeletic (care-giving) behaviour in Cetacea. pp. 755-89. In: Whales, Porpoises and Dolphins, K. S. Norris (ed), University of California Press, Berkeley, Los Angeles.

Caldwell, D. K. and D. S. Erdman 1963 The pilot whale in the West Indies. J. Mammal. 44:113-5.

Caldwell, D. K. M. C. Caldwell and D. W. Rice 1966 Behaviour of the sperm whale Physeter catodon L. pp. 677-717. In: Whales, Porpoises and Dolphins, K. S. Norris (ed), University of California Press, Berkeley, Los Angeles.

Canadian Broadcasting Commission. 1967 Short documentary film on killer whale in captivity at Vancouver.

Carl, G. C. 1946 A school of killer whales stranded at Estevan Point, Vancouver Island. Rept. Prov. Mus. nat. Hist. & Anthr. 1945:21-8.

Carrick, R. 1957 The Wildlife of Macquarie Island. Aust. Mus. Mag., 12:255-60.

Carrick, R. S. E. Csordas, S. E. Ingham and Keith, K. 1962 Studies on the Southern Elephant Seal Mirounga leonina (L); III The annual cycle in relation to age and sex. CSIRO Wildl. Res., 7:119-160.

Carrick, R. S. E. Csordas, and S. E. Ingham 1962 Studies on the Southern Elephant Seal Mirounga leonina (L); IV. Breeding and development. CSIRO Wildl. Res., 7: 161-197.

Carrick, R. and S. E. Ingham 1960 Ecological studies of the Southern Elephant Seal M.leonina (L.) at Macquarie Island and Heard Island. Mammalia, 24:325-42.

Carrick, R. and S. E. Ingham 1962a Studies on the Southern Elephant Seal M.leonina (L.); I. Introduction to the series. CSIRO Wildl. Res., 7:89-101.

Carrick, R. and S. E. Ingham 1962b Studies on the Southern Elephant Seal M.leonina (L.); II. Canine tooth structure in relation to function and age determination. CSIRO Wildl. Res., 7:102-118.

Carrick, R. and S. E. Ingham 1962c Studies on the Southern Elephant Seal M.leonina (L.); V. Population dynamics and utilisation. CSIRO Wildl. Res., 7:198-206.

Carte, A. and A. MacAlister 1868 On the anatomy of Balaenoptera rostrata. Phil. Trans. R. Soc., 158:201-61.

Carter, T. D. J. E. Hill & G. H. Tate 1945 Mammals of the Pacific World. Mac-Millan, New York, 227 pp.

Cave, A. J. E. & F. J. Aumonier 1961 The visceral histology of the primitive cetacean, Caperea (Neobalaena). J. Roy micr. Soc., 80:25-33.

Chapman, F. R. — 1893 — Notes on the depletion of the fur seal in the southern seas. Can. Rec. Sci. 1893: 446-59.

Chapman, D. G. K. R. Allen & S. J. Holt — 1964 — Reports of the Committee of Three Scientists on the Special Scientific Investigation of the Antarctic Whale Stocks. Intn. Whaling Comm. Rep., 14: 32-106.

Chapman, D. G. K. R. Allen, S. J. Holt & J. A. Gulland — 1965 — Reports of the Committee of Four Scientists on the Special Scientific Investigation of the Antarctic Whale Stocks. Intn. Whaling Comm. Rep., 15: 47-63.

Chittleborough, R. G. — 1953 — Aerial observations on the humpback whale, *Megaptera nodosa* (Bonnaterre), with notes on other species. Aust. J. mar. Freshwat. Res., 4:219-26.

Chittleborough, R. G. — 1954 — Studies on the ovaries of the humpback whale, *Megaptera nodosa* (Bonnaterre), on the Western Australian coast. Aust. J. mar. Freshwat. Res., 5:35-63.

Chittleborough, R. G. — 1955 — Aspects of reproduction of the male humpback whale. Aust. J. mar. Freshwat. Res., 5:1-29.

Chittleborough, R. G. — 1956 — Southern right whale in Australian waters. J. Mammal., 37:456-7.

Chittleborough, R. G. — 1958a — The breeding cycle of the female humpback whale, *Megaptera nodosa* (Bonnaterre). Aust. J. mar. Freshwat. Res., 9:1-18.

Chittleborough, R. G. — 1958b — An analysis of recent catches of humpback whales from the stocks in groups IV and V. Norsk Hvalfangsttid., 47: 109-37.

Chittleborough, R. G. — 1959a — Australian marking of humpback whales. Norsk Hvalfangsttid., 48:47-55.

Chittleborough, R. G. — 1959b — *Balaenoptera brydei* Olsen on the west coast of Australia. Norsk Hvalfangsttid., 48:62-6.

Chittleborough, R. G. — 1959c — Determination of age in the humpback whale, *Megaptera nodosa* (Bonnaterre). Aust. J. mar. Freshwat. Res., 10:125-43.

Chittleborough, R. G. — 1960 — Marked humpback whale of known age. Nature, 187:164.

Chittleborough, R. G. — 1962 — Australian catches of humpback whales, 1962. Rep. Div. Fish. Oceanogr. C.S.I.R.O. Aust., 34, 13 pp.

Chittleborough, R. G. — 1963 — Australian catches of humpback whales, 1962. Rep. Div. Fish. Oceanogr. C.S.I.R.O. Aust., 35, 5 pp.

Chittleborough, R. G. — 1965 — Dynamics of two populations of the humpback whale, *Megaptera novaeangliae* (Borowski). Aust. J. mar. Freshwat. Res., 16:33-128.

Chittleborough, R. G. & E. H. M. Ealey — 1953 — Seal marking at Heard Island. Interim Rep. Aust. natl. Antarct. Res. Exped., 1:1-23.

Christensen, G. — 1955 — The stocks of blue whales in the northern Atlantic. Norsk Hvalfangsttid., 44: 640-2.

Chuzhakina, E. S. — 1961 — (Morphological characterisation of the ovaries of the female sperm whale in connection with age determination.) Trudu. Inst. Morf. Zhivot., 34:33-53.

Claridge, G. C. — 1961 — Seal tracks in the Taylor Dry Valley. Nature, 190:559.

Clark, J. W. — 1873 — On the eared seals of the Auckland Islands. Proc. zool. Soc. Lond., 1873, pp. 750-60.

Clark, J. W. — 1875 — On the eared seals of the islands of St. Paul and Amsterdam, with a description of the fur-seal of New Zealand, and an attempt to distinguish and rearrange the New Zealand Otariidae. Proc. zool. Soc. Lond., 1875, pp. 650-77.

Clarke, G. L. — 1954 — Elements of Ecology. John Wiley & Sons, New York, 534 pp.

Clarke, M. R. — 1962 — The identification of cephalopod "beaks" and the relationship between beak size and total body weight. Bull. Br. Mus. nat. Hist., Zool., 8:421-80.

Clarke, R. — 1953 — Sperm whaling from open boats in the Azores. Norsk Hvalfangsttid., 42:265-77.

Clarke, R. — 1954 — Open boat whaling in the Azores: the history and present methods of a relic industry. 'Discovery' Rep., 26:281-354.

Clarke, R. — 1955a — A giant squid swallowed by a sperm whale. Norsk Hvalfangsttid., 44:353-7.

Clarke, R. — 1955b — The Antarctic whaling seasons of 1951-52, 1952-53 and 1953-54. Polar Rec., 7: 518-21.

Clarke, R. — 1956a — Sperm whales of the Azores. 'Discovery' Rep., 28:237-98.

Clarke, R. — 1956b — A biologia dos cachalotes capturados nos Azores. Notas e Estudos do Instituto de Biologia Maritima, Lisboa, 10:1-11.

Clarke, R. — 1962 — Whale observation and whale marking off the coast of Chile in 1958 and from Ecuador towards and beyond the Galapagos Islands in 1959. Norsk Hvalfangsttid., 51:265-87.

Clarke, R. — 1965 — Southern right whales on the coast of Chile. Norsk Hvalfangsttid., 54:121-28.

Clarke, R. & A. Aguayo L. — 1965 — Bryde's whale in the Southeast Pacific. Norsk Hvalfangsttid., 54:141-8.

Clarke, R. A. Aguayo L. & O. Paliza — 1964 — Progress report on sperm whale research in the southeast Pacific Ocean. Norsk Hvalfangsttid., 53:297-302.

Clarke, R. A. Aguayo L. & O. Paliza — 1968 — Sperm Whales of the Southeast Pacific. Part 1. Introduction. Part 2. Size range, external characters and teeth. Hvalrad. Skr., 51:1-80.

Coleman, E. — 1951 — Save the seals. Vict. Nat., 67:171-7.

Collett, R. — 1912 — Norges Pattedyr. H. Aschehoug & Co. Ltd. Kristiana. 744 pp.

Comrie, L. C. & A. B. Adam — 1938 — The female reproductive system and *Corpora lutea* of the false killer whale, *Pseudorca crassidens* Owens. Trans. R. Soc. Edinb., 59:521-32.

Courtenay—Latimer, M. — 1962 — Two rare seal records for South Africa. Ann. Cape Prov. Mus., 1:102.

Cowan, I. M. — 1939 — The sharp-headed finner whale of the eastern Pacific. J. Mammal., 20:215-25.

Cowan, I. M. — 1945 — A beaked whale stranded on the coast of British Columbia. J. Mammal., 26:93-4.

Cowan, I. M. & C. J. Guiguet — 1952 — Three Cetacean records from British Columbia. Murrelet, 33:10-1.

Cowan, I. M. & J. Hatter — 1940 — Two mammals new to the known fauna of British Columbia. Murrelet, 21:9.

Crile, G. C. & D. P. Quiring — 1940 — A comparison of the energy-releasing organs of the white whale and the thorough-bred horse. Growth, 4:291-298.

Cross, C. M. P. — 1928 — South African sealing industry. Fur J. 1928, Jan. pp. 24 & 37-9.

Crowther, W. L. — 1919 — Notes on Tasmanian whaling. Pap. Roy. Soc. Tasmania: 130-51.

Csordas, S. E. — 1958 — Breeding of the fur seal, *Arctocephalus forsteri* (Lesson), at Macquarie Island. Aust. J. Sci., 21:87.

Csordas, S. E. — 1962a — The Kerguelen fur seal on Macquarie Island. Vict. Nat., 79:226-9.

Csordas, S. E. — 1962b — Leopard seals on Macquarie Island. Vict. Nat., 79:358-62.

Csordas, S. E. — 1963a — Sealions on Macquarie Island. Vict. Nat., 80:32-5.

Csordas, S. E. 1963b The history of fur seals on Macquarie Island. Vict. Nat., 80:255-8.

Csordas, S. E. & S. E. Ingham 1965 The New Zealand fur seal *Arctocephalus forsteri* (Lesson) at Macquarie Island 1949-64. C.S.I.R.O. Wildl. Res., 10:83-99.

Dakin, W. J. 1934 Whalemen Adventurers. Angus & Robertson, Sydney. 263 pp.

Daniel, J. C. 1963 Stranding of a blue whale *Balaenoptera musculus* Linn. near Surat, Gujarat, with notes on earlier literature. J. Bombay nat. Hist. Soc., 60:252-4.

Daniel, R. J. 1925 Animal life in the sea. Hodder and Stoughton, London. 119 pp.

Daugherty, A. E. 1965 Marine mammals of California. California Dept. Fish Game, Sacramento. 87 pp.

Davies, J. L. 1958 The Pinnipedia: an essay in zoogeography. Geogr. Rev., 48:474-93.

Davies, J. L. 1960 The southern form of the pilot whale. J. Mammal., 41:29-34.

Davies, J. L. 1961 Birth of an elephant seal on the Tasmanian coast. J. Mammal., 42:113-4.

Davies, J. L. 1963a The antitropical factor in Cetacean speciation. Evolution, 17:107-16.

Davies, J. L. 1963b The Whales and Seals of Tasmania. Tasmanian Mus. and Art Gallery. 32 pp.

Davies, J. L. & E. R. Guiler 1957 A note on the pygmy right whale, *Caperea marginata* Gray. Proc. Zool. Soc. Lond., 129:579-89.

Davies, J. L. & E. R. Guiler 1958 A newborn piked whale in Tasmania. J. Mammal., 39:593-94.

Dawbin, W. H. 1954 Maori whaling. Norsk Hvalfangsttid. 43:269-81.

Dawbin, W. H. 1956a The migrations of humpback whales which pass the New Zealand coast. Trans. R. Soc. N.Z., 84:147-96.

Dawbin, W. H. 1956b Whale marking in South Pacific waters. Norsk Hvalfangsttid., 45:485-508.

Dawbin, W. H. 1959a New Zealand and South Pacific whale marking and recoveries to the end of 1958. Norsk Hvalfangsttid., 48:213-38.

Dawbin, W. H. 1959b Evidence on growth-rates obtained from two marked humpback whales. Nature, Lond., 183:1749-50.

Dawbin, W. H. 1960 An analysis of the New Zealand catches of humpback whales from 1947 to 1958. Norsk Hvalfangsttid., 49:61-75.

Dawbin, W. H. 1964 Movements of humpback whales marked in the Southwest Pacific Ocean 1952 to 1962. Norsk Hvalfangsttid:, 53:68-78.

Dawbin, W. H. 1966 The seasonal migratory cycle of humpback whales. In: K. S. Norris (Ed.) Whales, Dolphins and Porpoises, pp. 145-70. University of California Press, Los Angeles.

Dearborn, J. H. 1962 An unusual occurrence of the leopard seal at McMurdo Sound, Antarctica. J. Mammal., 43:273-5.

Degerbøl, M. 1956 Campbell Island—home of elephant seals and albatrosses: In: Report of the 'Galathea' Deep Sea Expedition. Macmillan Co., New York. pp. 257-71.

Dell, R. K. 1960 The New Zealand occurrences of the pygmy sperm whale, *Kogia breviceps*. Rec. Dom. Mus., Wellington, 3:229-34.

Dempsey, F. W. & G. B. Wislocki 1941 The structure of the ovary of the humpback whale *Megaptera nodosa*. Anat. Rec., 80:243-57.

Deraniyagala, P. E. P. 1962 Some southern temperate zone fishes, a bird and whales that enter the Ceylon area. Spolia Zeylan, 29:233-7.

Devereux, I. 1968 Oxygen isotope paleotemperatures from the Tertiary of New Zealand. Tuatara, 16:41-4.

DeVries, A. L. & D. E. Wohlschlag 1964 Diving depths of the Weddell seal. Science, 145:292.

Dickson, M. R. 1964 The skull and other remains of *Prosqualodon marplesi*, a new species of fossil whale. N.Z. Jl Geol. Geophys., 7: 626-35.

Dieffenbach, E. 1841 New Zealand and its native population. London.

Doi, T. & S. Ohsumi 1969 Fifth memorandum on results of Japanese stock assessment of whales in the North Pacific. Intn. Whaling Comm. Rep., 19: 123-9.

Dorsett, E. I. 1954 Hawaiian whaling days. Amer. Neptune, 14:42-6.

Doutch, H. F. 1952 Rookery Island (Macquarie). Geogrl. Mag., 25:224-31.

Downs, W. T. 1914 Pelorus Jack — tuni rangi. J. Polynesian Soc., 23:176.

Dreher, J. J. 1961 Linguistic considerations of porpoise sounds. J. Acoustical Soc. Amer., 33: 1799-1800.

Dreher, J. J. 1966 Cetacean communication: Small group experiment. pp. 529-43. In: Whales, Dolphins and Porpoises. K. S. Norris (Ed.), University of California Press, Berkeley.

Dreher, J. J. 1967 Bistatic target signatures and their acoustic recognition: A suggested animal model, pp. 317-38. In: Marine Bioacoustics Vol. 2. W. N. Tavolga (Ed.). Pergamon Press, New York. 353 pp.

Dreher, J. J. & W. E. Evans 1964 Cetacean communication. pp. 373-93. In: Marine Bio-acoustics Vol. 1. W. N. Tavolga (Ed.) Pergamon Press, New York. 413 pp.

Dudok van Heel, W. H. 1966 Navigation in Cetacea. pp. 597-606. In: Whales, Dolphins and Porpoises. K. S. Norris (Ed.). University of California Press, Berkeley.

Dunbabin, T. 1925 Bay whaling in the early days. Navy League J. 1926.

Dunstan, D. J. 1957 Caudal presentation at birth of a humpback whale *Megaptera nodosa* (Bonnaterre). Norsk Hvalfangsttid, 46:553-5.

Dziedzic, A. 1952 Etude d'un compteur d'impulsions acoustiques destiné à des meseres de rythmes d'activité biologique d'animaux marins. Cahiers de Biol. Marine, 3:417-31.

Edmondson, C. H. 1948 Records of *Kogia breviceps* from the Hawaiian Islands. J. Mammal., 29:76-7.

Eichelberger, L. L. Leiter & E. M. K. Geiling 1940 Water and electrolyte content of dolphin kidney and extraction of pressor substance. Proc. Soc. Exp. Biol., 44:356-359.

Eklund, C. R. 1960 Elephant seals in Antarctica. J. Mammal., 41:277.

Elliot, G. H. 1958 Politics and the Antarctic whale. Sea Frontiers, 4:171-81.

Elliott, H. F. I. 1953 The fauna of Tristan da Cunha. Oryx, 41-53.

Ellison, N. F. 1951 Lesser rorqual in river Mersey. Northwest. Nat., 23:1-4.

Ellison, N. F. 1954a Bottle-nosed whales in Mersey. Northwest. Nat. N.S., 2:460.

Ellison, N. F. 1954b Lesser rorqual on Lancashire coast. Northern Nat. N.S., 2:460.

Elsner, R. 1969 Cardiovascular adjustments to diving. pp. 117-45. In: The Biology of Marine Mammals, H. T. Andersen (ed), Academic Press, New York.

Engel, S. 1954 Respiratory tissue of the large whales. Nature, 173: 128-9.

Erdman, D. S. 1962a Stranding of a beaked whale, *Ziphius cavirostris* Cuvier on the south coast of Puerto Rico. J. Mammal., 43:276-7.

Erdman, D. S. 1962b New fish records and one whale record from Puerto Rico. Caribbean J. Sci., 1:39-40.

Essapian, F. S. 1955 Speed-induced skin folds in the bottle-nosed porpoise, *Tursiops truncatus*. Breviora Mus. Comp. Zool. (Harvard), 43:1-4.

Essapian, F. S. 1962 Courtship in captive saddle-backed porpoises, *Delphinus delphis* L. 1758. Z. Saeugetierk, 27:211-7.

Evans, W. E. 1967 Vocalization among marine mammals: pp. 159-86. In: Marine Bio-acoustics. Vol. 2. W. N. Tavolga (Ed.). Pergamon Press, New York. 353 pp.

Evans, W. E. & J. J. Dreher 1962 Observations on scouting behaviour and associated sound production by the Pacific bottle-nosed porpoise *(Tursiops gilli* Dall). Bull. So. Calif. Acad. Sci., 61:217-26.

Evans, W. E. & R. Haugen 1963 An experimental study of the echolocation ability of a California sea lion *Zalophus californianus* (Lesson). Bull. S. Calif. Acad. Sci., 62:165-75.

Evans, W. E. & J. H. Prescott 1962 Observations of the sound production capabilities of the bottlenose porpoise: A study of whistles and clicks. Zoologica, 47:121-8.

Evans, W. E. W. W. Sutherland & R. G. Beil 1964 The directional characteristics of Delphinoid sounds. pp. 353-72. In: Marine Bio-acoustics. Vol. 1. W. N. Tavolga (Ed.). Pergamon Press, New York. 413 pp.

Fairford, F. 1916 Whale hunting in the North Atlantic. Empire Rev. 1916 (June): 222-4.

Falla, R. A. 1953 Southern seals: population studies and conservation problems. Proc. 7th Pacif. Sci. Congr., 4:706.

Falla, R. A. 1962 Exploitation of seals, whales and penguins in New Zealand. Proc. N.Z. Ecol. Soc., 9:34-8.

Fanning, E. 1924 Voyages and discoveries in the South Seas 1792-1832. Marine Research Society, Salem, U.S.A.

Fay, F. H. 1957 History and present status of the Pacific walrus population. Trans. No. Amer. Wildl. Conf., 22:431-45.

Fay, F. H. & C. Ray 1966 The influence of climate on the distribution of walruses *Odobenus rosmarus* (Linnaeus). I. Evidence from Thermoregulatory Behaviour. Zoologica, 53:1-14.

Featherston, D. W. 1965 Some aspects of the biology of *Glandiocephalus perfoliatus* (Raillient and Henry, 1912); a pseudophyllidean cestode found in the bile duct of *Leptynochotes weddelli*. M.Sc. Thesis. University of Canterbury, Christchurch, New Zealand. 42 pp.

Felts, W. J. L. 1966 Some functional and structural characteristics of Cetacean flippers and flukes. pp. 255-76. In: K. S. Norris (Ed.). Whales, Porpoises and Dolphins. University of California Press, Los Angeles.

Fetcher, E. S. 1939 The water balance in marine mammals. Quart. Rev. Biol., 14:451-9.

Fetcher, E. S. 1940 Experiments on the water balance of the dolphin. Amer. J. Physiol., 133:274-275.

Fetcher, E. S. & G. W. Fetcher 1942 Experiments on the osmotic regulation of dolphins. J. Cell. comp. Physiol. 19: 123-30.

Ficus, C. H. & K. Niggol 1965 Observations of Cetaceans off California, Oregon and Washington. U.S. Fish. Wildl. Spec. Sci. Rep., 498:1-27.

Fish, P. A. 1898 The brain of the fur seal, *Callorhinus ursinus*; with a comparative description of those of *Zalophus californianus*, *Phoca vitulina*, *Ursus americanus* and *Monachus tropicalis*. J. comp. Neurol., 8: 58-91.

Fitch, J. E. & R. L. Brownell jnr. 1968 Fish otoliths in cetacean stomachs and their importance in interpreting feeding habits. J. Fish. Res. Bd. Canada, 25: 2561-74.

Flanigan, N. J. 1966 The anatomy of the spinal cord of the Pacific striped dolphin, *Lagenorhynchus obliquidens*. pp. 207-31. In: K. S. Norris (Ed.). Whales, Porpoises and Dolphins. University of California Press, Los Angeles.

Fleming, C. A. 1950 Some South Pacific sea-bird logs. Emu, 49:169-88.

Fleming, C. A. 1952 The White Island trench: a submarine graben in the Bay of Plenty. Proc. 7th Pac. Sci. Congr., 3:210-3.

Fleming, C. A. 1962 New Zealand biogeography: a palaeontologist's approach. Tuatara, 10:53-108.

Fleming, C. A. & J. J. Reed 1951 Mernoo Bank, east of Canterbury, New Zealand. N.Z. Jl Sci. Tech., B32:17-30.

Flower, W. H. 1872 On the recent ziphioid whales, with a description of the skeleton of *Berardius arnouxi*. Trans. zool. Soc. Lond., 8:203-34.

Flower, W. H. 1879 A further contribution to the knowledge of the existing ziphioid whales, genus *Mesoplodon*. Trans. zool. Soc. Lond., 10: 415-37.

Flower, W. H. 1880 On the external characters of two species of British dolphins (*Delphinus delphis* Linn. and *Delphinus tursio* Fabr.). Trans. zool. Soc. Lond., 11:1-5.

Flower, W. H. 1882 On the cranium of a new species of *Hyperoodon* from the Australian seas. Proc. zool. Soc. Lond., 1882:392-6.

Flower, W. H. 1883 On the characters and divisions of the family Delphinidae. Proc. zool. Soc. Lond., 1883:466-513.

Flyger, V. M. S. R. Smith, R. Damm & R. S. Peterson 1965 Effects of three immobilizing drugs on Weddell seals. J. Mammal., 46:345-7.

Forbes, H. O. 1893 Observations on the development of the rostrum in the Cetacean genus *Mesoplodon*, with remarks on some of the species. Proc. zool. Soc. Lond., 1893:216-36.

Fraser, F. C. 1934 Report on Cetacea stranded on the British coasts from 1927 to 1932. Brit. Mus. nat. Hist. London, 11:34-6.

Fraser, F. C. 1936 Vestigial teeth in specimens of Cuvier's whale *(Ziphius cavirostris)* stranded on the Scottish coast. Scott. Nat., 1936:153-7.

Fraser, F. C. 1937 Early Japanese whaling. Proc. Linn. Soc. Lond., 1937-8:19-20.

Fraser, F. C. 1940 Three anomalous dolphins from Blacksod Bay, Ireland. Proc. Roy. Irish. Acad., 45(B): 413-55.

Fraser, F. C. 1945 On a specimen of the southern bottle-nosed whale, *Hyperoodon planifrons*. "Discovery" Rep., 23:21-36.

Fraser, F. C. 1946 Report of Cetacea stranded on the British coasts from 1933 to 1937. Bull. Brit. Mus. (nat. Hist.), 12:49-51.

Fraser, F. C. 1950 Notes on a skull of Hector's beaked whale *Mesoplodon hectori* (Gray) from the Falkland Islands. Proc. Linn. Soc. Lond., 162:50-2.

Fraser, F. C.   1953   Report on Cetacea stranded on the British coasts from 1938 to 1947. Bull. Brit. Mus. (nat. Hist.), 13:1-48.

Fraser, F. C.   1955a   A skull of *Mesoplodon gervaisi* (Deslongchamps) from Trinidad, West Indies. Ann. Mag. nat. Hist., ser. 12, 8:624-30.

Fraser, F. C.   1955b   The southern right whale dolphin *Lissodelphis peroni* (Lacépède), external characters and distribution. Bull. Brit. Mus. (nat. Hist.) Zool., 2:339-46.

Fraser, F. C.   1966a   Comments on the Delphinoidea, pp. 7-30. In: K. S. Norris (Ed.). Whales, Dolphins and Porpoises. University of California Press, Los Angeles.

Fraser, F. C.   1966b   Guide for the identification and reporting of stranded whales, dolphins and porpoises on the British coasts. Bull. Brit. Museum (nat. Hist.), London: pp. 34.

Fraser, F. C. & P. E. Purves   1954   Hearing in Cetaceans. Bull. Brit. Mus. (nat. Hist.), 2:103-16.

Fraser, F. C. & P. E. Purves   1955   The 'blow' of whales. Nature, 176: 1221-2.

Fraser, F. C. & P. E. Purves   1960   Hearing in Cetaceans. Bull. Br. Mus. (nat. Hist.), Zool., 7:1-140.

Freeman, W. H. & B. Bracegirdle   1966   An Atlas of Histology. Heinemann Educ., London. p. 140.

Fujino, K.   1953   On the serological constitution of the sei-, fin-, blue-, and humpback whales (I). Scient. Rep. Whales Res. Inst., Tokyo, 8:103-25.

Fujino, K.   1954   On the body proportions of the fin whales *Balaenoptera physalus* (L.) caught in the northern Pacific Ocean. Scient. Rep. Whales Res. Inst., Tokyo, 9:121-63.

Fujino, K.   1960   Immunogenetic and marking approaches to identifying subpopulations of the North Pacific whales. Scient. Rep. Whales Res. Inst., Tokyo, 15:85-142.

Fujino, K.   1964   Fin whale subpopulations in the Antarctic whaling areas, II, III and IV. Scient. Rep. Whales Res. Inst., Tokyo, 18:1-28.

Galbreath, E. C.   1963   Three beaked whales stranded on the Midway Islands, Central Pacific Ocean. J. Mammal., 44:422-3.

Gambell, R.   1964   A pygmy blue whale at Durban. Norsk Hvalfangsttid., 53:66-8.

Gambell, R.   1967   Seasonal movements of sperm whales. Symp. zool. Soc. Lond., 19:237-54.

Gambell, R.   1968   Seasonal cycles and reproduction in sei whales of the southern hemisphere. Discovery Reps., 35:31-134.

Gambell, R.   1969   Interim report on the sperm whale stock off Durban. Intn. Whaling Comm. Rep., 19:76-83.

Garner, D. M.   1953   Physical characteristics of inshore surface waters between Cook Strait and Banks Peninsula, New Zealand. N.Z. Jl Sci. Tech., B35:239-46.

Garner, D. M.   1954   Sea surface temperature in the Southwest Pacific Ocean, from 1949 to 1952. N.Z. Jl Sci. Tech., B36:285-303.

Garner, D. M.   1955   Some hydrological features of the tropical South-west Pacific Ocean. N.Z Jl Sci. Tech., B37:39-46.

Garner, D. M.   1958   The Antarctic convergence south of New Zealand, N.Z. Jl Geol. Geophys., 1: 577-94.

Garner, D. M.   1959   The sub-tropical convergence in New Zealand surface waters. N.Z. Jl Geol. Geophys., 2:315-37.

Garner, D. M.   1961   Hydrology of New Zealand coastal waters, 1955. N.Z. Dept. sci. industr. Res. Bull., 138:1-85.

Garner, D. M. & N. M. Ridgway   1965   Hydrology of New Zealand offshore waters. N.Z. Dept. sci. industr. Res. Bull., 162:1-62.

Gaskin, D. E.   1963   Whale-marking cruises in New Zealand waters made between February and August 1963. Norsk Hvalfangsttid., 52: 307-21.

Gaskin, D. E.   1964a   Whale-marking cruises in New Zealand waters made between August and December 1963. Norsk Hvalfangsttid., 53:29-41.

Gaskin, D. E.   1964b   Recent observations in New Zealand waters on some aspects of behaviour of the sperm whale (*Physeter macrocephalus*). Tuatara, 12:106-14.

Gaskin, D. E.   1964c   Return of the southern right whale (*Eubalaena australis* Desm.) to New Zealand waters, 1963. Tuatara, 12:115-8.

Gaskin, D. E.   1965   New Zealand whaling and whale research 1962-4. N.Z. Sci. Rev., 23:19-22.

Gaskin, D. E.   1966   New records of the pigmy sperm whale *Kogia breviceps* Blainville 1838, from New Zealand, and a probable record from New Guinea. Norsk Hvalfangsttid. 55: 35-7.

Gaskin, D. E.   1967a   Report on the *Chiyoda Maru No. 5* expedition December 1966-March 1967. In: Report on whale observations in the Antarctic 1966-67. F.A.O. Fish. Circ., 111: i-v and 1-48. pp. 21-48.

Gaskin, D. E.   1967b   Luminescence in a squid *Moroteuthis* sp. (probably *ingens* Smith), and a possible feeding mechanism in the sperm whale *Physeter catodon* L. Tuatara, 15:86-88.

Gaskin, D. E.   1968a   The New Zealand Cetacea. Fish. Res. Bull. (N. Ser.), 1:1-92.

Gaskin, D. E.   1968b   Analysis of sightings and catches of sperm whales (*Physeter catodon* L.) in the Cook Strait area of New Zealand in 1963-4. N.Z. Jl mar. Freshwat. Res., 2:260-72.

Gaskin, D. E.   1968c   Distribution of Delphinidae (Cetacea) in relation to sea surface temperatures of eastern and southern New Zealand N.Z. Jl mar. Freshwat. Res., 2:527-34.

Gaskin, D. E.   1970a   The origins of the New Zealand fauna and flora: A review. Geogrl Revs. 60: 414-34.

Gaskin, D. E.   1970b   Composition of schools of sperm whales *Physeter catodon* L. east of New Zealand. N.Z. Jl mar. Freshwat. Res. 4:456-71.

Gaskin, D. E. in press₁, 1971   Distribution and movements of sperm whales *Physeter catodon* L. in the Cook Strait region of New Zealand. Nytt. mag. Zool. 1971.

Gaskin, D. E. in press₂ 1971   Distribution of beaked whales (Cetacea: Ziphiidae) related to sea surface temperatures off southern New Zealand. N.Z. Jl mar. Freshwat. Res. 5.1971.

Gaskin, D. E. & M. W. Cawthorn   1967a   Squid mandibles from the stomachs of sperm whales (*Physeter catodon* L.) captured in the Cook Strait region of New Zealand. N.Z. Jl mar. Freshwat. Res., 1:59-70.

Gaskin, D. E. & M. W. Cawthorn   1967b   Diet and feeding habits of the sperm whale (*Physeter catodon* L.) in the Cook Strait region of New Zealand N.Z. Jl mar. Freshwat. Res., 1:156-79.

Gates, D. J.   1963   Australian whaling since the war. Norsk Hvalfangsttid., 52:123-27.

Gates, D. J.   1964   Value of whale products falls. Fisheries News1., 23:16-17.

Geiling, E. M. K.   1935   The hypophysis cerebri of the finback and sperm whale. Bull. Johns Hopkins Hospital, 57:123-7.

Geraci, J. R. &   1966   Relationship of dietary histamine to
  K. E. Gerstmann   gastric ulcers in the dolphin. J. Amer.
                                  Vet. Med. Assoc. 149(7):884-90.

Gibbney, L. F.   1957   The seasonal reproductive cycle of the
                                  female elephant seal, Mirounga leonina
                                  (L.) at Heard Island. Interim Rep. Aust.
                                  natl. Antarct. Res. Exped., (B), 1:1-26.

Gilmore, R. M.   1959   On the mass strandings of sperm whales.
                                  Pacific Nat., 1 (10):9-16.

Gilmore, R. M.   1961   Whales, Porpoises and the U.S. Navy.
                                  Norsk Hvalfangsttid., 50: 89-108.

Glauert, L.   1947   The genus Mesoplodon in western Aus-
                                  tralian seas. Aust. Zool., 11:73-5.

Godfrey, K.   1964   Sperm whale chase — W.A. aerial survey.
                                  Fisheries Newsl., 23:14, 15, 17.

Goudappel, J. R.   1958   Microscopic structure of the lungs of
  & E. Slijper   the bottlenose whale. Nature, 182:479.

Graham, D. H.   1953   Mammals. In: A treasury of New Zealand
                                  fishes. A. H. and A. W. Reed, Welling-
                                  ton. 404 pp.

Gray, D.   1882   The bottle-nose whale (migrations).
                                  Proc. zool. Soc. Lond., 1882:726-31.

Gray, J. E.   1827   Description of the skulls of two appar-
                                  ently undescribed species of dolphins,
                                  which are in the British Museum. Phil-
                                  os. Mag., n.s., 2:375-6.

Gray, J. E.   1828-30   Spicilegia zoologica. Pt. 1:1-8. Pt. 2:9-12.
                                  Treuetel, Weurtz, London.

Gray, J. E.   1837   Description of some new or little known
                                  mammalia, principally in the British
                                  Museum collection. Mag. nat. Hist. n.s.
                                  1:577-87.

Gray, J. E.   1844   The seals of the southern hemisphere.
                                  In: The zoology of the voyage of HMS
                                  Erebus and Terror during the years 1839
                                  to 1843. London. pp. 1-8.

Gray, J. E.   1846   Mammalia: On the Cetaceous animals.
                                  In: J. Richardson and J. E. Gray (Eds.),
                                  The Zoology of the Voyage of HMS
                                  Erebus and Terror, under command of
                                  Captain Sir James Clark Ross, R.N.,
                                  F.R.S., during years 1839 to 1843. Vol.
                                  1, Janson, London.

Gray, J. E.   1865   Ziphius layardi from Cape of Good Hope.
                                  Proc. zool. Soc. Lond., 1865:358.

Gray, J. E.   1871a   Notes on the skull of Balaena marginata
                                  described in Transactions of the New
                                  Zealand Institute, Vol. II, p. 26, as the
                                  type of a new genus Neobalaena. Trans.
                                  Proc. N.Z. Inst., 3:123-4.

Gray, J. E.   1871b   Notes on the Berardius of New Zealand
                                  waters. Ann. Mag. nat. Hist., ser. 4:
                                  115-7.

Gray, J. E.   1874a   List of seals, whales and dolphins of
                                  New Zealand. Trans. Proc. N.Z. Inst.,
                                  6:87-9.

Gray, J. E.   1874b   On the skeleton of the New Zealand pike
                                  whale, Balaenoptera huttoni (Physalus
                                  antarcticus Hutton). Ann. Mag. nat. Hist.,
                                  ser. 4, 13:448-52.

Gray, J. E.   1874c   Description of the skull of a new species
                                  of dolphin (Feresa attenuata). Ann. Mag.
                                  nat. Hist., ser. 4, 14:238-9.

Greenshields, F.   1937   Bottle-nosed whales (Hyperoodon rostra-
                                  tus) in the river Tay. Scot. Nat. Edinb.:
                                  268-76.

Gregory, M. E.   1955   The composition of the milk of the blue
  S. K. Kon, S. J.   whale. J. Dairy Res., 72:108.
  Rowland & S. Y.
  Thompson

Gudkov, V. M.   1963   (On colour variations of sperm whales
                                  in the northeast Pacific.) Trudy Inst.
                                  Okeanol., 71:207-22.

Guiguet, C. J.   1954   A record of Baird's dolphin (Delphinus
                                  bairdii Dall) in British Columbia.
                                  Canad. Fld. Nat., 68:136.

Guiler, E. R.   1961   A pregnant female pygmy right whale.
                                  Aust. J. Sci., 24:297.

Gwynn, A. M.   1953a   The status of the leopard seal at Heard
                                  Island and Macquarie Island 1948-50.
                                  Interim Rep. Aust. natl. Antarct. Res.
                                  Exped., 3:1-33.

Gwynn, A. M.   1953b   Notes on the fur seals (and sea lion)
                                  at Macquarie Island and Heard Island.
                                  Interim Rep. Aust. natl. Antarct. Res.
                                  Exped., 4:1-16.

Haga, R.   1961   Birds and seals around Japanese Syowa
                                  base on Prince Harald coast Antarctica
                                  (preliminary). Antarct. Rec. Tokyo, 11:
                                  146-8.

Haldane, R. C.   1905   Whaling in Scotland I. Ann. Scot. nat.
                                  Hist., 54:65-72.

Haldane, R. C.   1908   Whaling in Scotland II. Ann. Scot. nat.
                                  Hist., 61: 10-15.

Haldane, R. C.   1910   Whaling in Scotland III. Ann. Scot. nat.
                                  Hist., 66:65-72.

Hale, H. M.   1932a   Beaked whales — Hyperoodon planifrons
                                  and Mesoplodon layardii — from South
                                  Australia. Rec. S. Aust. Mus., 4:291-311.

Hale, H. M.   1932b   The pygmy right whale (Neobalaena
                                  marginata) in South Australian waters.
                                  Rec. S. Aust. Mus., 4:314-9.

Hale, H. M.   1939   Rare whales in South Australia. S. Aust.
                                  Nat., 19(4):5-8.

Hale, H. M.   1947   The pigmy sperm whale (Kogia breviceps
                                  Blainville) on South Australian coasts.
                                  Rec. S. Aust. Mus., 8:531-46.

Hale, H. M.   1962a   The pigmy sperm whale (Kogia breviceps)
                                  on South Australian coasts. Part III.
                                  Rec. S. Aust. Mus., 14:197-229.

Hale, H. M.   1962b   Occurrence of the whale Berardius arnuxi
                                  in Southern Australia. Rec. S. Aust. Mus.,
                                  14:231-43.

Hall, E. R.   1959   The Mammals of North America. II.
  & K. R. Kelson   Ronald Press, New York. 1,083 pp.

Hall, T. S.   1903   Crabeater seals in Australian waters.
                                  Note in: Nature, 67:327-8.

Halstead, B. W.   1959   Dangerous Marine Animals. Cornell Mari-
                                  time Press, Cambridge, Maryland. pp.
                                  i-xiv, 1-146 pp.

Hamilton, J. E.   1934   The southern sea lion Otaria byronia (de
                                  Blainville). 'Discovery' Rep., 8:269-318.

Hamilton, J E.   1939a   The leopard seal Hydrurga leptonyx (de
                                  Blainville). 'Discovery' Rep., 18:239-64.

Hamilton, J. E.   1939b   A second report on the southern sea
                                  lion, Otaria byronia (de Blainville).
                                  'Discovery' Rep., 19:121-64.

Hamilton, J. E.   1940   On the history of the elephant seal,
                                  Mirounga leonina (Linn.). Proc. Linn.
                                  Soc. Lond., 1939, 40:33-7.

Hamilton, J .E.   1945   The Weddell seal in the Falkland Islands.
                                  Proc. zool. Soc. Lond., 114: 549.

Hamon, B. V.   1965   The East Australian Current, 1960-1964.
                                  Deep-Sea Res., 12:899-921.

Hancock, D.   1965   Killer whales kill and eat a minke whale.
                                  J. Mammal., 46:341-2.

Handley, C. O.   1966   A synopsis of the genus Kogia (pygmy
                                  sperm whales.) pp. 62-9. In: K. S. Norris
                                  (Ed.), Whales, Dolphins, and Porpoises.
                                  University of California Press, Los
                                  Angeles.

Hanstroem, B.   1944   Zur histologie und vergleichenden
                                  anatomie der hypophyse der Cetaceen.
                                  Acta Zool., 25:1-25.

Harboe, A. & A. Schrumpf — 1952 — The red blood cell diameter in the blue whale and humpback. Norsk Hvalfangsttid, 41:416-8.

Harmer, S. F. — 1924 — On *Mesoplodon* and other beaked whales. Proc. zool. Soc. Lond., 1924: 541-87.

Harmer, S. F. — 1927 — Report on Cetacea stranded on the British coasts from 1913 to 1926. Brit. Mus. nat. Hist., 10:1-91.

Harmer, S. F. — 1928 — History of whaling. Proc. Linn. Soc. Lond., 140:51-95.

Harmer, S. F. — 1929 — Southern whaling. Proc. Linn. Soc. Lond., 142:85-163.

Harrison, R. J. — 1949 — Observations on the female reproductive organs of the ca'aing whale, *Globicephala melaena* Traill. J. Anat. London, 83:238-53.

Harrison, R. J. — 1960 — Reproduction and reproductive organs in common seals *(Phoca vitulina)* in the Wash, East Anglia. Mammalia, Paris, 24:372-385.

Harrison, R. J. — 1969a — Reproduction and reproductive organs. pp. 253-348 In: The Biology of Marine Mammals, H. T. Andersen (ed), Academic Press, New York.

Harrison, R. J. — 1969b — Endocrine organs: Hypophysis, thyroid and adrenal. pp. 349-90 In: The Biology of Marine Mammals, H. T. Andersen (ed), Academic Press, New York.

Harrison, R. J., R. C. Boice, & R. L. Brownell jnr. — 1969 — Reproduction in wild and captive dolphins. Nature, 222:1143-7.

Harrison, R. J., F. R. Johnson and B. A. Young — 1970 — The oesophagus and stomach of dolphins *(Tursiops, Delphinus, Stenella).* J. Zool. Lond., 1970 160:377-90.

Harrison, R. J. & J. E. King — 1965 — Marine Mammals. Hutchinson University Library, London. 192 pp.

Harrison, R. J. C. H. Matthews, & J. M. Roberts — 1952 — Reproduction in some Pinnipedia. Trans. zool. Soc. Lond., 27:437-540.

Harrison, R. J. & J. D. W. Tomlinson — 1956 — Observations on the venous system in certain Pinnipedia and Cetacea. Proc. zool. Soc. Lond., 126:205-33.

Harrison, R. J. & J. D. W. Tomlinson — 1963 — Anatomical and physiological adaptations in diving mammals. Viewpoints in Biology, 2:115-162.

Hart, J. S. & L. Irving — 1959 — The energetics of harbor seals in air and in water with special consideration of seasonal changes. Canad. J. Zool., 37:447-57.

Haynes, F. & A. H. Laurie — 1937 — On: The histological structure of Cetacean lungs. Discovery Rep., 17:1-6.

Hector, J. — 1873a — On the whales and dolphins of the New Zealand seas. Trans. Proc. N.Z. Inst., 5:154-70.

Hector, J. — 1873b — Notes on the whales and dolphins of the New Zealand seas. Ann. Mag. nat. Hist., ser. 4, 11:104-12.

Hector, J. — 1875 — Notes on New Zealand whales. Trans. Proc. N.Z. Inst., 7:251-65.

Hector, J. — 1877 — Notes on New Zealand Cetacea. Trans. Proc. N.Z. Inst., 9:477-84.

Hector, J. — 1878 — Notes on the whales of the New Zealand seas. Trans. Proc. N.Z. Inst., 10:331-43.

Heine, A. J. — 1960 — Seals at White Island, Antarctica. Antarctic, 2: 272-3.

Heldt, J. H. — 1954 — *Tursiops truncatus* Montagu dans le Golfe de Tunis. Bull. Soc. Sci. nat. Tunis, 6:61-2.

Hennings, H. — 1950 — The whale hypophysis with special reference to its ACTH. content. Acta. Endrocrinologica, 5:376-86.

Hepburn, D. & D. Waterston — 1904 — A comparative study of the grey and white matter of the motor cell groups, and of the spinal accessory nerve in the spinal cord of the porpoise *(Phocoena communis).* J. Anat. Physiol., 38:105-218.

Hershkovitz, P. — 1966 — Catalog of living whales. U.S. Natl. Museum Bull., 246: viii and 1-259.

Higham, T. F. — 1960 — Nature note: dolphin riders. Ancient stories vindicated. Greece and Rome, 7:82-6.

Highley, E. — 1967 — Oceanic circulation patterns off the east coast of Australia. C.S.I.R.O. Div. Fish. Oceanogr. Tech. Pap., 23:1-19.

Hinton, M. A. C. — 1925 — Report on the papers left by the late Major Barrett-Hamilton, relating to the whales of South Georgia. Crown Colony Agents, London: 57-209.

Hjort, J. G. Jahn, & P. Ottestad — 1933 — The optimum catch. Hvalrad. Skr. 7: 92-127.

Hocken, T. M. — 1871 — Report on the whale fisheries. Trans. Proc. N.Z. Inst., 3:68-70.

Hoedemaker, N. J. ten Cate — 1935 — Mitteilung ueber eine reife Plazenta von *Phocaena phocaena* (Linnaeus). Arch. Néerl. zool., 1:330-8.

Holdgate, M. W. — 1957 — Gough Island — a possible sanctuary. Oryx, 4:168-76.

Holdgate, M. W. — 1963 — Observations of birds and seals at Anvers Island, Palmer Archipelago in 1955-57. Bull. Br. Antarc. Surv., 2:45-51.

Holdgate, M. W. R. W. LeMaitre, M. K. Swales & N. M. Wace — 1956 — The Gough Island scientific survey 1955-56. Nature, 178:234-6.

Holm, J. L. & A. Jonsgaard — 1959 — Occurrence of the sperm whale in the Antarctic and the possible influence of the moon. Norsk Hvalfangsttid., 48: 161-82.

Home, E. — 1812 — An account of some peculiarities in the structure of the organ of hearing in *Balaena mysticetus* of Linnaeus. Philos. Trans., 102:83-8.

Hopkins, G. H. E. — 1949 — The host-associations of the lice of mammals. Proc. zool. Soc. Lond., 119: 387-604.

Hosokawa, H. — 1951 — On the extrinsic eye muscles of the whale, with special remarks upon the ennervation and function of the musculus retractor bulbi. Scient. Rep. Whales Res. Inst., Tokyo, 6:1-34.

Houck, W. J. — 1958 — Cuvier's beaked whale from northern California. J. Mammal., 39:308-9.

Howard, P. — 1954 — A.N.A.R.E. bird banding and seal marking. Vict. Nat., Melbourne, 71:73-82.

Howell, A. B. — 1928 — Contribution to the comparative anatomy of the eared and earless seals (genera *Zalophus* and *Phoca*). Proc. U.S. Nat. Mus., 73:1-142.

Howell, A. B. — 1930 — Myology of the narwhal *(Monodon monoceros).* Amer. J. Anat., 46:187-216.

Hubbs, C. L. — 1946 — First records of two beaked whales, *Mesoplodon bowdoini* and *Ziphius cavirostris*, from the Pacific coast of the United States. J. Mammal., 27: 242-55.

Hubbs, C. L. — 1951a — Probable record of the beaked whale, *Ziphius cavirostris*, in Baja, California. J. Mammal., 32:365-6.

174

Hubbs, C. L. 1951b Eastern Pacific records and general distribution of the pygmy sperm whale. J. Mammal., 32:403-10.

Hubbs, C. L. 1953 Dolphin protecting dead young. J. Mammal., 34:498.

Hubbs, C. L. 1956 Back from oblivion. Pacific Discovery, 9:14-21.

Huber, E. 1934 Anatomical notes on Pinnipedia and Cetacea. Publ. Carnegie Inst. Wash., 447:105-36.

Hunter, J. 1787 Observations on the structure and oeconomy of whales. Philos. Trans., 77:371-450.

Hutton, Capt. F. W. & J. Drummond 1923 Seals. In: The Animals of New Zealand, pp. 43-8. Whitcombe and Tombs, Christchurch.

Hutton, J. 1950 On a Bahamas reef. Field, 195:751-2.

Ichihara, T. 1966a Criterion for determining age of fin whale with reference to ear plug and baleen plate. Scient. Rep. Whales Res. Inst., Tokyo, 20:17-82.

Ichihara, T. 1966b The pygmy blue whale, Balaenoptera musculus brevicauda, a new subspecies from the Antarctic. pp. 79-113. In: Whales, Dolphins and Porpoises, K. S. Norris (Ed.). University of California Press, Los Angeles.

Ichara, T. & T. Doi. 1934 Stock assessment of pygmy blue whales in the Antarctic, Norsk Hvalfangsttid., 53:145-67.

Ingebrigtsen, A. 1929 Whales caught in the North Atlantic and other seas. Rapp. Cons. Explor. Mer., 56:1-26.

Ingham, S. E. 1957 Elephant seals on the Antarctic continent. Nature, 180:1215-16.

Ingham, S. E. 1960 Status of seals (Pinnipedia) at Australian Antarctic stations. Mammalia, 24:422-30.

International Commision on Whaling 1969 Nineteenth report of the Commission, Office of the Commission, London, 148 pp.

International Whaling Statistics, Vol. 51 1964 Det Norske Hvalrads Statistiske Publikasjoner, Oslo.

International Whaling Statistics, Vol. 59 1967 Det Norske Hvalrads Statistiske Publikasjoner, Oslo.

International Whaling Statistics, Vol. 60 1968 Det Norske Hvalrads Statistiske Publikasjoner, Oslo.

International Whaling Statistics, Vol. 61 1968 Det Norske Hvalrads Statistiske Publikasjoner, Oslo.

Iredale, T. & E. leG. Troughton 1933 The correct generic names for the grampus or killer whale, and the so-called grampus or Risso's dolphin. Rec. Aust. Mus., Sydney, 19:28-36.

Iredale, T. & E. leG. Troughton 1934 A check-list of the mammals recorded from Australia. Mem. Aust. Mus., 6, 122 pp.

Irving, L. 1939 Respiration in diving mammals. Physiol. Rev., 19:122-34.

Irving, L. 1969 Temperature regulation in marine mammals. pp. 147-75 In: The Biology of Marine Mammals, H. T. Andersen (ed), Academic Press, New York.

Irving, L. & J. S. Hart 1957 The metabolism and insulation of seals as bare-skinned mammals in cold water. Canad. J. Zool., 35:497-511.

Irving, L. L. J. Peyton, C. H. Bahn & R. S. Peterson 1962 Regulation of temperature in fur seals. Physiol. Zool., 35:275-84.

Irving, L. P. F. Scholander & S. W. Grinnell 1941 The respiration of the porpoise Tursiops truncatus. J. Cell. comp. Physiol., 17:145-68.

Isachsen, G. 1927 Modern Norwegian whaling in the Antarctic. Geogr. Rev., 19:387-403.

Ishikawa, Y. & S. Tejima 1949 Protein digestive power of sperm whale pancreatic enzyme (1). Scient. Rep. Whales Res. Inst., Tokyo, 2:55-60.

Ivanova, E. I. 1955 (Body proportions of sperm whales.) Trudy Inst. Okeanol., 18:100-12.

Ivanova, E. I. 1962 (Morphology of Eubalaena sieboldi.) Trudy Inst. Morf. Zhiv., 34:216-25.

Ivashin, M. 1958 (On the systematic position of the humpback whale (Megaptera nodosa lalandi Fischer) of the southern hemisphere.) U.S.S.R. Arctic and Antarctic Sci. Res. Inst. Leningrad, 3:77-8.

Ivashin, M. V. 1962 (Marking humpback whales in the southern hemisphere.) Zool. Zh., 41:1848-58.

Jacobsen, A. P. 1941 Endocrinological studies on the blue whale (Balaenoptera musculus L.) Hvalrad. Skr., 24:1-84.

Jacobson, H. C. 1893 Tales of Bank's Peninsula. Christchurch, iv, 200 pp.

Jansen, J. 1951 The morphogenesis of the cetacean cerebellum. J. comp. Neurol., 93:341-400.

Jansen, J. 1953 Studies on the cetacean brain. The gross anatomy of the rhombencephalon of the fin whale (Balaenoptera physalus L.). Hvalrad. Skr., 37:1-35.

Jansen, J. jr. & J. Jansen 1953 A note on the amygdaloid complex in the fin whale (Balaenoptera physalus L.) Hvalrad. Skr., 39:1-14.

Jansen, J. & J. K. S. Jansen 1969 The nervous system of Cetacea. pp. 175-252 In: The Biology of Marine Mammals, H. T. Andersen (ed), Academic Press, New York.

Jellison, W. M. 1953 A beaked whale, Mesoplodon sp., from the Pribilofs. J. Mammal., 34:249-51.

Jenkins, D. G. 1968 Planktonic Foraminiferida as indicators of New Zealand Tertiary Paleotemperatures. Tuatara, 16:32-7.

Jenkins, J. T. 1921 A history of the whale fisheries from the Basque fisheries of the tenth century to the hunting of the finner whale at the present date. London. 336 pp.

Jenkins, J. T. 1932 Whales and modern whaling. London. 239 pp.

Joensen, J. S. 1962 (Pilot whales (Globicephalus melaena Traill) killed in Faroe 1940-62.) Frodskaparrit, 11:34-44.

Johnsen, A. O. 1960 Den moderne hvalfangsts historie. I. Oslo.

Johnston, T. H. & P. M. Mawson 1939 Internal parasites of the pigmy sperm whale. Rec. S. Aust. Mus., 6:263-74.

Jonsgaard, A. 1951 Studies on the little piked or minke whale (Balaenoptera acutorostrata Lacépède.) Norsk Hvalfangsttid, 40:209-32.

Jonsgaard, A. 1952 On the growth of the fin whale (Balaenoptera physalus) in different waters. Norsk Hvalfangsttid, 41:57-65.

Jonsgaard, A. 1953 Fin whale (Balaenoptera physalus) with six foetuses. Norsk Hvalfangsttid, 42:685-6.

Jonsgaard, A. 1955 The stocks of blue whales (Balaenoptera musculus) in the northern Atlantic Ocean and adjacent Arctic waters. Norsk Hvalfangsttid, 44:505-19.

Jonsgaard, A. 1962 Population studies on the minke whale *Balaenoptera acutorostrata* Lacépède. pp. 159-67. In: Le Cren, E. D. and M. W. Holdgate. The Exploitation of Natural Animal Populations. Oxford.

Jonsgaard, A. 1966 The distribution of Balaenopteridae in the North Atlantic Ocean. pp. 114-124. In: K. S. Norris (Ed.). Whales, Dolphins and Porpoises. University of California Press, Los Angeles.

Jonsgaard, A. 1969 Age determination of marine mammals. pp. 1-30, In: The Biology of Marine Mammals, H. T. Andersen (ed), Academic Press, New York.

Jonsgaard, A. & P. Hoidal 1957 Strandings of Sowerby's whale *(Mesoplodon bidens)* on the west coast of Norway. Norsk Hvalfangsttid, 46:507-12.

Jonsgaard, A. & E. J. Long 1959 Norway's small whales. Sea Frontiers, 5:168-74.

Jonsgaard, A. & P. B. Lyshoel 1970 A Contribution to the Knowledge of the Biology of the Killer Whale. Nytt mag. for Zool. 18:41-48.

Jonsgaard, A. & P. Oynes 1952 Om bottlenosen og spekhoggeren. Fauna, Oslo, 1:18 pp.

Jorpes, V. E. 1950 The insulin content of whale pancreas. Hvalrad. Skr., 35:1-15.

Juel, N. 1888 Hvalfangsten i Finmarken. Norsk Fiskgritidende, 1888:1-231.

Junge, G. C. A. 1950 On a specimen of the rare fin whale, *Balaenoptera edeni* Anderson, stranded on Puli Sugi, near Singapore. Zool. Verh., Leiden, 9, 26 pp.

Kanwisher, J. & H. Leivestad 1957 Thermal regulation in whales. Norsk Hvalfangsttid., 46:1-5.

Kanwisher, J. & A. Senft 1960 Physiological measurements on a live whale. Science, 131:1379-80.

Kanwisher, J. & G. Sundnes 1966 Thermal regulation in Cetaceans. pp. 397-409, In: K. S. Norris (Ed.). Whales, Dolphins and Porpoises. University of California Press, Los Angeles.

Kasuya, T. & T. Ichihara 1965 Some information on minke whales from the Antarctic. Scient. Rep. Whales Res. Inst., Tokyo, 19:37-43.

Kasuya, T. & S. Ohsumi 1966 A secondary sexual character of the sperm whale. Scient. Rep. Whales Res. Inst., Tokyo, 20:89-94.

Kellogg, R. 1922 Pinnipeds from Miocene and Pleistocene deposits of California. Bull. Dep. Geol. Univ. Calif. 13:23-132.

Kellogg, R. 1929 What is known of the migration of some of the whalebone whales. Rep. Smithson. Instn., 1928:467-94.

Kellogg, R. 1936 A review of the Archaeoceti. Publ. Carneg. Instn., 482:1-366.

Kellogg, R. 1940 Whales, giants of the sea. Nat. Geogr., 77:35-90.

Kellogg, R. 1942 Tertiary, Quaternary ,and Recent marine mammals of South America and the West Indies. Proc. 8th Amer. Sci. Congr., Washington, 1940, 3:445-73.

Kellogg, W. N. 1960 Auditory scanning in the dolphin. Psych. Rev., 10:25-7.

Kellogg, W. N. 1961 Porpoises and Sonar. University of Chicago Press, Chicago. 177 pp.

Kellogg, W. N. & C. E. Rice 1966 Visual discrimination and problem solving in a bottlenose dolphin. pp. 731-54. In: Whales, Dolphins and Porpoises, K. S. Norris (ed), University of California Press, Los Angeles.

Kenyon, K. W. 1952 A bottlenose dolphin from the California coast. J. Mammal., 33:385-7.

Kenyon, K. W. 1961 Cuvier beaked whales stranded in the Aleutian Islands. J. Mammal., 42:71-6.

Kenyon, K. W. & V. B. Scheffer 1955 The seals, sea-lions and sea otter of the Pacific coast. U.S. Dept. Interior, Fish and Wildl. Ser. Circ., 32:1-34.

Kenyon, K. W. V. B. Scheffer, & D. G. Chapman 1954 A population study of the Alaska fur-seal herd. U.S. Dept. Interior, Fish and Wildl. Spec. Sci. Rept. Wildl., 12:1-77.

Kenyon, K. W. & F. Wilke 1953 Migration of the northern fur seal, *Callorhinus ursinus*. J. Mammal., **34**: 86-98.

Kettlewell, H. B. D. & R. W. Rand 1955 Elephant seal cow and pup on South African coast. Nature, 175: 1000-1.

Keyes, I. W. 1968 Cenozoic marine temperatures indicated by the Scleractinian coral fauna of New Zealand. Tuatara, 16:21-25.

Kimura, S. 1957 The twinning in southern fin whales. Scient. Rep. Whales Res. Inst., Tokyo, 12:103-25.

Kimura, S. & T. Nemoto 1956 Note on a minke whale kept alive in aquarium. Scient. Rep. Whales Res. Inst., Tokyo, 11:181-9.

King, J. E. 1954 The otariid seals of the Pacific coast of America. Bull. Brit. (Nat. Hist.) Zool., 2:309-37.

King, J. E. 1957 On a pup of the crabeater seal *Lobodon carcinophagus*. Ann. Mag. nat. Hist., 10:619-24.

King, J. E. 1959a The northern and southern populations of *Arctocephalus gazella*. Mammalia, 23:19-40.

King, J. E. 1959b A note on the specific name of the Kerguelen fur seal. Mammalia, 23:381.

King, J. E. 1961 The feeding mechanism and jaws of the crabeater seal *(Lobodon carcinophagus)*. Mammalia, 25:462-6.

King, J. E. 1962 Some of the aquatic modifications of seals. Norsk Hvalfangsttid., 51:104-20.

King, J. E. 1964 Seals of the world. British Museum Press (London). 154 pp.

Kirpichnikov, A. A. 1950 (Observations on the distribution of large whale species in the Atlantic Ocean.) Priroda, 10:63-4.

Kleinenberg, S. E. 1956 (Mammals of the Black and Azov Seas.) Publ. Acad. Sci. USSR., Moscow.

Kleinenberg, S. E. 1959 On the origin of Cetacea. Proc. Int. Congr. zool., 15:445-7.

Klumov, S. K. 1954 (On the reproduction of the Black Sea dolphin) Trudy Inst. Okeanol., 8:206-19.

Klumov, S. K. 1962 (Right whale [Japanese] of the Pacific Ocean.) Trudy Inst. Okeanol., 58:202-97.

Knox, F. J. 1850 The whale and whaling. New Zealand Mag. 1850.

Knox, F. J. 1871 Observations on the Ziphidae, a family of the Cetacea. Trans. Proc. N.Z. Inst., 3:125-8.

Kojima, T. 1951 On the brain of the sperm whale *(Physeter catodon* L.). Scient. Rep. Whales Res. Inst., 6:49-72.

Kooyman, G. L. & H. T. Andersen 1969 Deep diving. pp. 65-94 In: The Biology of Marine Mammals, H. T. Andersen (ed), Academic Press, New York.

Korotkevich, E. S. 1958 (Antarctic mummies.) Sov. Ant. Exp., 1: 89-91.

Kotov, V. A. 1958 [On the distribution of *B. bairdi*.] Priroda, 8:119.

Kramer, G. 1954 Ueber relatives Wachstum bei Bartenwalen. Zool. Anz., 152:58-64.

Krefft, G. 1870 Notes on the skeleton of a rare whale probably identical with *Dioplodon seychellensis*. Proc. zool. Soc. Lond., 1870: 426-7.

Kristensen, L. 1896 Journal of the Right-whaling Cruise of the Norwegian Steamship "Antarctic" in the South Polar Seas under the Command of Capt. Christensen during the Years 1894-95. Trans. R. Geogr. Soc. Australia 1896.

Kruger, L. 1959 The thalamus of the dolphin (Tursiops truncatus) and comparison with the other mammals. J. Comp. Neurol., 111:133-94.

Kruger, L. 1966 Specialised features of the Cetacean brain. pp. 232-54. In: K. S. Norris (Ed.). Whales, Dolphins and Porpoises. University of California Press, Los Angeles.

Kuekenthal, W. 1886 (White whale fishery.) Deutsche Geographische Blaetter, 11:38.

Kuekenthal, W. 1887 (Bottlenose whaling.) Deutsche Geographische Blaetter, 12:6.

Kuekenthal, W. 1890 (Finmark whaling.) Deutsche Geographische Blaetter, 18:14.

Lami, R. 1955 Grampus griseus Cuv. dans la région Malouine. Bull. Lab. marit. Dinard, 41:60.

Lami, R. 1961 Le dauphin de Risso Grampus griseus Cuv. au Cap Frehel. Penn ar Bed N.S., 3:102.

Lang, T. G. 1966 Hydrodynamic analysis of Cetacean performance. pp. 410-432. In: Whales, Dolphins and Porpoises, K. S. Norris (ed.), University of California Press, Berkeley, Los Angeles.

Langworthy, O. R. 1931 Factors determining the differentiation of the cerebral cortex in sea-living mammals (the Cetacea): A study of the brain of the porpoise, Tursiops truncatus. Brain, 54:225-36.

Langworthy, O. R. 1932 A description of the central nervous system of the porpoise (Tursiops truncatus). J. comp. Neurol., 54:437-99.

Langworthy, O. R. 1935 The brain of the whalebone whale, Balaenoptera physalus. Bull. Johns Hopk. Hosp., 57:143-7.

Langworthy, O. R. F. H. Hesser, & L. C. Kolb 1938 A physiological study of the cerebral cortex of the Hair Seal (Phoca vitulina). J. comp. Neurol., 69:351-69.

Lauer, B. H. & B. E. Baker 1969 Whale milk. I. Fin whale (Balaenoptera physalus) and beluga whale (Delphinapterus leucas) milk: gross composition and fatty acid constitution. Can. J. Zool., 47:95-7.

Law, P. G. & T. Burstall 1953 Heard Island. Interim Rep. Aust. Natl. Antarct. Res. Exped., 7:1-32.

Laws, R. M. 1952a The elephant seal industry at South Georgia. Polar Rec., 6:746-54.

Laws, R. M. 1952b Seal marking methods. Polar Rec., 6:359-61.

Laws, R. M. 1952c A new method of age determination for mammals with special reference to the elephant seal. Nature, 169:972-4.

Laws, R. M. 1953a The seals of the Falkland Islands and Dependencies. Oryx, 2:87-97.

Laws, R. M. 1953b The elephant seal (Mirounga leonina Linn.) I. Growth and age. Falkland Is. Depend. Surv. Sci. Reps., 8:1-62.

Laws, R. M. 1956a The elephant seal (Mirounga leonina Linn.) II. General, social and reproductive behaviour. Falkland Is. Depend. Surv. Sci. Rep., 13:1-88.

Laws, R. M. 1956b The elephant seal (Mirounga leonina Linn.) III. The physiology of reproduction. Falkland Is. Depend. Surv. Sci. Rep., 15:1-66.

Laws, R. M. 1957 Polarity of whale ovaries. Nature, London, 179:1011-1012.

Laws, R. M. 1958 Growth rates and age of crabeater seals (Lobodon carcinophagus, Jacquinot and Pucheran). Proc. zool. Soc. Lond., 130:275-88.

Laws, R. M. 1959 The foetal growth rates of whales, with special reference to the fin whale, Balaenoptera physalus Linn. 'Discovery' Rep. 29:281-308.

Laws, R. M. 1960 The southern elephant seal. Norsk Hvalfangsttid., 49:466-476; 520-42.

Laws, R. M. 1961 Reproduction, growth and age of southern fin whales. 'Discovery' Rep., 31:327-486.

Laws, R. M. 1962 Age determination of pinnipeds with special reference to growth layers in the teeth. Z. Saeugetierk., 27:129-46.

Laws, R. M. & P. E. Purves 1956 The ear plug of the Mysticeti as an indicator of age, with special reference to the North Atlantic fin whale (Balaenoptera physalus Linn.) Norsk Hvalfangsttid., 45:413-25.

Laws, R. M. & R. J. F. Taylor 1957 A mass dying of crabeater seals, Lobodon carcinophagus (Gray). Proc. zool. Soc. Lond., 129:315-24.

Le Danois, E. 1912 Description d'un embryon de Grampus griseus Gray. Arch. Zool. exp. Gén., Ser. 5, 8:399-419.

Le Souef, A. S. 1929 Occurrence of the crabeating seal Lobodon carcinophaga, Hombron and Jacquinot, in New South Wales. Aust. J. Zool., 6:99.

Lenfant, C. 1969 Physiological properties of blood of marine mammals. pp. 95-116 In: The Biology of Marine Mammals, H. T. Andersen (ed), Academic Press, New York.

Lewis, F. 1942 Notes on Australian seals. Vic. Nat., Melbourne, 59:24-26.

Lillie, D. G. 1910 Observations on the anatomy and general biology of some members of the larger Cetacea. Proc. zool. Soc. Lond., 1910:769-92.

Lillie, D. G. 1915 Cetacea. Br. Antarct. 'Terra Nova' Exped. 1910, Nat. Hist. Rep. Zool., 1:85-124.

Lilly, J. C. 1961 Man and Dolphin. Doubleday, New York. 312 pp.

Lilly, J. C. 1962 Vocal behaviour of the bottlenose dolphin. Proc. Amer. Philosophical Soc., 106:520-9.

Lilly, J. C. 1963 Distress call of the bottlenose dolphin: stimuli and evoked behavioural responses. Science, 139:116-8.

Lilly, J. C. 1966 Sonic-ultrasonic emissions of the bottlenose dolphin. pp. 503-9. In: Whales, Porpoises and Dolphins, K. S. Norris (ed.), University of California Press, Los Angeles.

Lilly, J. C. & A. M. Miller 1961 Sounds emitted by the bottlenose dolphin. Science, 133:1689-93.

Lindsey, A. A. 1937 The Weddell seal in the Bay of Whales, Antarctica. J. Mammal., 18:127-44.

Lindsey, A. A. 1938 Notes on the crabeater seal. J. Mammal., 19:459-61.

Ling, J. K. 1965a Hair growth and moulting in the southern elephant seal, Mirounga leonina (Linn.). In: Biology of the Skin and Hair Growth. Ed. A. G. Lyne and B. F. Short, Angus and Robertson, Sydney. pp. 525-44.

Ling, J. K. 1965b Functional significance of sweat glands and sebaceous glands in seals. Nature, 208:560-2.

Ling, J. K. 1966 The skin and hair of the southern elephant seal, Mirounga leonina (Linn.) I. The facial vibrissae. Aust. J. Zool., 14:855-66.

Ling, J. K. 1968 The skin and hair of the southern elephant seal, *Mirounga leonina* (L.). III. Morphology of the adult integument. Aust. J. Zool., 16:629-45.

Ling, J. K. & D. G. Nicholls 1963 Immobilization of elephant seals using succinylcholine chloride. Nature, 200: 1021-2.

Ling, J. K. & C. D. B. Thomas 1967 The skin and hair of the southern elephant seal, *Mirounga leonina* (L.). II. Pre-natal and early post-natal development and moulting. Aust. J. Zool., 15: 349-65.

Littlepage, J. L. 1963 Diving behavior of a Weddell seal wintering in McMurdo Sound, Antarctica. Ecology, 44:775-7.

Littlepage, J. L. & J. S. Pearse 1962 Biological and oceanographic observations under an Antarctic ice shelf. Science, 137:689-91.

Liversidge, D. 1950 The elephant seals in the Antarctic. Discovery, 11:253-5.

Longman, H. A. 1926 New records of Cetacea, with a list of Queensland species. Mem. Qu. Mus., 8: 266-78.

Lubbock, B. 1937 The Arctic Whalers. Glasgow, i-xii, 1-483.

MacKay, J. J. 1886 The arteries of the head and neck and the rete mirabile of the porpoise. Proc. Phil. Soc. Glasgow, 17:366-376.

Mackintosh, N. A. 1942 The southern stocks of whalebone whales. 'Discovery' Rep., 22:197-300.

Mackintosh, N. A. 1959 Biological problems in the regulation of whaling. Norsk Hvalfangsttid., 48-395-404.

Mackintosh, N. A. 1965 The Stocks of Whales. Fishing News (Books) Ltd., London. 232 pp.

Mackintosh, N. A. 1966 The Distribution of Southern Blue and Fin Whales. pp. 125-44. In: Whales, Dolphins and Porpoises, K. S. Norris (ed.), University of California Press, Los Angeles.

Mackintosh, N. A. & J. F. G. Wheeler 1929 Southern Blue and Fin Whales. 'Discovery' Rep., 1:257-540.

Macy, O. 1835 The history of Nantucket; being a compendious account of the first settlement of the island by the English, together with the rise and progress of the whale fishery. Boston: i-xi; 1-300.

Mansfield, A. W. 1958 The breeding behaviour and reproductive cycle of the Weddell seal (*Leptonychotes weddelli* Lesson). Sci. Rep. Falkland Islands Dept. Survey, 18:1-44.

Manville, R. H. & R. P. Shanahan 1961 *Kogia* stranded in Maryland. J. Mammal., 42:269-70.

Marelli, C. A. 1953 Documentos incongraficos sobre cetaceos de las costas Argentinas. Falsa orca, delphin blanco, delphin de barard, tursion. An. Mus. Nahuel Huapi, Buenos Aires, 3:133-43.

Markham, C. R. 1881 On the whale fishery of the Basque Province of Spain. Proc. zool. Soc. Lond., 1881: 969-76.

Markowski, S. 1952 The cestodes of seals from the Antarctic. Bull. Brit. Mus. (Nat. Hist.), Zool., 1: 125-50.

Marlow, B. J. 1963 Rare whale washed up on Sydney Beach (*M. layardi*). Aust. Nat. Hist., 14:164.

Marples, B. J. 1949a Vertebrate palaeontology in New Zealand. Tuatara, 2:103-8.

Marples, B. J. 1949b Two endocranial casts of Cetaceans from the Oligocene of New Zealand. Amer. J. Sci., New Haven, 247:462-71.

Marples, B. J. 1956 Cetotheres (Cetacea) from the Oligocene of New Zealand. Proc. zool. Soc. Lond., 126:565-80.

Mathew, A. P. 1948 Stranding of a whale (*Megaptera nodosa*) on the Travancore coast in 1943. J. Bombay Nat. Hist. Soc., 47:732-3.

Matsushita, T. 1955 Daily rhythmic activity of the sperm whales in the Antarctic Ocean. Bull. Jap. Soc. Scient. Fish., 20:770-3.

Matsuura, Y. 1936a (Breeding habits of the sperm whale in the adjacent waters of Japan.) Zool. Mag. Japan, 48:260-6.

Matsuura, Y. 1936b (On the lesser rorqual found in the adjacent waters of Japan.) Bull. Jap. Soc. Scient. Fish., 4:325-30.

Matthews, L. H. 1929 The natural history of the elephant seal. 'Discovery' Rep., 1:234-55.

Matthews, L. H. 1932 Lobster-krill: Anomuran Crustacea that are the food of whales. 'Discovery' Rep., 5:467-84.

Matthews, L. H. 1937 The humpback whale, *Megaptera nodosa*. 'Discovery' Rep., 17:7-92.

Matthews, L. H. 1938a The Sperm whale, *Physeter catodon*. 'Discovery' Rep., 17:93-168.

Matthews, L. H. 1938b Notes on the Southern Right whale, *Eubalaena australis*. 'Discovery' Rep., 17: 169-81.

Matthews, L. H. 1938c The Sei whale, *Balaenoptera borealis*. 'Discovery' Rep., 17:183-290.

Matthews, L. H. 1948 Whales. Proc. phil. Soc. Glasgow, 73: 1-13.

Matthews, L. H. 1950 The male urogenital tract in *Stenella frontalis* (G. Cuvier). Atlantide Repts., 1:223-47.

Matthews, L. H. 1952 Sea Elephant. McGibbon and Kee, London. 1952. 190 pp.

Matthews, L. H. 1968 The Whale. Simon and Schuster, New York.

Matthiessen, L. 1893 Ueber den physikalisch-optischen Bau der Augen Knoelwal und Finnwal. Zeitschr. vergl. Augenheilk. 7:77.

Maxwell, G. 1967 Seals of the World. Constable, London.

Mayer, W. V. 1950 *Tursiops gillii*, the bottlenosed dolphin, a new record from the Gulf of California, with remarks on *Tursiops nuuanu*. Amer. Midl. Nat., 43:183-5.

McBride, A. F. & D. O. Hebb 1948 Behaviour of the captive bottlenose dolphin, *Tursiops truncatus*. J. comp. Physiol. Psychol., 41:111-23.

McBride, A. F. & Kritzler, H. 1951 Observations on pregnancy, parturition, and post-natal behaviour in the bottlenose dolphin. J. Mammal., 32:251-66.

McCann, C. 1961 The occurrence of the southern bottle-nosed whale, *Hyperoodon planifrons* Flower, in New Zealand waters. Rec. Dom. Mus., Wellington, 4:21-7.

McCann, C. 1962a Key to the Family Ziphiidae, beaked whales. Tuatara, 10:13-18.

McCann, C. 1962b The taxonomic status of the beaked whale, *Mesoplodon hectori* (Gray)—Cetacea. Rec. Dom. Mus., Wellington, 4:83-94.

McCann, C. 1962c The taxonomic status of the beaked whale, *Mesoplodon pacificus* Longman-Cetacea. Rec. Dom. Mus., Wellington, 4: 95-100.

McCann, C. 1964a The female reproductive organs of Layard's beaked-whale, *Mesoplodon layardi* (Gray). Rec. Dom. Mus., Wellington, 4:311-6.

McCann, C. 1964b A coincidental distribution pattern of some of the larger marine animals. Tuatara, 12:119-24.

McCann, C. 1964c Key to the seals (Pinnipedia) of New Zealand. Tuatara. 12:40-9.

McCann, C. & F. H. Talbot — 1963 — The occurrence of True's beaked whale (*Mesoplodon mirus* True) in South African waters, with a key to South African species of the genus. Proc. Linn. Soc. Lond., 175:137-44.

McGinitie, G. E. & N. McGinitie — 1949 — Natural History of Marine Animals. McGraw-Hill, Inc., New York. 523 pp.

McKenzie, R. A. — 1940 — Some marine records from Nova Scotia fishing waters. Proc. Nova Scotian Inst. Sci., Halifax, 20:42-6.

M. C-L — 1960 — Whales on our Coast. News Bull. zool. Soc. S. Afr., 2:23-4.

McLaren, I. A. — 1960 — Are the Pinnipedia biphyletic? Syst. Zool., 9:18-28.

McNab, R. — 1907 — Murihiku and the Southern Islands. William Smith, Invercargill, N.Z.

McNab, R. — 1913 — Old whaling days. A history of Southern New Zealand from 1830 to 1840. Melbourne. pp. i-xiii; 1-508.

McNab, R. — 1914 — From Tasman to Marsden. Wilke Press, Dunedin.

Medcof, J. C. — 1963 — Partial survey and critique of Ceylon's marine fisheries, 1953-55. Bull. Fish. Res. Sta. Ceylon, 16:29-118.

Meek, A. — 1918 — The reproductive organs of the Cetacea. J. Anat., 52:186-210.

Meschkat, A. — 1956 — Bei den Pelzrobben von Cape Cross. Kosmos Stuttgart, 52:461-7.

Miller, G. S. — 1920 — American records of whales of the genus *Pseudorca*. Proc. U.S. natn. Mus., 57:205-7.

Mitchell, E. D. — 1961 — A new walrus from the Imperial Pliocene of Southern California: With notes on odobenid and otariid humeri. Los Angeles Contrib. Sci. County Mus., 44:1-28.

Mitchell, F. D. — 1962 — A walrus and a sea lion from the Pliocene Purisima formation at Santa Cruz, California, with remarks on the type, locality and geological age of the sea lion *Dusignathus santacruzensis* Kellogg. Los Angeles Contrib. Sci. County Mus. 56:3-24.

Mitchell, E. D. — 1965 — Evidence for mass strandings of the False Killer whale (*Pseudorca crassidens*) in the Eastern North Pacific Ocean. Norsk Hvalfangsttid., 54:172-7.

Mitchell, E. D. — 1966 — Faunal succession of extinct North Pacific marine mammals. Norsk Hvalfangsttid., 55:47-60.

Mitchell, E. D. — 1967 — Controversy over Diphyly in Pinnipeds. Syst. Zool., 16:350-1.

Mitchell, E. D. — 1968a — The Mio-Pliocene Pinniped *Imagotaria*. J. Fish. Res. Bd. Canada, 25:1843-1900.

Mitchell, E. D. — 1968b — Northeast Pacific stranding, distribution and seasonality of Cuvier's beaked whale, *Ziphius cavirostris*. Can. J. Zool., 46:265-79.

Mitchell, E. D. — 1968c — North Atlantic whale research. Ann. Rev. Fisheries Council of Canada Reprint 1968:45, 47-8.

Mitchell, E. D. & W. J. Houck — 1967 — Cuvier's beaked whale (*Ziphius cavirostris*) stranded in Northern California. J. Fish. Res. Bd. Canada, 24:2503-13.

Mohl, B. — 1964 — Preliminary studies on hearing in seals. Vidensk. Medd. Fra Dansk Naturn. Foren., 127:283-94.

Mohr, E. — 1923 — Die Saugetiere der Suedsee-Expedition der Hamburgischen wissenschaftlichen Stiftung 1908-1909. Mitt. zool. Staatsinst. Zool. Mus. Hamburg, 40:7-18.

Moore, J. C. — 1953 — Distribution of marine mammals to Florida waters. Am. Midl. Nat., 49:117-58.

Moore, J. C. — 1958 — A beaked whale from the Bahama Islands and comments on the distribution of *Mesoplodon densirostris*. Amer. Mus. Novitates., 1897:1-12.

Moore, J. C. — 1960 — New records of the Gulf-Stream beaked whale, *Mesoplodon gervaisi*, and some taxonomic considerations. Amer. Mus. Novitates, 1993:1-35.

Moore, J. C. — 1963a — Recognizing certain species of beaked whales of the Pacific Ocean. Am. Midl. Nat., 70:396-428.

Moore, J. C. — 1963b — The Weddell seal. Chicago Nat. Mus. Bull., 34:6-7.

Moore, J. C. — 1966 — Diagnoses and distributions of beaked whales of the genus *Mesoplodon* known from North American waters. pp. 32-61. In: Whales, Dolphins and Porpoises, K. S. Norris (ed.), University of California Press, Los Angeles.

Moore, J. C. — 1968 — Relationships among the living genera of beaked whales with classifications, diagnoses and keys. *Fieldiana zool.* 53(4):209-98.

Moore, J. C. & R. S. Palmer — 1955 — More piked whales from southern North Atlantic. J. Mammal., 36:429-33.

Moore, J. C. & F. G. Wood — 1957 — Differences between the beaked whales *Mesoplodon mirus* and *Mesoplodon gervaisi*. Am. Mus. Novitates, 1831:1-25.

Morrell, W. P. — 1935 — New Zealand. The Modern World Series. pp. i-xv; 1-365.

Morrison, P. — 1962 — Body temperatures in some Australian mammals. III. Cetacea (*Megaptera*). Biol. Bull., 123:154-69.

Morton, W. R. M. & H. C. Mulholland — 1961 — The placenta of the pilot whale, *Globicephala melaena*. J. Anat., London, 95:605.

Moses, S. T. — 1948 — Notes on two whales stranded in Baroda State. Proc. Indian Sci. Congr. 34th, 1948:1-3.

Munroe, H. — 1853 — Statistics relative to the Northern Whale Fisheries from 1772 to 1852. J. Statist. Soc. Lond., 17:34.

Murata, T. — 1951 — Histological studies on the respiratory portions of the lungs of Cetacea. Scient. Rep. Whales Res. Inst., Tokyo, 6:35-47.

Murdaugh, H. V., J. K. Brennan, W. W. Pyron & J. W. Wood — 1962 — Function of the inferior vena cava valve of the harbour seal. Nature, 194:700-1.

Murie, J. — 1871 — On Risso's grampus, *G. rissoanus* (Desm.) J. Anat. Physiol., Lond., 5:118.

Murray, M. D. & D. G. Nicholls — 1965 — Studies on the ectoparasites of seals and penguins. I. The ecology of the louse *Lepidophthirus macrorhini* Enderlein on the southern elephant seal *Mirounga leonina* (L.). Aust. J. Zool., 13:437-54.

Murray, M. D., M. S. R. Smith & Z. Soucek — 1965 — Studies on the ectoparasites of seals and penguins. II. The ecology of the louse *Antarctophthirus ogmorhini* Enderlein on the Weddell seal, *Leptonychotes weddelli* Lesson. Aust. J. Zool., 13:761-71.

Nasu, K. — 1963 — Oceanography and whaling ground in the subarctic region of the Pacific Ocean. Scient. Rep. Whales Res. Inst., Tokyo, 17:105-155.

Naumov, S. P. — 1933 — (The Seals of the U.S.S.R. pp. 1-105. In: Economically Exploited Animals of the U.S.S.R.) All-Union Cop. United Publ. House, Moscow.

Nemoto, T. — 1962 — Food of Baleen whales collected in recent Japanese Antarctic whaling expeditions. Scient. Rep. Whales Res. Inst., Tokyo, 16:89-103.

Nemoto, T. 1963 New records of sperm whales with protruded rudimentary hind limbs. Scient. Rep. Whale Res. Inst., Tokyo, 17:79-81.

Nemoto, T. 1964 School of Baleen whales in the feeding areas. Sci. Rep. Whales Res. Inst., Tokyo, 18:89-110.

Nemoto, T. & K. Nasu 1958 *Thysanoeessa macrura* as a food of baleen whales in the Antarctic. Sci. Rep. Whales Res. Inst., Tokyo, 13:193-9.

Nemoto, T. & K. Nasu 1963 Stones and other aliens in the stomachs of Sperm whales in the Bering Sea. Scient. Rep. Whales Res. Inst., Tokyo, 17:83-91.

Neuville, H. 1936 Le Pancréas des Cétacés. Livre Jubilaire E. Bouvier, 1936:19-23.

Newell, B. S. 1966 Seasonal changes in the hydrological and biological environments off Port Hacking, Sydney, N.S.W. Aust. J. mar. Freshwat. Res., 17:77-91.

Newman, M. A. & P. L. McGeer 1966 The capture and care of a killer whale, *Orcinus orca*, in British Columbia. Zoologica, 51:59-70.

Nikolov, D 1963 La chasse aux dauphins et la répartition de leurs troupes au large du littoral bulgare de la Mer Noire. Izv. tsentr. nauch. izst. Inst. Rib. Varna, 3:183-98.

Nikulin, P. G. 1937 (Sea lions of the Okhotsk Sea, their utilisation.) Bull. Pacif. Sci. Inst. Fish, Vladivostok, 10:35-48.

Nippgen, J. 1921 L'industrie de la Baleine aux Isles Falkland. La Géographie, 36:3.

Nishida, S. & J. Amemiya 1962 Morphological notes of internal organs of Crabeating seal. Antarctic Rec., 14:97-101.

Nishiwaki, M. 1950 Determination of the age of Antarctic blue and fin whales by the colour changes in crystalline lens. Scient. Rep. Whales Res. Inst., Tokyo, 4:115-61.

Nishiwaki, M. 1952 On the age-determination of Mystacoceti, chiefly blue and fin whales. Scient. Rep. Whales Res. Inst., Tokyo, 7:87-119.

Nishiwaki, M 1955 On the sexual maturity of the Antarctic male sperm whale (*Physeter catodon* L.) Scient. Rep. Whales Res. Inst., Tokyo, 10:143-9.

Nishiwaki, M. 1959 Humpback whales in Ryukyuan waters. Scient. Rep. Whales Res. Inst., Tokyo, 14:49-87.

Nishiwaki, M. 1960 Ryukyuan humpback whaling in 1960. Scient. Rep. Whales Res. Inst., Tokyo, 15:1-15.

Nishiwaki, M. 1962a *Mesoplodon bowdoini* stranded at Akita Beach, Sea of Japan. Scient. Rep. Whales Res. Inst., Tokyo, 16:61-77.

Nishiwaki, M. 1962b Observation on two mandibles of *Mesoplodon*. Scient. Rep. Whales Res. Inst., Tokyo, 16:79-82.

Nishiwaki, M. 1963 Taxonomical consideration on genera of Delphinidae. Scient. Rep. Whales Res. Inst., Tokyo, 17:93-103.

Nishiwaki, M. 1964 Revision of the article "Taxonomical consideration on genera of Delphinidae" in No. 17. Scient. Rep. Whales Res. Inst., Tokyo, 18:171-2.

Nishiwaki, M. 1966a Distribution and migration of the larger Cetaceans in the North Pacific as shown by Japanese whaling results. pp. 171-91. In: Whales, Dolphins and Porpoises, K. S. Norris (ed.). University of California Press, Los Angeles.

Nishiwaki, M. 1966b A discussion of rarities among the smaller cetaceans caught in Japanese waters. pp. 192-204. In: Whales, Dolphins and Porpoises, K. S. Norris (ed.). University of California Press, Los Angeles.

Nishiwaki, M. & C. Handa 1958 Killer whales caught in the coastal waters off Japan for the recent 10 years. Scient. Rep. Whales Res. Inst., Tokyo, 13:85-96.

Nishiwaki, M. & K. Hayashi 1950 Biological survey of fin and blue whales taken in the Antarctic season 1947-48 by the Japanese fleet. Scient. Rep. Whales Res. Inst., Tokyo, 3:132-90.

Nishiwaki, M. & T. Hibiya 1951 On the sexual maturity of the sperm whale (*Physeter catodon*) found in the adjacent waters of Japan (I). Scient. Rep. Whales Res. Inst., Tokyo, 6:153-65.

Nishiwaki, M. & T. Hibiya 1952 On the sexual maturity of the Sperm whale (*Physeter catodon*) found in the adjacent waters of Japan (II). Scient. Rep. Whales Res. Inst., Tokyo, 7:121-4.

Nishiwaki, M. T. Hibiya, & S. Ohsumi 1958 Age study of Sperm whale based on reading of tooth laminations. Scient. Rep. Whales Res. Inst., Tokyo, 13:135-53.

Nishiwaki, M. T. Ichihara & S. Ohsumi 1958 Age studies of fin whale based on ear plug. Scient. Rep. Whales Res. Inst., Tokyo, 13:155-69.

Nishiwaki, M. & T. Kamiya 1958 A beaked whale *Mesoplodon* stranded at Oiso Beach, Japan. Scient. Rep. Whales Res. Inst., Tokyo, 13:53-83.

Nishiwaki, M. & T. Kamiya 1959 *Mesoplodon stejnegeri* from the coast of Japan. Scient. Rep. Whales Res. Inst., Tokyo, 14:35-48.

Nishiwaki, M. & K. S. Norris 1966 A new genus, *Peponocephala*, for the Odontocete Cetacean species *Electra electra*. Scient. Rep. Whales Res. Inst., Tokyo, 20:95-100.

Nishiwaki, M. S. Ohsumi & T. Kasuya 1961 Age characteristics in the Sperm whale mandible. Norsk Hvalfangsttid., 50:499-507.

Nishiwaki, M. & T. Ohe 1951 Biological investigation on blue whales (*Balaenoptera musculus*) and fin whales (*Balaenoptera physalus*) caught by the Japanese Antarctic whaling fleets. Scient. Rep. Whales Res. Inst., Tokyo, 5:91-167.

Nishiwaki, M. & H. C. Yang 1961 A curiously tailed dolphin caught in Formosa. Norsk Hvalfangsttid., 50:507-12.

Norman, J. R. & F. C. Fraser 1937 Giant Fishes, Whales and Dolphins. Putnam, London. 361 pp.

Norris, K. S. (Ed.) 1961 Standardized methods for measuring and recording data on the smaller cetaceans. J. Mammal., 42:471-6.

Norris, K. S. 1964 Some problems of echolocation in Cetaceans. pp. 317-36. In: Marine Bioacoustics, Vol. I, W. N. Tavolga (ed.). Pergamon Press, New York, 413 pp.

Norris, K. S. 1965 Trained porpoise released in the open sea. Science, 147:1048-50.

Norris, K. S. (Ed.) 1966 Whales, Porpoises and Dolphins. University of California Press, Los Angeles.

Norris, K. S. 1968 The Evolution of Acoustic Mechanisms in Odontocete Cetaceans. pp. 297-324. In: Evolution and Environment, E. T. Drake (ed.). Yale University Press, New Haven.

Norris, K. S. 1969 The echolocation of marine mammals. pp. 391-423 In: The Biology of Marine Mammals, H. T. Andersen (ed) Academic Press, New York.

Norris, K. S. H. A. Baldwin & D. J. Samson 1965 Open ocean diving test with a trained porpoise. Deep Sea Res., 12:505-9.

Norris, K. S. & W. E. Evans 1967 Directionality of echo-location clicks in the rough-tooth porpoise, *Steno bredanensis* (Lesson). pp. 305-16. In: Marine Bio-acoustics Vol. 2. W. N. Tavolga (Ed.). Pergamon Press, New York.

Norris, K. S. & J. H. Prescott 1961 Observations on Pacific Cetaceans of Californian and Mexican waters. Univ. Calif. Publs. Zool., 63:291-402.

Norris, K. S.
J. H. Prescott,
P. V. Asa-Dorian
& P. Perkins
1961
An experimental demonstration of echo-location behaviour in the porpoise, *Tursiops truncatus* (Montagu). Biol. Bull., 120:163-76.

Nowell, P. T. 1956 The collection of foetal whale thymus glands on FF *Balaena* for medical research. Norsk Hvalfangsttid., 45:165-171.

Oesau, W. 1955 Hamburgs Groenlandfahrt auf Walfisch-fang und Robbenschlag von 17-19 Jahrundert. Gluckstadt. Hamburg. (Verlag J. J. Augustin): 1-316.

Ogawa, T. 1935 (Studien ueber die Zahnwale in Japan.) Pt. I. Botany and Zool., Tokyo, 4:1150-7.

O'Gorman, F. A. 1961 Fur seals breeding in the Falkland Islands Dependencies. Nature, 192:914-6.

O'Gorman, F. A. 1962a Observations on terrestrial locomotion in Antarctic seals. Proc. zoo. Soc. Lond., 141:837-51.

O'Gorman, F. A. 1962b The return of the Antarctic fur seal. New Scient., 20:374-6.

Ohlin, A. 1893 Some remarks on the bottlenose whale (*Hyperoodon*). K. fysiogr. Saellsk. Lund. Handlingar 4.

Ohno, M.
& K. Fujino
1952
Biological investigation on the whales caught in the Japanese Antarctic whaling fleets, season 1950-51. Scient. Rep. Whales Res. Inst., Tokyo, 7:125-88.

Ohsumi, S. 1960 Relative growth of the fin whale, *Balaenoptera physalus* (Linn.). Scient. Rep. Whales Res. Inst., Tokyo, 15:17-84.

Ohsumi, S. 1964a Examination on age determination of the fin whale. Scient. Rep. Whales Res. Inst., Tokyo, 18:49-88.

Ohsumi, S. 1964b Comparison of maturity and accumulation rate of corpora albicantia between the left and right ovaries in Cetacea. Scient. Rep. Whales Res. Inst., Tokyo, 18:123-48.

Ohsumi, S. 1965 Reproduction of the Sperm whale in the Northwest Pacific. Scient. Rep. Whales Res. Inst., Tokyo, 19:1-35.

Ohsumi, S. 1966 Sexual segregation of the sperm whale. Scient. Rep. Whales Res. Inst., Tokyo, 20:1-16.

Ohsumi, S.
T. Kasuya
& M. Nishiwaki
1963
The accumulation rate of dentinal growth layers in the maxillary tooth of the Sperm whale. Scient. Rep. Whales Res. Inst., Tokyo, 17:15-35.

Ohsumi, S.
M. Nishiwaki
& T. Hibiya
1958
Growth of Fin whale in the Northern Pacific. Scient. Rep. Whales Res. Inst., Tokyo, 13:97-133.

Ohta, K.,
T. Watarai,
T. Oishi,
Y. Ueshiba,
S. Hirose,
T. Yoshizawa,
Y. Akikusa,
M. Sato,
& H. Okano
1955
Composition of Fin whale milk. Scient. Rep. Whales Res. Inst., Tokyo, 10:151-67.

Okada, Y. 1938 A catalogue of vertebrates of Japan. Maruzen Press, Tokyo. 412 pp.

Okutani, T.
& T. Nemoto
1964
Squids as the food of Sperm whales in the Bering Sea and Alaskan Gulf. Scient. Rep. Whales Res. Int., Tokyo, 18:111-22.

Oliver, W. R. B. 1921 The crabeating seal in New Zealand. Trans. N.Z. Inst., 53:360.

Oliver, W. R. B. 1922a The Paremata whale. N.Z. Jl Sci. Technol., 5:125-6.

Oliver, W. R. B. 1922b The whales and dolphins of New Zealand. N.Z. Jl Sci. Technol., 5:129-41.

Oliver, W. R. B. 1922c A review of the Cetacea of the New Zealand seas. Proc. zool. Soc. Lond., 1922:557-85.

Oliver, W. R. B. 1924a Strap-toothed whale at Kaitawa Point, entrance to Porirua Harbour. N.Z. Jl Sci. Technol., 7:187-8.

Oliver, W. R. B. 1924b Stranded blackfish at Marsden Point. N.Z. Jl. Sci. Technol., 7:188-9.

Oliver, W. R. B. 1937 *Tasmacetus shepherdi*: A new genus and species of beaked whale from New Zealand. Proc. zool. Soc. Lond. 107, ser. B, 371-81.

Oliver, W. R. B. 1946 A pied variety of the coastal porpoise. Dom. Mus. Rec. Zool., 1:1-4.

Olsen, O. 1913 On the external characters and biology of Bryde's whale (*Balaenoptera brydei*), a new rorqual from the coast of South Africa. Proc. zool. Soc. Lond., 1913: 1073-90.

Olsen, O. 1915 Hvaler og Hvalfangsti Sydafrika. Bergens Mus. Aarbok, 1914-5, 5:1-56.

Ommanney, F. D. 1932 The urino-genital system of the Fin whale (*Balaenoptera physalus*). 'Discovery' Rep., 5:363.

Ommanney, F. D. 1933 Whaling in the Dominion of New Zealand. 'Discovery' Rep., 7:239-52.

Omura, H. 1950 Diatom infection on Blue and Fin whales in the Antarctic whaling area V (Ross Sea area). Sci. Rep. Whales Res. Inst., Tokyo, 4:14-26.

Omura, H. 1953 Biological study on humpback whales in the Antarctic whaling areas IV and V. Scient. Rep. Whales Res. Inst. Tokyo, 8:81-102.

Omura, H. 1955 Whales in the northern part of the North Pacific. Norsk Hvalfangsttid:, 44:323-42.

Omura, H. 1957 Osteological study of the little piked whale from the coast of Japan. Scient. Rep. Whales Res. Inst., Tokyo, 12:1-21.

Omura, H. 1958a North Pacific Right whale. Scient. Rep. Whales Res. Inst., Tokyo, 13:1-52.

Omura, H. 1958b Note on embryo of Baird's beaked whale. Scient. Rep. Whales Res. Inst., Tokyo, 13:213-4.

Omura, H. 1959 Bryde's whales from the coast of Japan. Scient. Rep. Whales Res. Inst., Tokyo, 14:1-33.

Omura, H. 1962a Bryde's whale occurs on the coast of Brazil. Scient. Rep. Whales Res. Inst., Tokyo, 16:1-5.

Omura, H. 1962b Further information on Bryde's whale from the coast of Japan. Scient. Rep. Whales Res. Inst., Tokyo, 16:7-18.

Omura, H. 1966 Bryde's whale in the Northwest Pacific. pp. 70-78. In: Whales, Dolphins and Porpoises, K. S. Norris (ed.). University of California Press, Los Angeles.

Omura, H.
& K. Fujino
1954
Sei whales in the adjacent waters of Japan. II. Further studies on the external characters. Scient. Rep. Whales Res. Inst, Tokyo, 9:89-103.

Omura, H.
K. Fujino
& S. Kimura
1955
Beaked whale *Berardius bairdi* of Japan, with notes on *Ziphius cavirostris*. Scient. Rep. Whales Res. Inst., Tokyo, 10:89-132.

Omura, H.
& T. Nemoto
1955
Sei whales in the adjacent waters of Japan. III. Relation between movement and water temperature of the sea. Scient. Rep. Whales Res. Inst., Tokyo, 10:79-87.

Omura, H.
& H. Sakiura
1956
Studies on the little piked whale from the coast of Japan. Scient. Rep. Whales Res. Inst., Tokyo, 11:1-37.

Orr, R. T. 1950 Rarity of the deep. Pacif. Disc., 3: 13-15.

Orr, R. T. 1951 Cetacean records from the Pacific coast of North America. Wasmann J. Biol., 9: 147-8.

Orr, R. T. 1953 Beaked whale *(Mesoplodon)* from California with comments on taxonomy. J. Mammal., 34:239-49.

Orr, R. T. 1965 The rough-toothed dolphin in the Galapagos Archipelago. J. Mammal., 46:101.

Orr, R. T. 1966 Cuvier's beaked whale in the Gulf of California. J. Mammal., 47:339.

Orr, R. T. & T. C. Poulter 1965 The Pinniped population of the Ano Nuevo Island, California. Proc. Calif. Acad. Sci., ser. 4, 32:377-404.

Osgood, W. H. 1943 The mammals of Chile. Publ. Field Mus., zool. ser., 30:1-268.

Ottestad, P. 1950 On age and growth of Blue whales. Hvalrad. Skr., 33:67-72.

Ottestad, P. 1956 On the size of the stock of Antarctic fin whales relative to the size of the catch. Norsk Hvalfangsttid., 45:298-308.

Owen, R. 1846 A History of British Fossil Mammals. London.

Owen, R. 1866 On some Indian Cetacea collected by Walter Elliot. Trans. zool. Soc. Lond., 6:17-47.

Parker, T. J. 1885 Notes on the skeleton and baleen of a fin whale *(Balaenoptera musculus?)* recently acquired by the Otago University Museum. Trans. Proc. N.Z. Inst., 17:3-13.

Parker, W. R. 1933 Pelorus Jack. Proc. Linn. Soc. London, 1933, 34:2-3.

Parry, D. A. 1949a The anatomical basis of swimming in whales. Proc. zool. Soc. Lond., 119:49-60.

Parry, D. A. 1949b The swimming of whales and a discussion of Gray's paradox. J. Exptl. Biol., 26:24-34.

Parry, D. A. 1949c The structure of whale blubber and its thermal properties. Quart. J. Microbiol. Sci., 90:13-26.

Paul, L. J 1968 Some seasonal water temperature patterns in the Hauraki Gulf, New Zealand. N.Z. Jl mar. Freshwat. Res., 2:535-58.

Paulian, P. 1953 Pinnipèdes, Cétacés, oiseaux des Iles Kerguelen et Amsterdam. Mission Kerguelen 1951. Mém. Inst. Sci. Madagascar, 8A:111-234.

Paulian, P. 1955 Sur l'âge et la croissance du Léopard de Mer *Hydrurga leptonyx* (de Blainville). Mammalia, 19:347-56.

Paulian, P. 1957a Note sur les phoques des Iles Amsterdam. de l'otarie de'Ile Amsterdam. Mammalia, Paris, 21:9-14.

Paulian, P. 1957b Note sur les phoques des Iles Amsterdam. Mammalia, 21:211-25.

Paulian, P. 1960 Sur la taille à la naissance et la durée de l'allaitement chez le léopard de Mer *(Hydrurga leptonyx* de Blainville). Mammalia, 24:468-75.

Paulian, P. 1964 Contribution à l'étude de l'otarie de l'Ile Amsterdam. Mammalia, 28:1-146.

Paulus, M. 1962 Etude ostéographique et ostéométrique sur un *Ziphius cavirostris* G. Cuvier, 1823, échoué à Marseille-Estaque en 1879 (Collection du Muséum de Marseille). Bull. Mus. Hist. nat. Marseille, 22:17-48.

Pearson, J. 1936 The whales and dolphins of Tasmania. Part I. External characters and habits. Pap. Proc. R. Soc. Tasm., 1935:163-92.

Pedersen, T. 1952 The milk fat of the Sperm whale. Norsk Hvalfangsttid., 41:300.

Perkins, J. 1945 Biology at Little America, III. The west base of the United States Antarctic Service Expedition 1939-1941. Proc. Amer. Phil. Soc., 89:270-84.

Peters, W. C. H. .1875 Ueber eine neue Art von Seebaeren, *Arctophoca gazella,* von der Kergueleninseln. Monatsber. K. P. Akad. Wissensch. Berlin, 1875:393-9.

Peterson, R. S. 1965 Drugs for handling fur seals. J.. Wildl. Man., 29:688-93.

Peterson, R. S. 1968 Social Behaviour in Pinnipeds. pp. 3-53. In: The Behaviour and Physiology of Pinnipeds, R. J. Harrison, R. C. Hubbard, R. S. Peterson, C. E. Rice and R. J. Schusterman (Eds.). Appleton Century Crofts Ltd., New York.

Peterson, R. S. & G. A. Bartholomew 1967 The natural history and behaviour of the California sea lion. Am. Soc. Mammal. Spec. Pub., 1:1-79.

Peterson, R. S. & W. G. Reeder 1966 Multiple births in the northern fur seal. Sonderdruck Aus. Z. f. Saeugetierkunde Bd., 31:52-6.

Péwé, T. R. N. R. Rivard & G. A. Llano 1959 Mummified seal carcasses in the McMurdo Sound region, Antarctica. Science, 130:716.

Philip, E. J. 1935 Whaling ways of Hobart Town. Hobart, Tasmania.

Pike, G. C. 1950 Stomach contents of whales caught off the coast of British Columbia. Proc. Rep. Pacif. Cst Stns., 83:27-8.

Pike, G. C. 1951 What do whales eat? Canadian Fisherman, 37:11.

Pike, G. C. 1953a Preliminary report on the growth of finback whales from the coast of British Columbia. Norsk Hvalfangsttid., 42:11-15.

Pike, G. C. 1953b Colour pattern of humpback whales from the coast of British Columbia. J. Fish. Res. Bd. Canada, 10:320-5.

Ping, C. 1926 On some parts of the visceral anatomy of the porpoise, *Neomeris phocoenoides.* Anat. Rec. Philadelphia, 33:13-28.

Polack, J. S. 1838 New Zealand; being a narrative of travels and adventures during a residence in that country between 1831 and 1837. 2 vols. London.

Polkey, W. & W. N. Bonner 1966 The pelage of the Ross seal. Br. Antarct. Surv. Bull., 8:93-6.

Postel, E. 1956 Echouage d'un baleinoptère aux Iles Kerkennah. Bull. Sta. oceanogr. Salammbo,. 53:75-6.

Postel, E. & Mayrat, A. 1956 Un souffleur s'échoué à Kherredine. Bull. Sta. oceanogr. Salammbo., 53:75.

Poulter, T. C. 1963 Sonar signals of the sea lion. Science, 139:753-4.

Poulter, T. C 1968 Marine Mammals. 78 pp. Ch. 18. In: Animal Communication, T. A. Sebeok (ed.). Indiana Press. (Photostat ms.)

Pratt, W. T 1877 Colonial experiences; or, incidents and reminiscences of thirty-four years in New Zealand. London.

Pressey, H. E. & S. Cobb 1928 Observations on the spinal cord of *Phocaena.* J. comp. Neurol. Philad., 47:75-89.

Preuss, F. 1954 Die tunica albuginea penis und ihre Trabekel bei Pferd und Rind. Anat. Anz., 101:64-83.

Purves, P. E. 1955 The wax plug in the external auditory meatus of the Mysticeti. 'Discovery' Rep., 27:293-302.

Purves, P. E. 1963 Locomotion in whales. Nature, 197:334-7.

Purves, P. E. 1966 Anatomy and Physiology of the Outer and Middle Ear in Cetaceans. pp. 320-80. In: Whales, Porpoises and Dolphins, K. S. Norris (ed.). University of California Press, Los Angeles.

Purves, P. E. & M. D. Mountford 1959 Ear plug laminations in relation to the age composition of a population of Fin whales (*Balaenoptera physalus*). Bull. Br. Mus. nat. Hist., Zool., 5:125-61.

Purves, P. E. & W. L. van Utrecht 1963 The anatomy and function of the ear of the bottlenosed dolphin, *Tursiops truncatus*. Beaufortia, 9:241-56.

Pycraft, W. P. 1932 On the genital organs of a female common dolphin (*Delphinus delphis*). Proc. zool. Soc. London, 1932 pp. 807-811.

Pyper, J. 1929 History of the whale fisheries of Aberdeen. Scot. Nat., 1929:39-50, 69-80, 103-8.

Quay, W. B. 1954 The blood cells of Cetacea with particular reference to the Beluga *Delphinapterus leucas* Pallas, 1776. Saeugetirkund. Mitteilungen, 2:49-54.

Quay, W. B. 1957 Pancreatic weight and histology in the White whale. J. Mammal., 38:185-192.

Rabot, C. 1919 The Norwegians in Spitsbergen. Geogr. Rev., 8:209-76.

Rand, R. W. 1949 Studies on the Cape fur seal. (*Arctocephalus pusillus* Schreber).
1. Age grouping in the female, i+16.
2. Attendance in the rookery, i+19.
3. Age grouping in the male, i+23.
Cape Town Dept. Agric. Govt. Guano Island Admin. Progr. Rep., 1949:1-58.

Rand, R. W. 1955a Reproduction in the female Cape fur seal, *Arctocephalus pusillus* (Schreber). Proc. zool. Soc. Lond., 124:717-40.

Rand, R. W. 1955b Marion Islands — home of South Africa's Elephant seal. Afr. Wild Life, 9:7-9.

Rand, R. W. 1956a Notes on the Marion Island Fur seal. Proc. zool. Soc. Lond., 126:65-82.

Rand, R. W. 1956b The Cape Fur seal (*Arctocephalus pusillus* (Schreber), its general characteristics and moult. Investl. Rep. Div. Fish. S. Afr., 21:1-52.

Rand, R. W. 1959 The Cape Fur seal (*Arctocephalus pusillus*). Distribution, abundance and feeding habits of the southwestern coast of the Cape Province. Investl. Rep. Div. Fish. S. Afr., 34:1-75.

Rand, R. W. 1962 Elephant seals on Marion Island. Afr. Wildlife, 16:191-8.

Rand, R .W. 1967 The Cape Fur seal (*Arctocephalus pusillus*) 3. General behaviour on land and at sea. S. Afr. Div. Sea Fish. Invesl. Rep., 60:1-39.

Rankin, J. J. 1953 First record of the rare beaked whale, *Mesoplodon europaeus* Gervais from the West Indies. Nature, London, 172:873.

Rankin, J. J. 1956 The structure of the skull of the Beaked whale, *Mesoplodon gervaisi* Deslongchamps. J. Morphol., 99:329-58.

Rankin, J. J. 1961 The bursa ovarica of the Beaked whale, *Mesoplodon gervaisi* Deslongchamps. Anat. Record, 139:379-85.

Rapp, W. 1836 Bemergungen ueber die Gehoerwerkzeuge der Cetacean. Frociep. Notizen, 49:116-21.

Ratcliffe, H. L. 1942 Autopsy of a male pigmy Sperm whale (*Kogia breviceps*). Notul. Nat., 112:1-4.

Raven, H. C. 1937 Notes on the taxonomy and osteology of two species of *Mesoplodon* (*M. europaeus* Gervais *M. mirus* True). Am. Museum Novitates, 905:1-30.

Raven, H. C. & W. K. Gregory 1933 The spermaceti organ and nasal passages of the Sperm whale (*Physeter catodon*) and other odontocetes. Amer. Mus. Novitates, 677:1-18.

Ray, C. 1960 *Trichecodon huxleyi* (Mammalia: Odobenidae) in the Pleistocene of southeastern United States. Bull. Mus. Comp. zool., Harvard, 12:129-142.

Ray, C. 1965 Physiological ecology of marine mammals at McMurdo Sound, Antarctica. Bioscience, April 1965: 274-7.

Ray, C. 1966 Snooping on seals for science in Antarctica. Animal Kingdom (June): 66-75.

Ray, C. & D. Lavellee 1964 Self-contained diving operations in McMurdo Sound, Antarctica: Observations on the sub-ice environment of the Weddell seal *Leptonychotes weddelli* (Lesson). Zoologica: 121-36.

Ray, C. & M. S. R. Smith 1968 Thermoregulation of the pup and adult Weddell seal. *Leptonychotes weddelli* (Lesson), in Antarctica. Zoologica, 53: 33-46.

Rayner, G. W. 1939 *Globicephala leucosagmaphora*, a new species of the genus Globicephala. Ann. Mag. nat. Hist., 4:543-44.

Renard, P. 1954 Notes sur un exemplaire d'*Hyperoodon rostratus* échoué à sallenelles, Calvadus, le 17 Février 1953. Bull. Soc. Linn. Norm., 7: 74-9.

Reysenbach de Haan, F. W. 1957 Hearing in whales. Acta Otolaryngol., Suppl., 134:1-114.

Rice, D. W. 1960 Distribution of the bottle-nosed dolphin in the Leeward Hawaiian Islands. J. Mammal., 41:407-8.

Rice, D. W. 1963a The whale marking cruise of the Sioux City off California and Baja California. Norsk Hvalfangsttid., 52:153-60.

Rice, D. W. 1963b Progress report on biological studies of the larger Cetacea in the waters off California. Norsk Hvalfangsttid., 52: 181-7.

Rice, D. W. K. W. Kenyon & D. B. Lluch 1965 Pinniped populations at Islas Guadalupe, San Benito and Cedros, Baja California, 1965. Trans. San Diego Soc. Nat. Hist., 14:73-84.

Rice, D. W. & V. B. Scheffer 1968 A List of the Marine Mammals of the World. U.S. Fish and Wildlife Service, Special Scientific Report, Fisheries No. 579, pages 1-16. A revision of an earlier list under the same title. Scheffer and Rice, 1963.

Richards, L. P. 1952 Cuvier's Beaked whale from Hawaii. J. Mammal., 33:255.

Rickard, L. S. 1965 The Whaling Trade in Old New Zealand. Minerva Press, Auckland, 163 pp.

Ricker, W. E. 1958 Handbook of computations for biological statistics of fish populations. Bull. Fish. Res. Bd. Canada, 119:1-300.

Ridgway, N. M. 1960 Surface water movements in Hawke Bay, New Zealand. N.Z. Jl. Geol. Geophys., 3:253-61.

Ridgway, S. H. 1965 Medical care of marine mammals. J. Vet. Med. Assn., 147:1077-86.

Riese, F. A. & O. R. Langworthy 1937 A study of the surface structure of the brain of the whale (*Balaenoptera physalus* and *Physeter catodon*). J. Comp. Neurol., 68:1-36.

Risting, S. 1922 Av Hvalfangstens Historie. Chr. Christensens Hvalfangstmuseum. Publ. No. 2, Oslo.

Roberts, A. 1951 The mammals of South Africa. S. Africa, 1951. pp. xlviii + 700 pp.

Robins, J. P. 1954 Ovulation and pregnancy corpora lutea in the ovaries of the humpback whale. Nature, 173:201-3.

Roe, H. S. J. 1967 Seasonal formation of laminae in the ear plug of the Fin whale. 'Discovery' Rep., 35:1-30.

Roest, A. I. R. M. Storm & P. C. Dumas 1953 Cuvier's beaked whale (*Ziphius cavirostris*) from Oregon. J. Mammal. 34:251-2.

Roest, A. I. 1964 *Physeter* and *Mesoplodon* strandings on the central California coast. J. Mammal., 45:129-36.

Roppel, A. Y. & S. P. Davey 1965 Evolution of Fur seal management on the Pribilof Islands. J. Wildl. Man., 29:448-63.

Rudmose-Brown, R. N. 1913 The seals of the Weddell sea: Notes on their habits and distribution. Scottish Natl. Antarctic Exped. Sci. Res. Voyage "Scotia", 1902-4. 4 (Zool.): 181-98.

Ruud, J. T. 1937 Bottlenosen, *Hyperoodon rostratus* (Mueller). Norsk Hvalfangsttid., 26:456-8.

Ruud, J. T 1940 The surface structure of the baleen plates as a possible clue to age in whales. Hvalrad. Skr., 23:1-23.

Ruud, J. T 1945 Further studies on the structure of the baleen plates and their application to age determination. Hvalrad. Skr., 29:1-69.

Ruud, J. T. 1949 Two whales stranded on Tunisian coast. Bull. Soc. Hist. Nat. Tunisié, 2:35-6.

Ruud, J. T. 1952 Catches of Bryde-whale off French Equatorial Africa. Norsk Hvalfangsttid., 41:662-3.

Ruud, J. T. 1956 International regulation of whaling, a critical survey. Norsk Hvalfangsttid., 45:374-87.

Ruud, J. T. 1959 The use of baleen plates in age determination of whales. Proc. XVth Int. Congr. Zool.:302-3.

Ruud, J. T. A. Jonsgaard & P. Ottestad 1950 Age studies on Blue whales taken in Antarctic seasons 1945-46, 1946-47 and 1947-48. Hvalrad. Skr., 33:1-66.

Saetersdal, G. J. Mejia & P. Ramirez 1963 La Caza de Cachalotes en el Peru. Bol. Inst. Invest. Los Recursos Mar., 1:45-84.

Sala de Castellarnua, P. I. 1945 Un extrano cetaceo en las costas de Valencia, *Ziphius cavirostris Cuvier*. An. Assoc. Esp. Progr.Cienc., 10:576-83.

Salvesen, T. E. 1912 The whaling industry of today. J. R. Soc. Arts, London, 60:515-23.

Sanderson, I. T. 1956 Follow the whale. Little, Brown and Company. Boston.

Sanford, F. C. 1884 Notes upon the history of the American whale fishery. Rept. U.S. Fish. Comm., 1882:205-20.

Santiago Carrara, I. 1952 Lobos marinos, Pingueinos y guaneras de Las Costas del litoral Maritimo e Islas Adyacentes de La Republica Argentina. Universidad Nac. de La Plata, Facultad de Ciencias Veterinarias, Catedra de Higiene e Industrias. Publicacion Especial, 189 pp.

Sapin-Jaloustre, J. 1953a Les Phoques de Terre Adélie (1). Mammalia, 16:179-212.

Sapin-Jaloustre, J. 1953b Les Phoques de Terre Adélie (11). Mammalia, 17:1-20.

Sapin-Jaloustre, J. 1953c L'Identification des Cétacés Antarctiques à la Mer. Mammalia, 17:221-59.

Scattergood, L. W. 1949 Notes on the Little Piked whale. Murrelet, 30:3-16.

Schaefer, M. B. 1953 Fisheries dynamics and the concept of maximum equilibrium catch. Proc. Gulf Caribbean Fish Inst. 6th Session: 1-11.

Scharff, R. F. 1910 The Irish whale fishery. Irish Nat. 1910: 229-33.

Scheffer, V. B. 1942 A list of the marine mammals of the west coast of North America. Murrelet, 23:42-7.

Scheffer, V. B. 1949 Notes on three Beaked whales from the Aleutian Islands. Pacific Sci., 3:353.

Scheffer, V. B. 1950 The striped dolphin *Lagenorhynchus obliquidens*, Gill, 1865 on the coast of North America. Am. Midl. Nat., 44:750-8.

Scheffer, V. B. 1953 Measurements and stomach contents of eleven delphinids from the North-East Pacific. Murrelet, 34:27-30.

Scheffer, V. B. 1958 Seals, Sea Lions and Walruses. A Review of the Pinnipedia. Stanford University Press, California. 179 pp.

Scheffer, V. B. (ed.) 1967 Standard measurements of seals. J. Mammal., 48:459-62.

Scheffer, V. B. & D. W. Rice 1963 A list of the marine mammals of the world. U.S. Fish and Wildlife Ser. Spec. Sci. Rept. Fisheries, 431:1-12.

Scheffer, V. B. & J. W. Slipp 1948 The whales and dolphins of Washington State, with a key to the cetaceans of the west coast of North America. Amer. Midl. Natl., 39:257-337.

Schevill, W. E. 1964 Underwater sounds of cetaceans. pp. 307-16 In: Marine Bioacoustics Vol. 1, W. N. Tavolga (ed), Pergamon Press, New York, 413 pp.

Schevill, W. E. C. Ray, K. W. Kenyon, R. T. Orr & R. C. van Gelder 1967 Immobilizing drugs lethal to swimming mammals. Science, 157:631-2.

Schevill, W. E. & W. A. Watkins 1965 Underwater calls of *Leptonychotes* (Weddell seal). Zoologica, 50:45-6.

Scholander, P. F. 1940 Experimental investigations on the respiratory function in diving mammals and birds. Hvalrad. Skr., 22:1-131.

Scholander, P. F. L. Irving & S. W. Grinnell 1942 On the temperature and metabolism of the seal while diving. J. Cell. comp. Physiol., 19:67-78.

Scholander, P. F. & W. E. Schevill 1955 Counter-current vascular heat exchange in the fins of whales. J. Appl. Physiol., 8:279-82.

Schubert, K. 1951 Das Pottwalverkommen an der Peru Kueste Fischereiwelt, 3:130-1.

Schubert, K. 1955 Der Wolfgang der Gegenwart. Stuttgart, 1955: 175 pp.

Sclater, W. L. 1901 The Mammals of South Africa. II. Porter & Co., London, 241 pp.

Scoresby, W. 1820 An account of the Arctic regions, with a history and description of the northern whale fishery. A. Constable, Edinburgh. 551 pp.

Scoresby, W. 1849 The northern whale fishery. London. B.M. 4420 f.

Scott, A. 1960 A bottle-nosed dolphin, *Tursiops truncatus*, on Co. Antrim shore. Irish Nat. J., 1960:183-4.

Scott, E. O. G. 1942 Records of Tasmanian Cetacea: No. II. A large school of the Pacific pilot whale, *Globicephalus melas* (Traill, 1809), stranded at Stanley, north western Tasmania, in October, 1935. Rec. Queen Vict. Mus., 1:5-34.

Scott, H. H. & C. E. Lord 1920a Studies of Tasmanian Cetacea. Part I. Pap. Proc. R. Soc. Tasm., 1919:1-17.

Scott, H. H. & C. E. Lord 1920b Studies of Tasmanian Cetacea. Part II. Pap. Proc. R. Soc. Tasm., 1920:23-32.

Scott, H. H. & C. E. Lord 1921 Studies of Tasmanian Cetacea Part IV. Pap. Proc. R. Soc. Tasm., 1920:1-10.

Scott, H. H. & C. Lord 1926 The eared seals of Tasmania. Pap. Proc. Roy. Soc. Tasm., 1925:75-78; 187-94.

Scott, J. H. & T. J. Parker 1889 On a specimen of *Ziphius* recently obtained near Dunedin. Trans. zool. Soc. Lond., 12:241-8.

Scott Johnson C. 1967 Sound Detection Thresholds in Marine Mammals. pp. 247-60. In: Marine Bioacoustics, Vol. 2, W. N. Tavolga (ed.). Pergamon Press, New York. 353 pp.

Sebeok, T. A. 1965 Animal communication. Science, 147:1006-14.

Serene, R. 1934 Sur un échouage de *Kogia breviceps* Gray à proximité de l'Institut Océanographique de Nhatrang (Annam). Bull. Mus. Hist. nat., Paris, ser. 2, 6:398-9.

Sergeant, D. E. 1958 Dolphins in Newfoundland waters. Canadian Field-Nat., 72:156-9.

Sergeant, D. E. 1959 Age determination in Odontocete whales from dentinal growth layers. Norsk Hvalfangsttid., 48:273-88.

Sergeant, D. E. 1962a The biology of the pilot or pot-head whale *Globicephala melaena* (Traill) in Newfoundland waters. Bull. Fish. Res. Bd. Canada, 132:1-84.

Sergeant, D. E. 1962b On the external characters of the Blackfish or Pilot whales (genus *Globicephala*). J. Mammal., 43:395-413.

Sergeant, D. E. 1963 Minke whales, *Balaenoptera acutorostrata* Lacépède, of the western North Atlantic. J. Fish. Res. Bd. Canada, 20:1489-1504.

Sergeant, D. E. & H. D. Fisher 1957 The smaller Cetacea of eastern Canadian waters. J. Fish. Res. Bd. Canada, 14:83-115.

Serventy, D. L. 1948 A record of the Leopard seal in western Australia. W. Aust. Nat., Perth, 1:155.

Sheppard, T. 1911 The evolution of Kingston-upon-Hull. Hull, 203 pp.

Sherrin, R. A. A. & J. H. Wallace 1890 Early history of New Zealand. Auckland.

Siebenaler, J. B. & D. K. Caldwell 1956 Co-operation among adult dolphins. J. Mammal., 37:126-8.

Silas, E. G. & C. K. Pillay 1962 The stranding of two false Killer whales (*Pseudorca crassidens* (Owen)) at Pozhikara, North of Cape Comorin. J. Mar. biol. Ass. India, 2:268-71.

Simpson, G. C. 1945 The Principles of Classification and a Classification of Mammals. Bull. Amer. Mus. Nat. Hist., 85:1-350.

Sivertsen, E. 1954 A survey of the eared seals (Family Otariidae) with remarks on the Antarctic seals collected by M/K "Norvegica" in 1928-1929. Det Norsk Videnskaps-Akademi, Oslo: 36 (1954): 1-76.

Sleeman, J. L. 1921 Whaling in New Zealand waters. Field, London 1921 (Dec.): 34-7.

Sleptsov, M. M. 1940 (Determination of age in *Delphinus delphis* L.) Byull. Mosk. Obshch. Ispyt. Prir., 49:43-51.

Slijper, E. J. 1936 Die Cetaceen. Diss. Utrecht 1936, Capita Zoologica No. 7, 590 pp.

Slijper, E. J. 1938 Vergleichend mikroskopische-anatomishe Untersuchungen ueber das Corpus cavernosum Penis der Cetaceen. Arch. Néerl. Zool. Suppl 1938, pp. 205-18.

Slijper, E. J. 1939 *Pseudorca crassidens* (Owens). Zool. Meded., Leiden, 21:241-366.

Slijper, E. J. 1946 Die physiologische Anatomie der verdauungsorgane bei den Vertebraten. Digestion. Tabulae Biologicae 23, 1:1-128.

Slijper, E. J. 1948 On the thickness of the layer of blubber in Antarctic Blue and Fin whales. Proc. Kon. Ned. Akad. Wet., 51:1033-45; 1114-24; 1310-16.

Slijper, E. J. 1958 Walvissen. D. B. Centen: Uilgeversmaatschappij. Amsterdam. 524 pp.

Slijper, E. J. 1961 Locomotion and locomotory organs in whales and dolphins (Cetacea). Symp. zool. Soc. Lond., 5:77-96.

Slijper, E. J. 1962 Whales. Hutchinson, London, 475 pp.

Slijper, E. J. 1966 Functional morphology of the reproductive system in Cetacea. pp. 277-319. In: Whales, Porpoises and Dolphins, K. S. Norris (ed.). University of California Press, Los Angeles.

Slipp, J. W. & F. Wilke 1953 The Beaked whale *Berardius* on the Washington coast. J. Mammal., 34:105-13.

Smith, M. S. R. 1965 Seasonal movements of the Weddell seal in McMurdo Sound, Antarctica. J. Wildl. Man., 29:464-70.

Smith, M. S. R. 1966a Studies on the Weddell seal (*Leptonychotes weddelli*) in McMurdo Sound, Antarctica. Ph.D. Thesis. University of Canterbury, Christchurch, New Zealand. 161 pp.

Smith, M. S. R. 1966b Injuries as an indicator of social behaviour in the Weddell seal (*Leptonychotes weddelli*). Mammalia, 30:241-6.

Sokolov, V. E. 1961 (Determination of the studies of mature cycle in females of *Lagenorhynchus acutus* Gray by method of vaginal smear.) Trudy Sovesh. Ichtiol. Kom., 12:68-71.

Soot-Ryen, T. 1961 On a Bryde's whale stranded on Curaçao. Norsk Hvalfangsttid., 50:323-32.

Sorensen, J. H. 1940 *Tasmacetus shepherdi*. History and description of specimens cast ashore on Mason's Bay, Stewart Island, in February, 1933. Trans. R. Soc. N.Z., 70:200-4.

Sorensen, J. H. 1950 Elephant seals of Campbell Island. Cape Exped. Ser. Bull., 6:5-31.

Sorensen, J. H. 1951 Wild Life in the Subantarctic. Whitcombe and Tombs, Christchurch, 85 pp. illus.

Sorensen, J. H. 1969a New Zealand Seals with Special Reference to the Fur Seal. Fish. tech. Rep. N.Z. Mar. Dep. 39:1-35.

Sorensen, J. H. (Ed.) 1969b New Zealand Fur Seals with Special Reference to the 1949 Open Season. Fish. tech. Rep. N.Z. Mar. Dep. 42:1-80.

Southwell, T. 1882 The bottlenose whale fishery in the North Atlantic Ocean. Rept. U.S. Fish. Comm., 1882.

Southwell, T. 1904 On the whale fishery from Scotland. Ann. Scot. Nat. Hist., 1904:77.

Southwell, T. 1905 Some results of the North Atlantic Fin whale fishery. Ann. Mag. Nat. Hist., 16:403-21.

Stager, K. E. & W. Reeder 1951 Occurrence of the False Killer whale *Pseudorca* on the California coast. Bull. S. Calif. Acad. Sci., 50:14-20.

Stannius, A. 1841 Ueber den Verlaug der Arterien bei *Delphinus phocaena*. Arch. Anat. Physiol. Wiss. Med. von J. Mueller: 380-402.

Starbuck, A. 1878 History of the American whale fishery from its earliest inception to the year 1876. Waltham, Mass. pp. 1-768.

Starks, E. C. 1922 History of California shore whaling. Calif. Fish Game Comm. Bull., 6:1-38.

Starr, R. B. 1961 A thermal profile and sea surface observations across the Southern Tasman Sea. N.Z. Jl. Geol. Geophys., 4:125-31.

Starrett, A. & P. Starrett 1955 Observations on young blackfish, *Globicephala*. J. Mammal., 36:424-9.

Stejneger, L. 1884 Contributions to the history of the Commander Islands. No. 1 — Notes on the natural history including descriptions of new Cetaceans. Proc. U.S. Natl. Mus., 6:58-89.

Stephen, A. C. 1932 Sowerby's whale (Mesoplodon bidens) at Inverness. Scot. Naturalist, 197:133.

Stephens, W. M. 1963 The Killer. Sea Frontiers 9:262-73.

Stephenson, W. 1951 The lesser rorqual in British waters. Rep. Dov. Mar. Lab (1949) 1951:7-48.

Stirling, I. 1966a The seals at White Island. A hypothesis on their origin. Antarctic, 4:310-13.

Stirling, I. 1966b A technique for handling live seals. J. Mammal., 47:543-4.

Stirling, I. 1966c Seal marking in McMurdo Sound. Antarctic, 4:363-4.

Stirling, I. 1967 Population studies on the Weddell seal. Tuatara, 15:133-41.

Stirling, I. 1968 Diurnal movements of the New Zealand fur seal at Kaikoura. N.Z. Jl mar. Freshwat. Res., 2:375-7.

Stirling, I. 1970 Observations on the behaviour of the New Zealand Fur seal. (Arctocephalus forsteri). J. Mammal., 51:766-78.

Stirling, I. & E. D. Rudolf 1968 Inland record of a live Crabeater seal in Antarctica. J. Mammal., 49:161-2.

Stonehouse, B. 1965a Birds and Mammals (of Antarctica). pp. 153-86. In: Antarctica, ed. T. Hatherton. Frederick A. Praeger Press, New York. 511 pp. illus.

Stonehouse, B. 1965b Marine birds and mammals at Kaikoura. Proc. N.Z. ecol. Soc., 12:13-20.

Street, R. J 1964 Feeding habits of the New Zealand fur seal, Arctocephalus forsteri. Fish. tech. Rep. N.S. mar. Dep., 9:1-20.

Stump, C. W. J. P. Robins & M. L. Garde 1960 The development of the embryo and membranes of the Humpback whale, Megaptera nodosa (Bonnaterre). Aust. J. mar. Freshwat. Res., 11:365-86.

Sverdrup, A. & K. Arnesen 1952 Investigations on the anterior lobe of the hypophysis of the Finback whale. Hvalrad. Skr., 36:1-15.

Swales, M. K. 1956 Vertebrate Zoology. The Gough Island Scientific Survey, 1955-56. Nature. 178:236.

Symons, H. W. 1955a Do Bryde whales migrate to the Antarctic? Norsk Hvalfangsttid., 44:64-7.

Symons, H. W. 1955b The foetal growth rate of whales. Norsk Hvalfangsttid., 44:519-25.

Talbot, F. H. 1960 True's Beaked whale from the south-east coast of South Africa. Nature, London, 186:406.

Tamino, G. 1953a Rinvenimento di una Globicephala sul lido di Castelporziano (Roma). Boll. Zool., 20:13-16.

Tamino, G. 1953b Ricupero di un Grampus griseus (Cuv.) arento sul lido di Ladispoli (Roma) il 5 aprile del 1953. Boll. Zool., 20:45-8.

Tamino, G. 1953c Ricupero di una Balenottera arenata sul lido di Salerno il 10 Febbraio 1953. Boll. Zool., 20:51-4.

Tamino, G. 1954 Nota sui Cetacei VI. Sullo scheletro di una Globicephala del Museo di Roma, con particolari considerazioni sul cranio di questa e di altre specie di Cetacei e di diversi mammiferi. Att. Soc. Ital. Mus. Civ. Stor. Nat. Milano, 93:489-537.

Tamino, G. 1957 Nota sui Cetacei Italiani-Rinvenimento di uno Ziphius cavirostris Cuv. presso il lido di Fregene (Roma) il giorno 9 111 1957. Atti. Soc. Ital. Sci. nat., 96:203-10.

Tarasevich, M. N. 1963 (Feeding of Sperm whales in the Northern Kuriles.) Trudy Inst. Okeanol., 71:195-206.

Tarawa, T. 1951 On the respiratory pigments of whale. Scient. Rep. Whales Res. Inst., Tokyo, 3:96-101.

Tavolga, M. C. 1966 Behavior of the bottlenosed dolphin (Tursiops truncatus): social interactions in a captive colony. pp. 718-30. In: Whales, Dolphins and Porpoises, K. S. Norris (ed.). University of California Press, Los Angeles.

Tavolga, M. C. & F. S. Essapian 1957 The behavior of the bottle-nosed dolphin (Tursiops truncatus): mating, pregnancy, parturition and mother-infant behavior. Zoologica, N.Y., 42:11-31.

Taylor, F. H. C. M. Fujinaga & F. Wilke 1955 Distribution and food habits of the Fur seals of the North Pacific Ocean. U.S. Dept. Interior, Fish and Wildl. Ser., 86 pp.

Taylor, R. J. F. 1957 An unusual record of three species of whale being restricted to pools in Antarctic sea ice. Proc. zool. Soc. Lond., 129:325-32.

Thompson, D'A. W. 1918 On whales landed at Scottish whaling stations, especially during the years 1908-14. Scott. Nat. 1918 (Sept.): 197-208.

Thomson, A. S. 1859 The story of New Zealand past and present. London, 2 vols.

Thomson, G. M. 1921 Seals. In: Wild Life of New Zealand, Vol. I, Mammalia, Chapter 10. Marcus F. Marks, Govt. Printer, Wellington.

Thorpe, M. R. 1938 Notes on the osteology of a Beaked whale. J. Mammal., 19:354-62.

Tomilin, A. G. 1947 (New view on the 'blow' of whales.) C.R. Acad. Sci., URSS, 55:85-7.

Tomilin, A. G. 1951 (On thermoregulation in Cetacea.) Priroda, Mosc. 1951:55-8.

Tomilin, A. G. 1955 (On the behaviour and sonic signalling in Cetaceans.) Trud. Inst. Okeanol., 18:28-47.

Tomlinson, J. D. W. & R. J. Harrison 1961 Venous modifications in two rare marine mammals. J. Anat., 95:453.

Tortonese, E. 1957 Il Cetaceo odontocete Ziphius cavirostris G. Cuvier nel Golfo di Genova. Doriana, 2:1-7.

Townsend, C. A. 1930 Twentieth century whaling. Bull. N.Y. zool. Soc., 33:1-31.

Townsend, C. H. 1935 The distribution of certain whales as shown by logbook records of American whaleships. Zoologica, N.Y., 19:1-50.

Troughton, E. le G. 1951 Furred animals of Australia. 4th Edition. Angus and Robertson, Sydney. 376 pp.

True, F. W. 1885 Contributions to the history of the Commander Islands. No. 5. Description of a new species of Mesoplodon, M. stejnegeri, obtained by Dr. Leonard Stejneger in Bering Island. Proc. U.S. Natl. Mus., 8:584-5.

True, F. W. 1889 Contributions to the natural history of the Cetaceans: a review of the Family Delphinidae. Bull. U.S. Nat. Mus., 36:1-191.

True, F. W 1906 Description of a new genus and species of fossil seal from the Miocene of Maryland. Proc. U.S. Nat. Mus., 30:835-40.

True, F. W. 1910 An account of the Beaked whales of the Family Ziphiidae in the collection of the United States National Museum, with remarks on some specimens in other Museums Bull. U.S. nat. Mus. 73:1-89.

Turbott, E. G. 1949 Observations on the occurrence of the Weddell seal in New Zealand. Rec. Auckland Inst. Mus., 3:377-9.

Turbott, E. G. 1952 Seals of the Southern Ocean. pp. 195-215. In: The Antarctic Today, Wellington, N.Z.

Turner, R. N. 1964 Methodological Problems in the Study of Cetacean Behaviour. pp. 337-51. In: Marine Bio-acoustics Vol. 1, W. N. Tavolga (ed.). Pergamon Press, New York. 413 pp.

Turner, W. M. 1860 Upon the thyroid glands and the thymus in Cetacea. Trans. R. Soc. Edinb., 22: 319-23.

Turner, W. M. 1872 On the occurrence of *Ziphius cavirostris* in the Shetland Seas, and a comparison of its skull with that of Sowerby's whale (*Mesoplodon sowerbyi*). Proc. R. Soc. Edinb., 7:760.

Turner, W. M. 1889 On the occurrence of Sowerby's whale (*M. bidens*) in the Firth of Forth. Proc. Roy. Soc. Edinburgh, 10:5-13.

Turner, W. M. 1891 The lesser rorqual (*Balaenoptera rostrata*) in the Scottish seas, with observations on its anatomy. Proc. R. Soc. Edinb., 19:36-75.

Twist, T. F. & D. M. Twist 1956 Record of a Beaked whale from Balandra. J. Trinidad Field Nat. Club: 26-7.

Uda, M. 1959 The fisheries of Japan. Fisheries Research Board of Canada, Nanaimo Biological Station. 96 pp.

Uda, M. 1954 Studies of the relation between the whaling grounds and the hydrographic conditions (I). Scient. Rep. Whales Res. Inst., Tokyo, 9:179-87.

Uda, M. 1962 Subarctic oceanography in relation to whaling and salmon fisheries. Scient. Rep. Whales Res. Inst., Tokyo, 16: 105-19.

Uda, M. & A. Dairokuno 1957 Studies of the relation between the whaling grounds and the hydrographic conditions. (II). A study of the relation between the whaling grounds of Kinkazan and the boundary of water masses. Scient. Rep. Whales Res. Inst., Tokyo, 12:209-24.

Uda, M. & K. Nasu 1956 Studies of the whaling grounds in the northern sea-region of the Pacific Ocean in relation to the meteorological and oceanographic conditions (I). Scient. Rep. Whales Res. Inst., Tokyo, 11:163-79.

Uda, M. & N. Suzuki 1958 Studies of the relation between the whaling grounds and the hydrographic conditions. III. The averaged conditions of the whaling grounds and their trends of variation during 1946-55. Scient. Rep. Whales Res. Inst., Tokyo, 13:215-29.

Ulmer, F. A. Jr 1941 *Mesoplodon mirus* in New Jersey with additional notes on the New Jersey *M. densirostris* and a list and key to the ziphioid whales of the Atlantic coast of North America. Proc. Acad. Nat. Sci. Phila., 93:107-122.

Uriarte, L. B. 1943 Noticia sobre Cetaceos. An. Assoc. Esp. Progr.-Cienc., 8:281-4.

Valso, J. 1938 Biochemical studies of whaling problems. I. The hypophysis of the Blue whale (*Balaenoptera musculus* L.). Macroscopical and microscopical anatomy and hormone content. Hvalrad. Skr., 16:5-30.

Van Beneden, P. J. 1888 Les ziphioides des mers d'Europe. Bull. Acad. Roy. Belgique, 41:1-119.

Van Bree, P. J. H. & J. Cadenat 1968 On a skull of *Peponocephala electra* (Gray, 1846) (Cetacea, Globicephalinae) from Sénégal. Beaufortia, 177:193-202.

Van Der Spoel, S. 1963 The vascular system in the kidneys of the common porpoise (*Phocaena phocoena* L.). Bijdr. Dierk., 33:71-81.

Van Gelder, R. G. 1960 Results of the Puritan-American Museum of Natural History Expedition to Western Mexico. 10. Marine mammals from the coasts of Baja, California and the Tres Marias Islands, Mexico. Amer. Mus. Novitates, 1992:1-27.

Van Lennep, E. W. 1950 Histology of the corpora lutea in Blue and Fin whale ovaries. Proc. Kon. Ned. Akad. Wet., 53:593-9.

Van Utrecht, W. L. 1966 On the growth of the baleen plate of the Fin whale and the Blue whale. Bijdr. Dierk., 35:1-38.

Van Utrecht, W. L. 1968 Notes on some aspects of the mammary glands in the Fin whale, *Balaenoptera physalus* (L.) with regard to the criterion 'lactating'. Norsk Hvalfangsttid., 57: 1-13.

Van Utrecht-Cock, C. N. 1966 Age determination and reproduction of female Fin whales, *Balaenoptera physalus* (Linnaeus, 1758) with special regard to baleen plates and ovaries. Bijdr. Dierk., 35:39-100.

Vaz Ferreira, R. 1950 Observaciones sobre la Isla de Lobos. Univ. Repub. Uruguay Rev. Fac. Human. Cienc., Montevideo, 5:145-76.

Vaz Ferreira, R. 1956a Caracteristicas generales de las Islas Uruguayas habitadas por lobos marinos. Servicio Oceanografico y de Pesca, trabajos sobre Islas de Lobos y Lobos Marinos, Montevideo, 1:1-21.

Vaz Ferreira, R. 1956b Etologia terrestre de *Arctocephalus australis* (Zimmerman) (lobo fino) en las islas Uruguayas. Servicio Oceanografico y de Pesca, Trabajos sobre Islas de Lobos y Lobos Marinos, Montevideo, 2:1-22.

Vinson, J. 1956 Sur la présence de l'éléphant de mer aux Mascareignes. Proc. Roy. Soc. Arts. Sci. Mauritius, 1:313-8.

Vincent, F. 1959 Etudes préliminaires de certaines émissions acoustiques de *Delphinus delphis* L. en captivité. Bull. Inst. Oceangr. Monaco 57:1-23.

Von Haast, J. 1870 Preliminary notice of a Ziphid whale probably *Berardius arnuxii*, stranded on the 16th of December 1868, on the sea beach near New Brighton, Canterbury. Trans. Proc. N.Z. Inst., 2:190-2.

Von Haast, J. 1876 Further notes on *Oulodon*, a new genus of Ziphioid whales from the New Zealand seas. Proc. zool. Soc. Lond. 1876: 457-8.

Von Haast, J. 1877a Notes on a skeleton of *Epiodon novaezealandiae*. Trans. Proc. N.Z. Inst., 9: 430-42.

Von Haast, J. 1877b Notes on *Mesoplodon floweri*. Trans. Proc. N.Z. Inst., 9:442-50.

Von Haast, J. 1877c On *Oulodon*: a new genus of Ziphioid whales. Trans. Proc. N.Z. Inst., 9:450-7.

Von Haast, J. 1880 Notes on *Ziphius (Epiodon) novaezealandiae* von Haast — goose-beaked whale. Trans. Proc. N.Z. Inst., 12:241-6.

Von Haast, J. 1881 Notes on *Balaenoptera rostrata* Fabricius (*B. huttoni* Gray). Trans. Proc. N.Z. Inst., 13:169-75.

Von Haast, J. 1883 Notes on the skeleton of *Balaenoptera australis* Desmoulins, the great southern rorqual or "sulphur-bottom" of whalers. Proc. zool. Soc. Lond., 1883-592-4.

Waite, E. R. 1912 Guide to the Whales and Dolphins of New Zealand. Canterbury College, Christchurch, 21 pp.

Waite, E. R. 1913 A supposed occurrence of the bottle-nosed whale *(Hyperoodon)* in New Zealand. Rec. Canterbury Mus., 2:23-6.

Wakefield, N. A. 1963 Leopard seals on Victorian beaches. Vict. Nat. Melbourne, 80:97.

Wakefield, N. A. 1968 Whales and dolphins recorded for Victoria. Vic. Nat., 84:273-81.

Walmsley, R. 1938 Some observations on the vascular system of a female fetal finback *(Balaenoptera physalus* L.). Contr. Embryol. Carneg. Inst., 27:107-78.

Warneke, R. M. 1963 A record of the beaked whale *(Mesoplodon layardii)* in Victorian waters. Victoria Nat., 80:68-73.

Warneke, R. M. 1966 Seals of Westernport. Victoria's Resources. Australia, 8:44-6.

Wassif, K. 1956 *Pseudorca crassidens* from Mediterranean shores of Egypt. J. Mammal., 37:456.

Webb, J. S. 1871 Note on coastal whaling stations, and the probability of their being instrumental in the destruction of the young of the whale. Trans. Proc. N.Z. Inst., 3.

Weber, M. 1923 Die Cetaceen der Siboga—Expedition. Siboga Exped., 58:1-38.

Webermen, E. C. 1914 The whale fishery in Russia. Pt. I. Moscow 1-312.

Wheeler, J. F. G. 1930 The age of Fin whales at physical maturity with a note on multiple ovulations. 'Discovery Rep.', 2:403-34.

Wilke, F. T. Taniwaki & N. Kuroda 1953 *Phocoenoides* and *Lagenorhynchus* in Japan, with notes on hunting. J. Mammal., 34:488-97.

Williamson, G. R. 1961 Two kinds of minke whale in the Antarctic. Norsk Hvalfangsttid., 50:133-41.

Wilson, E. A. 1907 Mammalia (Whales and Seals), National Antarctic Exped., 1901-1904, Natural History. British Museum, London. 2:1-66.

Wilson, R. B. 1933 The anatomy of the brain of the whale *(Balaenoptera sulfurea)*. J. comp. Neurol., 58. 419-80.

Wislocki, G. B. 1929 On the structure of the lungs of the porpoise *(Tursiops truncatus)*. Amer. J. Anat., 44:47-77.

Wislocki, G. B. 1933 On the placentation of the harbour porpoise. Biol. Bull., 65:80-98.

Wislocki, G. B. 1942 The lungs of the Cetacea with special reference to the harbour porpoise *(Phocaena phocaena* L.). Anat. Rec., 84: 117-123.

Wislocki, G. B. & L. F. Bélanger 1940 The lungs of the larger Cetacea compared with those of smaller species. Biol. Bull., 78:289-297.

Wislocki, G. B. & R. K. Enders 1941 The placentation of the bottle-nosed porpoise. Amer. J. Anat., 68:97.

Wislocki, G. B. & E. M. K. Geiling 1936 The anatomy of the hypophysis of whales. Anat. Rec. 66:17-41.

Wood, F. G. 1954 Underwater sound production and concurrent behaviour of captive porpoises *Tursiops truncatus* and *Stenella plagiodon*. Bull. Mar. Sci. Gulf. & Carib., 3: 120-33.

Wood Jones, F. 1925a Part III (Conclusion) containing the Monodelphia. pp. 271-449. In: The Mammals of South Australia. Government Printer, Adelaide.

Wood Jones, F. 1925b The Eared seals of South Australia. Rec. S. Aust. Mus., 3:9-16.

Woollen, W. W. 1921 Whales and whale fisheries of the North Pacific. Proc. Indiana Acad. Sci., 1919: 50-8.

Wyrtki, K. 1960 The surface circulation of the Tasman and Coral seas. CSIRO Div. Fish. Oceanogr. Tech. Pap. 8.

Yablokov, A. V. 1961 On the rate of development of the male womb viewed as a taxonomic character in whales and its significance. Bull. Soc. Nat. Moskou, Sec. Biol., 66: 149-50.

Yablokov, A. V. 1962 (The key to a biological riddle — the whale at a depth of 2,000 metres.) Priroda, 4:95-98.

Yamada, M. 1953 Contribution to the anatomy of the organ of hearing in whales. Scient. Rep. Whales Res. Inst., Tokyo, 8:1-79.

Yamada, M. 1954 Some remarks on the pygmy sperm whale, *Kogia*. Scient. Rep. Whales Res. Inst., Tokyo, 9:37-58.

Zemsky, V. A. 1956 Methods of establishing traces of corpora lutea of pregnancy and ovulation on the ovaries of female whalebone whales. Bull. Soc. Nat. Moskou., Sec. Biol., 61:5.

Zenkovich, B. A. 1962 Sea mammals as observed by the Round-the-World expedition of the Academy of Sciences of the USSR in 1957-58. Norsk Hvalfangsttid., 51:198-210.

Acknowledgments

If this book meets with any success it will owe much to the willing cooperation of a large number of people, both scientists and laymen, who have contributed information and photographs.

Grateful thanks are due to: Mr M. W. Cawthorn, now of the Fisheries Research Board of Canada, Ste Anne de Bellevue, Quebec, for gathering much basic bibliographic information for the pinniped section while he was working for me in Wellington; Miss J. E. King of the Department of Zoology, British Museum of Natural History, Mr R. M. Warneke of the Victorian Department of Fisheries and Wildlife, Melbourne, and Dr W. C. Clark of the Department of Zoology, Canterbury University, New Zealand, all of whom read and gave useful comments on the original drafts of some chapters; my wife Maureen for much help with checking the bibliography; Mr George E. Gale of the University of Guelph, Ontario, Canada, for his work in preparing the first index of the manuscript; Mr J. W. Brodie, formerly Acting Director of the Fisheries Research Division, Wellington, for permission to reproduce data obtained during my work for the Division in the service of the New Zealand Marine Department; Dr R. A. Falla, formerly Director of the Dominion Museum, Wellington, and Mr J. Moreland, Ichthyologist of the Dominion Museum, for permission to use data from the museum archives, and for much assistance with this data; the Alexander Turnbull Library, Wellington, for extracting information on early New Zealand sealing and obtaining photocopies of several otherwise inaccessible manuscripts; Messrs A. C. Kaberry and J. H. Sorensen of the New Zealand Marine Department, Wellington; Mr A. Ellis, District Fisheries Officer and Inspector for the Wellington Province; Mr R. J. Street of the Marine Department, Dunedin; Mr C. McCann of the Oceanographic Institute, Wellington; Directors and staff of the Marineland of New Zealand, Napier, especially the mayor Mr Peter Tait, town clerk Mr Pat Ryan, and trainers Mr Alex Dobbins, Mr Frank Robson and Mr Bruce Robson; Mr John H. Roake of the Mount Maunganui Marineland, Tauranga; Cpt. Gordon Henry, formerly of the Meteorological Office, Wellington, who maintained a whale-spotting service through coastal vessels for three years at the cost of much labour and no reward; Messrs Gilbert and Joseph Perano, Mr A. Krummel and Mr Trevor Norton, all formerly of the Tory Channel Whaling Company, for their kindly cooperation with the data and specimen collecting activities of my team; Dr Bernard Stonehouse of the Department of Zoology, Canterbury University, New Zealand; Mr L. Gurr of the Department of Zoology, Massey University, New Zealand; the Royal New Zealand Air Force for their cooperation on aerial surveys, and especially Flt. Lt. W. Willis, formerly of Defence Headquarters; the Royal New Zealand Navy, especially Lt. Commanders J. Beauchamp and D. Davies, and Lts. L. Merton, F. Arnott and S. F. Teagle, all onetime commanders of the Fisheries Protection vessels *Paea, Manga* and *Mako,* used extensively for whale survey work; Mr F. Newman of the *Ida Marian* (Greymouth) and Mr Ted Forbes and the late Mr Tom Garbes, both of Kaikoura; Messrs Charles and John Ashton for helpful liaison between my team and the whaling company;

officers of the Holm, Union Steamship Companies for reporting whales to me for a period of several years; Marine Department lighthouse keepers for regular whale sighting reports; the late Professor Dr E. J. Slijper of Amsterdam; Mr P. B. Best of the Division of Sea Fisheries, Capetown; Dr E. R. Guiler of the Zoology Department, University of Tasmania; Mr J. L. Bannister and Dr W. D. L. Ride of the Western Australian Museum; Dr R. G. Chittleborough of CSIRO, Cronulla, Australia; Dr F. C. Fraser of the Department of Zoology, British Museum of Natural History; Mr J. C. Hozack of Nuku'alofa, Tonga; Mr S. J. Holt of the FAO Fisheries Division, Rome; Mr Michael Watt, Department of Agriculture, Western Samoa; Mr Lorne Hume, formerly manager of the Western Canada Whaling Company; my erstwhile colleagues of the Fisheries Research Division, Messrs A. G. York, L. J. Paul, I. McCallum, C. Humphrey, P. Fraser, B. Tunbridge, and Miss M. McKenzie; personnel of the Raoul Island and Campbell Island weather stations, especially Messrs Allan Wright and Peter Ingram; Capt. S. B. Brown and Capt. S. Smith of Suva, Fiji; Mr Bruce Palmer of the Fiji Museum, Suva, Fiji; Monsieur C. Roger, of Orstom, Noumea, New Caledonia; Mr G. Svenne of the South Pacific Fishing Co., New Hebrides; Mr A. V. Atkins, Assistant Administrator, Agricultural Branch, Darwin; Mr L. W. C. Filewood, of Department of Agriculture, Stock and Fisheries, Konedobu, Papua; Mr B. J. Marlow, Curator of Mammals, Australian Museum, Sydney; Mr W. F. Ellis, Director of the Queen Victoria Museum, Launceston, Tasmania; Mr R. T. Archbold, Librarian of the South Australian Museum, Adelaide; Mr J. McNally of the National Museum of Victoria, Melbourne; Mr C. G. Setter, Department of Primary Industry, Canberra; Mr L. Grigg of Green Island Marineland, Cairns, Queensland; and Mr C. G. LeC. Eggleston of the Marineland of Australia Queensland.

Photographs and Line Drawings

I am particularly grateful to Mr M. W. Cawthorn for his work in illustrating the cetaceans noted in this book and to Mr G. D. Waugh, Director of the Fisheries Research Division, Wellington, for permission to reproduce these here, some in a form modified from the originals published in my bulletin *The New Zealand Cetacea*. All the seal drawings were produced by my wife.

Special thanks are due to Mr S. E. Csordas, now of the Central Chest Clinic, Melbourne, for the use of his excellent photographs from Macquarie Island.

I am very grateful for other photographs obtained from the following: The Marineland of New Zealand; Mr John Logan of Logan Print Co. Ltd, Gisborne; Professor Warren J. Houck of the Division of Biological Sciences, Humboldt State College, Arcata, California; Mr M. W. Cawthorn; Dr Hideo Omura, Director of the Whales Research Institute, Tokyo; Dr R. A. Falla; Mr R. J. Street; Mr J. Moreland; Mr Alex Dobbins; Mr J. W. D. Hall of Orewa; Mr A. S. Chamberlin of Motuora Island, Auckland; the *Northern Advocate* of Whangarei; the South Australian Museum, Adelaide; Mr L. Gurr of Massey University, Palmerston North; the British Antarctic Survey; the Royal New Zealand Air Force; the Sharon Whaling Museum, Massachusetts; Mr J. Johnson, Dept. Fisheries, Chester, Nova Scotia; Dr D. K. Caldwell of the Marineland of Florida; Dr I. Stirling, Canadian Wildlife Service; and numerous personnel of the Zoology Department, University of Guelph, Ontario.

Index

PART I: Author Index

Schroeder, R. J. 26
Schrumpf, A. 13
Schuler, W. 9
Schwartz, H. M. 157
Sclater, W. L. 103
Scoresby, W. 36
Scott, A. 129
Scott, E. O. G. 116, 117
Scott, H. H. 112, 129, 134, 156, 157
Scott-Johnson, C. 27
Sebeok, T. A. 152
Senft, A. 14
Serene, R. 103
Sergeant, D. E. 18, 21, 25, 66, 79, 80, 81, 97, 107, 116, 117, 124, 126, 129, 131, 134, 136
Serventy, D. L. 145
Shanahan, R. P. 102
Sheppard, T. 36
Sherrin, R. A. A. 30
Siebenaler, J. B. 130, 131
Silas, E. G. 122
Simpson, G. C. 1
Sivertsen, E. 143, 145, 154
Sleeman, J. L. 32
Sleptsov, M. M. 134, 136, 137
Slijper, E. J. 3, 9, 10, 11, 12, 13, 14, 16, 17, 18, 24, 35, 68, 70, 77, 78, 82, 89, 100, 113, 114, 119, 120, 121, 124, 125, 129, 130, 131, 133, 190
Slipp, J. W. 107, 110, 118, 124, 133, 134
Smith, F. 47
Smith, G. J. D. 12, 17
Smith, M. S. R. 14, 26, 143, 144
Smith, S. 190
Sokolov, V. E. 18, 126
Soot-Ryen, T. 77
Sopite, F. 35
Sorensen, J. H. 49, 50, 86, 106, 145, 149, 155, 157, 158, 159, 161, 162, 189
Sorlle, P. 41
Soucek, Z. 143, 144
Southwell, T. 36, 39, 114
Spoel, S. van der 11
Stager, K. E. 122
Stannius, A. 8
Starbuck, A. 29, 38
Starks, E. C. 38
Starr, R. B. 53
Starrett, A. 117

Starrett, P. 117
Stejneger, L. 110, 112
Stephen, A. C. 107
Stephens, W. M. 121
Stephenson, W. 80
Stirling, I. 25, 141, 143, 154, 157, 159, 160, 161, 162, 190
Stonehouse, B. 145, 159, 189
Storm, R. M. 112
Street, R. J. 155, 159, 161, 162, 189
Stump, C. W. 18, 82
Sundnes, G. 14, 130
Sutherland, W. W. 27
Suzuki, N. 56
Svenne, G. 190
Sverdrup, A. 9
Swales, M. K. 148
Symons, H. W. 71, 78
Tait, P. 189
Talbot, F. H. 107
Tamino, G. 70, 112, 115, 132
Taniwaki, T. 126
Tarasevich, M. N. 100
Tarawa, T. 13
Tate, G. H. H. 143
Tavolga, M. C. 130
Taylor, F. H. C. 142, 157
Taylor, R. J. F. 80, 110
Teagle, S. F. 33, 189
Tejima, S. 10
Terhune, J. 140
Thomas, C. D. B. 14, 151
Thompson, D. A. 36
Thompson, S. Y. 10
Thomson, A. S. 30
Thomson, G. M. 145, 149
Thorpe, M. R. 107
Tomilin, A. G. 14, 130, 137
Tomlinson, J. D. W. 11, 13, 18, 127
Tortonese, E. 112
Townsend, C. A. 41
Townsend, C. H. 51, 82, 85, 92, 93, 96
Troughton, E. le G. 107, 118, 123, 129, 143, 145, 154
True, F. W. 3, 106, 107, 115, 128
Turbott, E. G. 143, 145, 149
Tunbridge, B. R. 159, 190
Turner, R. N. 27
Turner, W. M. 10, 80, 106, 112

Twist, D. M. 107
Twist, T. F. 107
Uda, M. 56
Ulmer, F. A. Jr 107
Uriate, L. B. 112
Utrecht, W. L. van 8, 10, 72, 130
Utrecht-Cock, C. N. van 18, 72
Valso, J. 9
Vincent, F. 28
Vinson, J. 149
Wace, N. M. 148
Waite, E. R. 108, 112, 114, 131
Wakefield, N. A. 80, 122, 134, 145
Wallace, J. H. 30
Walmsley, R. 13
Warneke, R. M. 107, 149, 153, 157, 189
Wassif, K. 122
Waterston, D. 7
Watkins, W. A. 143, 144
Watt, M. 190
Waugh, G. D. 190
Webb, J. S. 31
Weber, M. 116, 118, 129, 134
Weberman, E. C. 36
Wettstein, A. 9
Wheeler, J. F. G. 20, 68, 71, 72
Wilke, F. 14, 46, 50, 110, 126, 157
Williamson, G. R. 79
Willis, W. 189
Wilson, E. A. 141, 142, 143, 145
Wilson, R. B. 6
Wislocki, G. B. 9, 14, 18, 130
Wohlschlag, D. E. 143, 144
Wood, F. G. 107, 130
Wood, J. W. 13
Wood-Jones, F. 134, 141, 155
Woollen, W. W. 38
Wright, A. 71, 190
Wright, D. 48
Wyrtki, K. 53
Yablokov, A. V. 18, 100
Yamada, M. 8, 102
Yang, H. C. 122
Yepes, J. 116, 118, 143, 145, 148
Yensen, R. 4
York, A. G. 190
Young, B. A. 10, 12
Zemsky, V. A. 18
Zenkovitch, B. A. 86

PART II: Subject Index